K. Nagata, K. Takeyasu (Eds.)

Nuclear Dynamics

Molecular Biology and Visualization of the Nucleus

K. Nagata, K. Takeyasu (Eds.)

Nuclear Dynamics

Molecular Biology and Visualization of the Nucleus

With 48 Figures, Including 34 in Color

Springer

Kyosuke Nagata, Ph.D.
Professor
Department of Infection Biology
Graduate School of Comprehensive Human Sciences
University of Tsukuba
Tennohdai, Tsukuba, Ibaraki 305-8575, Japan

Kunio Takeyasu, Dr.Sci.
Professor
Laboratory of Plasma Membrane and Nuclear Signaling
Kyoto University Graduate School of Biostudies
Sakyo-ku, Kitashirakawa-Oiwake-cho, Kyoto 606-8502, Japan

Library of Congress Control Number: 2006938399

ISBN-10 4-431-30054-6 Springer Tokyo Berlin Heidelberg New York
ISBN-13 978-4-431-30054-0 Springer Tokyo Berlin Heidelberg New York

Springer is a part of Springer Science+Business Media
springer.com

Printed in Japan

Typesetting: Camera-ready by the editors and authors
Printing and binding: Shinano, Japan

Printed on acid-free paper

Preface: Overview of Nuclear Organization and Nuclear Dynamics

Kunio Takeyasu

Laboratory of Plasma Membrane and Nuclear Signaling, Kyoto University Graduate School of Biostudies, Kitashirakawa Oiwake-cho, Sakyo-ku, Kyoto 606-8502, Japan

In contrast to the fact that prokaryotes, which do not have cell nuclei, can live only as unicellular organisms, eukaryotes bearing cell nuclei exist as a huge variety of organisms from single cell to higher multicellular organisms having higher-level functions. The reason for this is that cell nuclei of eukaryotes have a combination of mechanisms to store, retain, and transfer the very long genome stably and to produce cells carrying different functions from the identical genome information. Such functions of cell nuclei are supported by the intranuclear higher-order architecture. To understand vital activities closely related to genome function, such as cell proliferation, environment responsiveness, reproduction, development, differentiation, and aging, it is necessary to know the fundamental structure and construction of the cell nucleus, "a container" of genome (shown above). In this book, we investigate the dynamics of the nuclear structure as the basis supporting genome function, thereby comprehensively understanding how the higher-order structure of the nucleus is established and how it correlates with the expression of a variety of vital activities, including cell proliferation and ontogenesis, by combining biochemical and molecular genetic methods with the latest imaging techniques.

Nuclear Organization

A cell nucleus (hereafter referred to as nucleus) is an organelle that is compartmentalized by a special membrane structure, the nuclear envelope. A nucleus has a number of structural characteristics that other organelles do not have. One of these is internal storage of genomic DNA, a chemically single and physically very long supermolecule. In the case of human cells, DNA approximately 2 m long is stored in a tiny nucleus with a diameter of ~5 μm. Accurate replication and distribution of such a long

DNA molecule is required for the cell cycle regulation, and selective transcription of appropriate genes is necessary for transcriptional regulation. These facts conjure up an image of a highly orderly structure being required in a nucleus to keep and to express genome functions. The second characteristic of a nucleus is that it provides "space" for molecules such as proteins and RNAs to dissociate and distribute easily, since it is a giant organelle and lacks an internal membrane structure as a partition. In fact, many functional complexes (several kinds of nuclear bodies, such as nuclear speckle, Cajal body, and promyelocytic leukemia oncoprotein [PML] body, as well as nucleolus) exist in the nucleus (shown above), and its structure and function show dynamic alteration depending on the physiological conditions. Such functional complexes are designated as intranuclear compartments and/or intranuclear domains according to their functions and sizes. The third characteristic is a nuclear envelope structure, which exists as a boundary between the nucleus and the cytoplasm. Nuclearpore complexes exist to link two lipid bilayers that constitute a nuclear envelope, and regulate the bidirectional transport between the nucleus and the cytoplasm. A mesh-like structure, known as the nuclear lamina, in the inner layer of the nuclear envelope provide the scaffolding for chromosome functions in animal cells. Thus, the nuclear envelope, including nuclear pores, supports genome function of a nucleus structurally and in terms of substance transport.

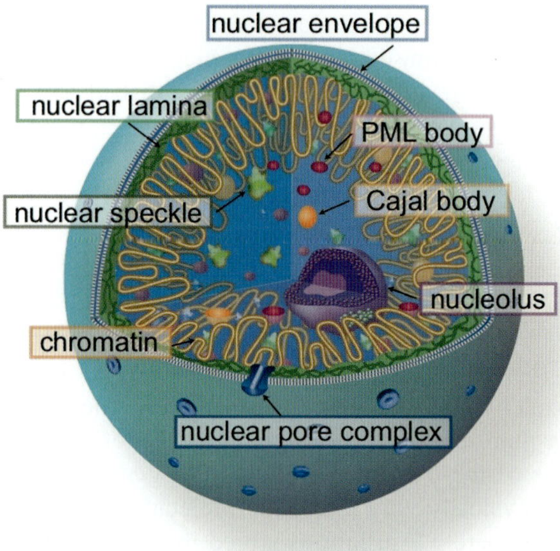

Fundamental Problems to be Solved

Intensive investigations of individual events in the nucleus, such as transcription of genes, DNA replication, repair, recombination, and RNA processing, have been conducted to date. Biological factors involved in the raw process of individual vital functions have been discovered by biochemical, molecular, biological, and genetic methods, and their molecular bases and molecular mechanisms have been analyzed. Nevertheless, how individual reactions are regulated temporally and spatially has not been investigated so intensively in spite of its importance, since most of the findings were based on *in vitro* reactions.

The amount of intranuclear proteins is huge (several hundred mg proteins/ml, assuming that they are distributed evenly in the nucleus). These proteins function without forming precipitations in the living cells by forming highly orderly complexes with chromosomal DNAs and RNAs. However, they have not been well-investigated biochemically, since most nuclear proteins form highly insoluble precipitations as a result of biochemical treatment, including DNA degradation. Although such insoluble proteins, called nuclear matrix proteins, may potentially affect gene expression as a nuclear architecture, the function of these has not been clarified thus far due to difficulties in the biochemical analysis.

Trend of New Nucleus Research

Genomic sequences of a variety of living organisms have been read, and genes coding for proteins, components of organisms, are being unraveled comprehensively. The presence and function of novel proteins can be speculated on and searched for using such genome information. In addition, thanks to the development of proteomics technology, the protein profile of a number of organelles in a nucleus, including the nucleolus, is now being clarified inclusively. Under the circumstances, single molecular dynamics in living cells and visualization of intermolecular binding are now enabled by great advances in the development of several fluorescent proteins and imaging techniques as independent technologies. By expressing target proteins fused to fluorescent proteins in cells, membrane proteins as well as nuclear proteins, which have thus far been difficult to handle biochemically, can be dynamically investigated in living cells. Based on the genome information accumulated as mentioned above, it becomes possible to map the dynamics of individual proteins in cells in a three-dimensional structure, over time, by a combination of biochemical and molecular genetic methods with molecular imaging techniques.

Based on the dynamic viewpoint of time, unforeseen and unexpected findings are being revealed, including the fact that importin β, which transports nuclear proteins into the nucleus in the intermittent phase, is also involved in spindle formation, that components of the nuclear pore complex are present in the centromere in the mitotic phase and are involved in chromosome disjunction, and that RNAs produced as transcriptional products transform DNA in their transcriptional regions into inactive heterochromatin. In other words, spatiotemporal changes of cellular molecules and structures in vital activity can now be captured.

In addition to the progress in the technologies investigating molecular dynamics, techniques derived to analyze the genome are also advancing markedly, and it has now been found that abnormalities of nuclear proteins induce several diseases, thanks to identification of genes responsible for these diseases. These include "nuclear membrane diseases," which are triggered by abnormalities in nuclear envelope proteins. For instance, a number of diseases, such as Emery-Dreifuss muscular dystrophy, familial lipodystrophy, progeria syndrome, and Pelger–Huët malformation, occur as a result of deletions and/or mutations of proteins of the inner layer of the nuclear envelope as well as nuclear lamina. The fact that nuclear envelope proteins are responsible for several diseases showing different symptoms suggests that the nuclear envelope architecture can regulate the expression of genome information dynamically, depending on the developmental and environmental status.

Hope for Fruitful Research on Nuclear Dynamics

Many authors of chapters in this book participated in the International Workshop on Nuclear Dynamics: Approaches from Biochemistry, Molecular Biology and Visual Biology at Yokohama, Japan on December 5, 2002. All of us agreed that, in order to understand vital activity closely related to genome function, it is prerequisite to understand the inner structure of the cell nucleus that supports the genome function. Since the structure is not always stable but varies from hour to hour, it is necessary to capture dynamic changes of the structure and function from the viewpoint of molecular dynamics. Since dynamic structural changes might be related to unexpected functions, researchers from a broad range of specialized areas including nuclear envelope, nuclear transport, chromosome structure, transcription, RNA processing, and nuclear domain structures should promote studies by collaborating organically with researchers specializing in imaging and nanotechnology. From this point of view, this book was written. It took over 3 years to complete, with the tremendous efforts of individual authors.

We hope that the outcomes of this field are prerequisite for the understanding of gene expression, DNA replication, and chromatin structure, and are likely to open new avenues for the elucidation of genes responsible for "nuclear membrane diseases" and abnormalities in chromosomal distribution, as well as mechanisms involved in nuclear reprogramming. Study results may also support research activities aiming at applications in regenerative medicine, which include the development of artificial chromosomes used in gene therapy as well as artificial cells that have specialized nuclear functions. In this sense, nuclear dynamics research could open up an extremely important field in terms of understanding the basics of life.

Contents

1. Visual Biology of Nuclear Dynamics: From Micro- to Nano-dynamics of Nuclear Components

Shige H. Yoshimura

Graduate School of Biostudies, Kyoto University, Kitashira-kawa-oiwake-cho, Sakyo-ku, Kyoto 606-8502, Japan

1.1 Cell Nucleus and Inner Nuclear Compartments

When you look at an interphase nucleus in a living cell through a light microscope, you will see a round, static organelle separated from the cytoplasm. If you continue the live-cell observation, you will easily learn that the cell nucleus does not undergo any significant morphological changes until it reaches the mitosis, where the nuclear envelope and the chromosomes show dynamic structural changes. Because of these morphological properties, the cell nucleus had previously been considered a "container" of genome that provides an enclosed space for genomic events to be carried out. However, recent progress in molecular and cellular biological approaches has led to the revelation that the cell nucleus is composed of various kinds of different "compartments," each of which is supposed to have a distinct "structure" and "function." These include promeyelocytic leukemia (PML) bodies, Cajal bodies, nucleolus, nuclear speckles, and nuclear foci (see figure in the Preface). Recent developments in various fluo-rescence observation techniques have revealed that these compartments are moving within a nucleus and there is a constant flow of proteins between nucleoplasm and these compartments (Fig. 1). In this chapter, therecent progress in various "visualization techniques" will be reviewed, and how these techniques have been utilized to visualize the structures and the dynamics of the inner nuclear compartments and chromosomes will be described.

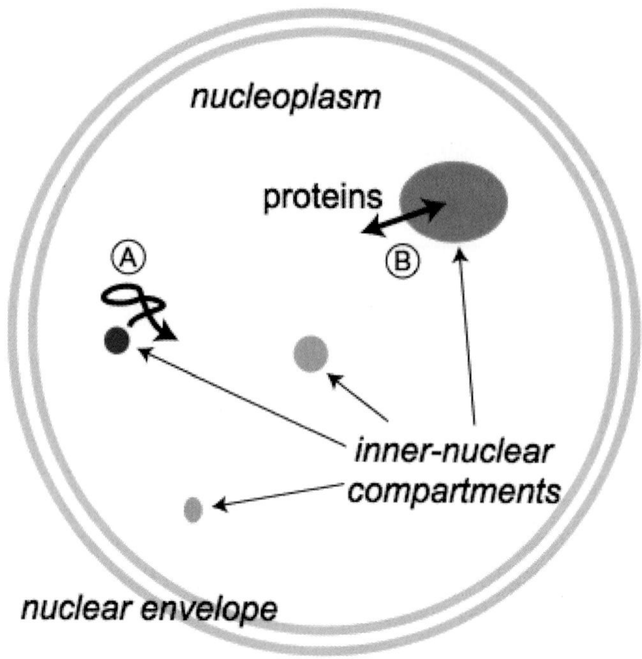

Fig. 1. Dynamics of inner nuclear compartments. The cell nucleus is composed of various inner nuclear compartments, such as nucleolus, promyelocytic leukemia (PML) body, Cajal body, and nuclear speckles. Two types of dynamics were investigated by fluorescence-based observation techniques. *A* The movement of the compartment itself can be chased by time-lapse imaging of a living cell expressing the green fluorescence protein (GFP)-fused marker protein. *B* The movement of the protein between the compartment and the nucleoplasm can be monitored by a photobleaching technique, such as fluorescence recovery after photobleaching (see later sections).

1.2 Dynamics of Inner Nuclear Compartments

Fluorescence-Based Imaging Techniques Enable Real-Time Visualization of DNA and Proteins in Nucleus

The recent progress that has come about in fluorescence imaging tech-niques has been one of the biggest achievements in cell biology. Such advancement relies on the progress both in software and hardware development for fluorescence microscopy, i.e., the development of a variety of fluorescent probes and a series of fluorescent proteins, and a great improvement in microscope setups including lasers and cameras. Useful fluorescent probes for observing nuclear dynamics include intercalators that specifically bind to DNA (DAPI, Hoechst, YOYO-1 (Invitrogen), and others; Fig. 2), a variety of ion-specific probes (Fura2 for Ca^{2+}, BCECF and SNARF for H^{+}), and phospholipid-specific probes (DiO and others). The advantages of these chemical probes are: (i) most of them have little effect on cell viability, (ii) they are so small that they can easily cross the plasma membrane and diffuse into the cell, and (iii) they can be easily applied to live-cell imaging. Where protein dynamics is concerned, a series of fluorescent proteins (green fluorescent protein [GFP], and its derivatives, Fig. 3) have been the most useful tool. Recent progress in laser-scanning confocal microscopy and high-speed video microscopy has enabled four-dimensional imaging of these fusion proteins in a living cell.

Real-time imaging of GFP-tagged protein has provided a large amount of information that could not be obtained otherwise, e.g., snapshot images of fixed cells. Dynamic morphological changes of the nuclear envelope breakdown (NEBD) was recently visualized and analyzed by the group of Ellenberg. The time-lapse observations of fluorescent protein-fused lamin and other nuclear envelope proteins, combined with photobleaching techniques, revealed that microtubules tear the nuclear envelope to facilitate the NEBD (Beaudouin et al. 2002). The involvement of dynein in NEBD was also reported (Salina et al. 2002). In mitosis, the nuclear lamina completely disappears, and the nuclear envelope is vesiculated and diffused into the cytoplasm. Many nuclear pore complex (NPC) subunits are also released into the cytosol. Reassemblies of the NPC, the nuclear envelope, and the nuclear lamina at the end of mitosis have also been studied by fluorescence-based four-dimensional microscopy, together with electron microscopy (for review see Vasu and Forbes 2001; Burke and Ellenberg 2002; Suntharalingam and Wente 2003). Some of the NPC subunits bind to the chromatin, then vesiculated envelopes are recruited to the surface of the chromatin and fused together to cover the surface. The formation of the nuclear lamina occurs after the nuclear envelope completely surrounds the chromatin and forms a round nucleus. Free lamins are transported into the nucleus through newly formed NPC and then form the lamina structure.

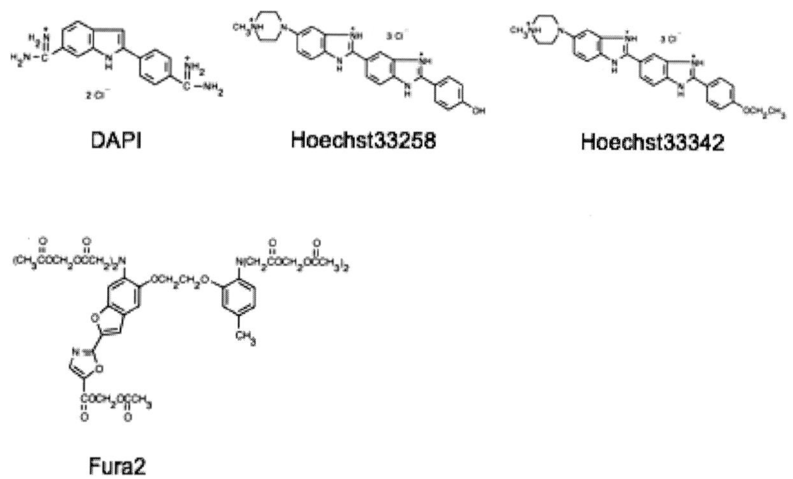

Fig. 2. Various fluorescence probes for studying nuclear dynamics. **Top** DNA-specific probes. DAPI (4′,6-diamidino-2-phenylindole, dihydrochloride, λ_{ex}=358 nm, λ_{em}=461 nm) and Hoechst are commonly used fluorescent dyes for DNA. Both of them intercalate between bases of double-stranded DNA but have slightly different properties for living cells (Haraguchi et al. 1999). Hoechst 33342 (λ_{ex}=350 nm, λ_{em}=461 nm) has better membrane permeability, better DNA specificity and better retention property than Hoechst 33258 (λ_{ex}=352 nm, λ_{em}=461 nm) and DAPI, and thus is more suitable for living cells. One simply adds Hoechst 33342 to the culture medium and then it specifically binds to DNA in all kinds of cells, and can be detected by UV light irradiation. Because detection of the dye requires UV excitation, long-time live imaging may cause defects in the genome, and microscopes that are not equipped with UV lasers cannot be used for this imaging.

Fig. 2 ***(continued)*** **Bottom** Cation-specific probe. Fura2 is a Ca^{2+}-specific fluorescent probe designed especially for intracellular Ca^{2+} imaging. Fura2 carries five carboxyl groups covalently attached through ester bonds. Since this esterified Fura2 is hydrophobic, it can penetrate the plasma membrane and can be taken up by the cell. Once Fura2 enters the cytoplasm, ester bonds are hydrolyzed by cytoplasmic esterase and Fura2 moiety is released. Since hydrolyzed Fura2 is very hydrophilic, it can no longer penetrate the membrane and remains in the cytoplasm. A single Fura2 specifically binds to one Ca^{2+} in a certain concentration range and thus can be used to monitor the Ca^{2+} concentration in a cell. Not only cytoplasmic Ca^{2+}, but also Ca^{2+} dynamics in the nucleus, have been extensively investigated utilizing Fura2 and similar dyes. The Ca^{2+} concentration in the nucleus is controlled independently of the cytoplasmic Ca^{2+}, and Ca^{2+} cannot diffuse through the nuclear pore complex (NPC) into or out of the nucleus. H^{+}-specific fluorescent probes (e.g. BCECF, SNARF, etc.) are also available to monitor intracellular pH.

Four-Dimensional Fluorescence Microscopy Reveals the Dynamics of Inner Nuclear Compartments

In contrast to its dynamic morphological changes in mitosis, the nucleus seems to be static throughout the interphase. Many proteins in inner nuclear compartments (PML bodies, nucleolus, etc.) were found in the insoluble fraction; i.e., when cultured cells were successively treated with detergent, high-salt solution and DNase I to remove all soluble proteins and chromosomes, these proteins remained in the insoluble fraction (nuclear matrix fraction), suggesting that these compartments are relatively immobile structures. Such a nuclear matrix was supposed to function as an immobile scaffold, to which chromosomes and other protein complexes were attached. However, recent fluorescence imaging techniques revealed that some of the inner nuclear architectures are slowly but constantly moving while changing their sizes and shapes.

The time-lapse observation of GFP-fused coilin, one of the protein components of a Cajal body, demonstrated that Cajal bodies are mobile (Platani et al. 2000). Two different types of Cajal body (large and small) have been known to exhibit different mean velocities; 0.48 and 0.65 μm/min for large and small bodies, respectively. Occasionally, two Cajal bodies join together or a single body separates into two individual bodies (Platani et al. 2000). A similar behavior of the Cajal body was observed when fluorescent protein-tagged SMN (Survival of Motor Neuron proteins) was used as a marker (Sleeman et al. 2003). The kinetic analysis of the movement suggested that the Cajal bodies move in the nucleoplasm mostly by anomalous diffusion, i.e., the movement of the particle is governed both by free diffusion and by a constraint to some immobile fractions. The Cajal body was known to associate with chromatin in an adenosine triphosphate (ATP)- and transcription-dependent manner, and its mobility was drastically reduced when it was associated with chromatin (Platani et al. 2002).

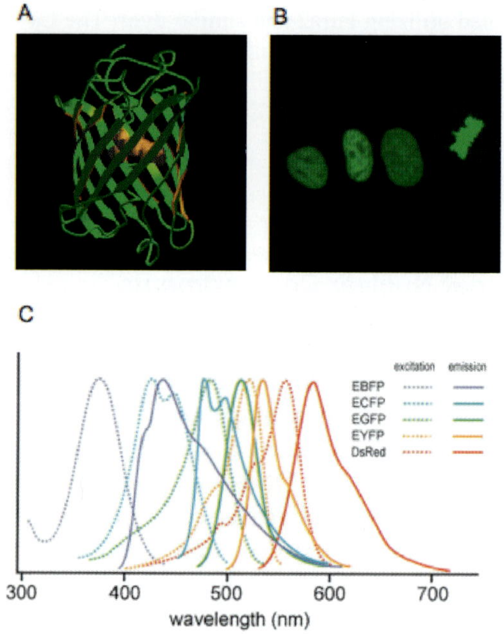

Fig. 3 Fluorescent proteins. Green fluorescent protein (GFP) was firstly discovered in a jellyfish (*Aequorea victoria*) and is composed of 238 amino acids. Since the fluorescence property of GFP is, in most cases, not affected by fusing with other proteins, it has been a useful tool for the construction of fluorescent proteins. The expression of such chimeric GFP protein in a cell is directly and easily confirmed by fluorescence microscopy without fixation or staining of the cell. GFP can be functionally expressed in a wide variety of cells and animals, from bacteria to the higher eukaryotes. **A** The structure of GFP. **B** A fluorescence image of HeLa cells expressing histone H3 fused with enhanced GFP (*EGFP*). Since the addition of GFP occasionally changes the intracellular localization of the target protein, the localization of GFP fusion proteins should be carefully compared to that of the native protein. **C** Intensive mutation work has developed a series of useful variants, which have different emission and excitation spectra.

Fig. 3. ***(continued)*** The combination of these fluorescent proteins enables the simultaneous live imaging of several different proteins. *EBFP*, enhanced blue fluorescent protein; *ECFP*, enhanced cyan fluorescent protein; *EYFP*, enhanced yellow fluorescent protein.

Dynamic movements of other nuclear compartments have also been observed. Pre-mRNA splicing factor (SF2/ASF) fused with GFP showed a transcription-coupled dynamic movement in the nucleoplasm (0.2 μm/min) (Eils et al. 2000). Similar to the Cajal bodies, some of these speckles underwent separation (budding off), producing two small separate speckles. PML bodies also turned out to be dynamic compartments (Muratani et al. 2002). Time-lapse imaging of EYFP-fused Sp100 protein (a component of the PML body) demonstrated that although ~25% of the PML bodies were immobile, ~63% showed restricted movement which was similar to that of the Cajal bodies. Interestingly, a small population of the PML bodies (~12%) exhibited rapid and large movement (4.0~7.2 μm/min), which could be abolished by ATP depletion (Muratani et al. 2002). These observations suggest that the inner nuclear compartments are always moving by simple diffusion within a constrained area, but occasionally show energy-dependent rapid movements.

FRAP and FLIP Visualize Dynamic Protein Flow into and out of the Compartments

The relatively slow movement of the inner nuclear architectures does not necessarily mean that the protein components stably remain in the compartment and have a low turnover rate. FRAP (fluorescence recovery after photobleaching), also referred to as fluorescence microphotolysis, is an effective method of analyzing protein mobilities in living cells. A small region of interest (ROI) with fluorescently tagged protein is irradiated by a strong laser pulse to irreversibly bleach the fluorophore, and the subsequent recovery of the fluorescence signal is monitored by time-lapse microscopy (Fig. 4A) (see Chapter 8 by Kimura). FRAP data are often analyzed with a kinetic model of protein diffusion and provide useful information on the diffusion coefficient. A number of FRAP analyses of various nuclear proteins, together with FLIP (fluorescence loss in photobleaching, Fig. 4B) analyses, have revealed that the fluorescence recovery time of nuclear proteins varies from several seconds to hours. Interestingly, even in the nuclear compartment that is not apparently physically moving, proteins are rapidly moving between the compartment and the nucleoplasm on the order of a few seconds.

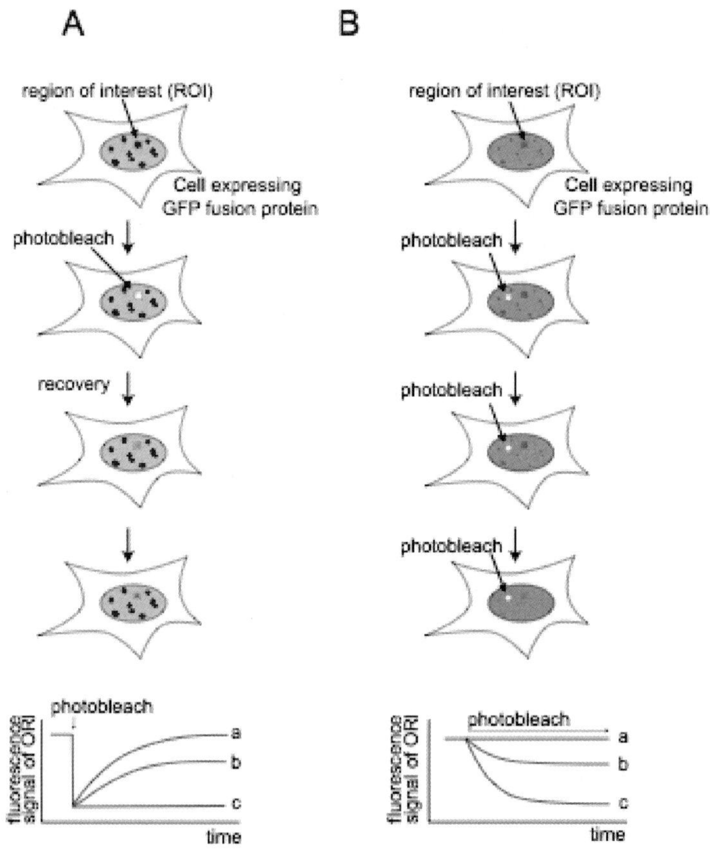

Fig. 4. Fluorescence recovery after photobleaching (FRAP) and fluorescence loss in photobleaching (FLIP). FRAP and FLIP are two major techniques used to measure the mobility of molecules in a cell. These techniques are easily done with conventional confocal laser scanning microscopy (CLSM). **A** In FRAP, a small region of interest (*ROI*) with fluorescently tagged protein is irradiated by a strong laser pulse to irreversibly bleach the fluorophore. The subsequent recovery of the fluorescence signal, which is caused by the influx of unbleached molecules from other areas, is monitored by time-lapse observation. **B** In FLIP, a small region outside the ROI is constantly photobleached and the fluorescence signal of ROI is monitored as a function of time until the signal has completely disappeared.

The relatively fast-moving proteins include nucleolar proteins (fibrillarin, nucleolin, B23) (Dundr et al. 2000; Phair and Misteli 2000; Chen and Huang 2001), proteins in the perinucleolar compartment (Huang et al. 1998), splicing factors (ACF, SF2/ASF, SC35) (Kruhlak et al. 2000; Phair and Misteli 2000), and proteins in the Cajal bodies (coilin, fibrillarin, SMN) (Handwerger et al. 2003; Sleeman et al. 2003), which were found in the insoluble fraction of the nucleus and thus had been supposed to be relatively static compartments. Nucleolar proteins involved in rDNA transcription (UBF1), pre-rRNA processing (fibrillalin, nucleolin, and Rpp29), and ribosome assembly (B23 and nucleolin) move relatively fast between nucleolus and nucleoplasm, with diffusion coefficients varying from 0.020 to 0.14 $\mu m^2/s$, whereas the ribosomal proteins (S5 and L9) show relatively slow mobility (diffusion coefficient 0.019 $\mu m^2/s$) (Phair and Misteli 2000; Snaar et al. 2000; Chen and Huang 2001). The mobilities of these proteins are differentially reduced when RNA polymerase I, which plays a major role in rDNA transcription, is selectively inhibited by actinomycin D (Chen and Huang 2001). These protein exchanges between nucleolus and nucleoplasm could be critical in the regulation of rRNA synthesis. Coilin and SMN, both of which are localized in the Cajal bodies, showed different recovery speeds after photobleaching; coilin is 50 times faster than SMN, indicating that even in the same compartment, each protein component has a different exchange rate.

The small diffusion coefficients of these nuclear proteins indicate that they are not "completely freely" diffusing in a nucleus. The mobility of nuclear proteins lies in between "completely immobile" and "freely diffused." Based on the model of Misteli and others (Misteli 2001), the slower mobility might result from the repeated transient interaction between the nuclear proteins and more immobile nuclear components (for review see Pederson 2000, 2001; Shopland and Lawrence 2000; Carmo-Fonseca et al. 2002). The movements of GFP-tagged nuclear proteins in the nucleoplasm were much slower than those of GFP itself and also than fluorescence-labeled dextran molecules that were artificially directed into the nucleus (Seksek et al. 1997). These facts suggest the existence of "immobile" fractions in a nucleus. However, the question of whether such "immobile fractions" really exist in relation to what we know as the "nuclear matrix" remains to be addressed

In clear contrast to these mobile proteins, components of nuclear lamina (lamin A, B, C), and NPC subunits (POM121, Nup153) are less mobile in the interphase (Moira et al. 2000; Daigle et al. 2001; Griffis et al. 2002), indicating the stabilities of these structures. However, recent studies using FRAP revealed that an individual NPC subunit has different recovery rate ranging from several seconds to hours (Griffis et al. 2002). This suggests that the entire NPC, which is composed of 30-50 different polypeptides in vertebrates, is not a static complex and some of the subunits are rapidly moving in and out of the complex, whereas other subunits are less mobile and might be involved in constructing a fundamental architecture of the pore. Interestingly, Nup98, one of the relatively mobile NPC subunits involved in the export of mRNA, is found not only in the NPC but also in the nucleoplasm (GLFG body), and the mobility differs depending on the ongoing transcription by RNA polymerase I or II (Griffis et al. 2002). The mobile NPC subunits may not be directly involved in the construction of the NPC architecture, but rather has regulatory roles in the transportation process.

The mobility of chromosomal proteins also varies. The least mobile are core histones, which are one of the most fundamental structural proteins of the chromosome. FRAP analysis of GFP-tagged histone H2B revealed that the recovery takes several hours or more (Lever et al. 2000), and that the recovery rate was independent of the chromosomal location (for example, euchromatin versus heterochromatin) (Misteli et al. 2000). A clear contrast to the core histones was seen in the linker histones (Lever et al. 2000; Misteli et al. 2000) and HMG proteins (Phair and Misteli 2000) (t1/2 of several seconds). The chromosomal proteins involved in higher-order folding of chromatin fibers (H1, HMG) are more mobile than the core histones. A FRAP experiment on a mitotic chromosome demonstrated that the recovery of fluorescence was observed on a spatially separated chro-mosome in a mitotic cell (Lever et al. 2000). This result indicates that physical interaction between chromatin fibers is not necessary for histone H1 exchange, and this exchange may occur through dissociation from a chromosome, diffusion through the nucleoplasm, and reassociation to another chromosome.

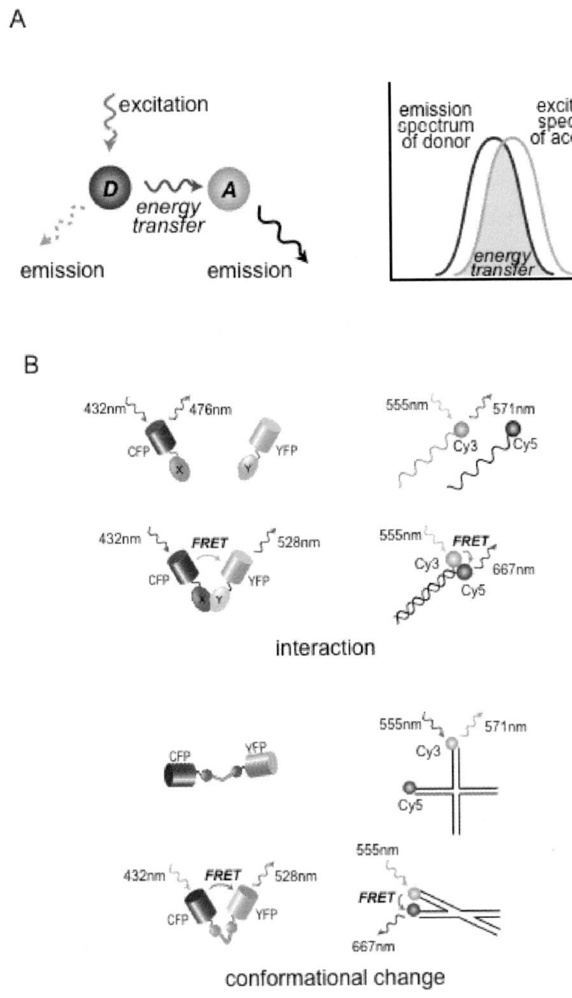

Fig. 5 Fluorescent resonance energy transfer (FRET). **A** FRET can occur when the emission spectrum of the donor fluorophore is overlapped with the excitation spectrum of the acceptor. The FRET efficiency largely depends on the spatial distance between the donor and the acceptor. Only when they are close to each other FRET occurs. **B** FRET can be used to detect protein–protein (*top left*), DNA–DNA (*top right*), and DNA–protein interactions and conformational changes in proteins (*bottom left*) and DNA (*bottom right*).

Visualization of Protein-Protein Interactions (FRET)

Fluorescent resonance energy transfer (FRET) is a recently developed fluorescence technique to visualize protein-protein and protein-DNA in-teractions. It can also be used to detect conformational changes of proteins, DNAs, and RNAs (Mergny and Maurizot 2001). It utilizes the transfer of fluorescence energy from a donor to an acceptor molecule; when the excitation spectrum of the acceptor overlaps the emission spectrum of the donor, the emission energy from the donor is transferred to the acceptor for excitation, before it is emitted as fluorescence (Fig. 5). When the acceptor is also a fluorophore, the fluorescence from the acceptor can be detected. FRET can be widely used both *in vivo* and *in vitro*. In the case of *in vivo* FRET experiments using living cells, CFP and YFP are often used as a donor and an acceptor, respectively. FRET has often been utilized to elucidate protein-protein interactions in the NPC (Damelin and Silver 2000, 2002), nuclear lamina (Moir et al. 2000; Kalab et al. 2002; Macara 2002; Plafker and Macara 2002; Shimi et al. 2004), the PML bodies (Marcello et al. 2001) and other nuclear components (Marcello et al. 2001; Day et al. 2001) (see Chapter 4 by Haraguchi).

The development of high-speed, highly sensitive cameras enabled the visualization of a single-molecule fluorescence in both *in vivo* and *in vitro* experiments. Single-molecule methods, such as single particle tracking and single-molecule FRET, have better space and time resolutions than the mass fluorescence methods such as FRAP and FLIP. Total internal reflection fluorescence microscopy (TIRFM) is now used to visualize fluorescence signals with low background and high contrast (for detail, see Fig. 6). Because of the limited reach of excitation light, TIRFM has often been used for *in vitro* imaging of various cellular components (Tokunaga et al. 1997; Harada et al. 1999; Greene and Mizuuchi 2004). *In vivo* (in situ) imaging of cytoskeletal components (Rappoport et al. 2003; Bretschneider et al. 2004) and of nuclear events (Jones et al. 2004) have also been reported.

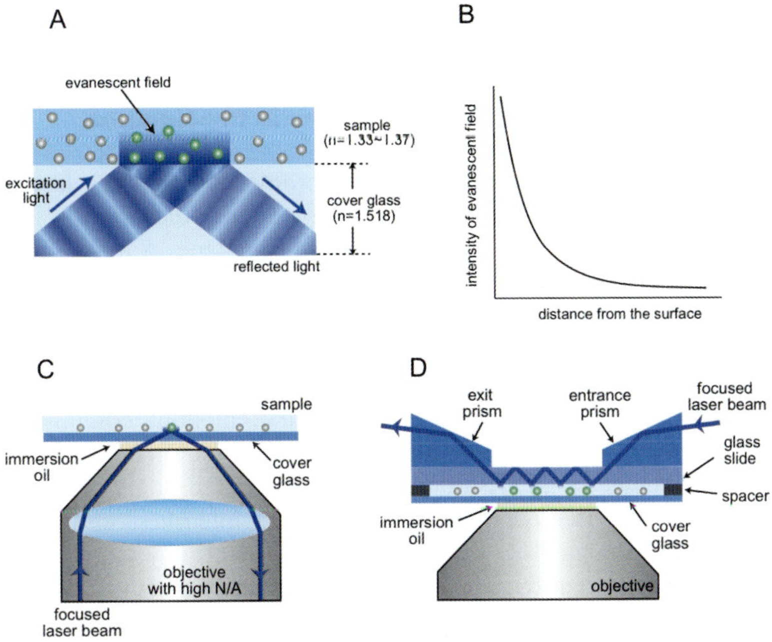

Fig. 6. Total internal reflection fluorescence microscopy (TIRFM). **A** TIRFM requires an excitation light beam traveling at a high incident angle through the glass cover slide (or plastic tissue culture container), where the proteins or the cells adhere. Refractive index differences between the glass and water phases regulate how light refracts or reflects at the interface as a function of incident angle. At a specific critical angle, light totally reflects at the glass/water interface, rather than passing through and refracting. The reflection generates a very thin electromagnetic field (evanescent field, usually less than 200 nm) in the aqueous medium, which has an identical frequency to that of the incident light. TIRFM utilizes this evanescent light for excitation of the fluorophore. **B** The intensity of evanescent light decays very rapidly depending on the distance from the reflection surface, enabling the excitation of fluorophores only close to the glass surface. **C, D** In conventional microscopes, two different systems are available to obtain the total reflection at the surface of the sample The objective-based system utilizes an immersion objective with a high numerical aperture to produce excitation illumination at supercritical angles and to retrieve fluorescence information emitted by the specimen (**C**). The prism-based system is composed of a prism attached to the slide glass and a focusing lens to obtain excitation beam light at a supercritical angle (**D**). The fluorescence is retrieved through an objective placed on the opposite side of the sample. Because of its simple equipment, the objective-based system is now more commonly used.

1.3 Chromosome Dynamics

FISH Reveals Chromosome Territories in the Nucleus

The most commonly used technique for chromosome labeling is fluorescence in situ hybridization (FISH). FISH utilizes fluorescently labeled DNA probe to visualize a specific locus in the genome. A large variety of FISH techniques are available now and a number of different sets of fluorescent probes can be purchased from many manufacturers. For example, SKY™ FISH (Spectral Karyotyping FISH, Applied Spectral Imaging, [Vista, California, U.S.]) can distinguish ("paint") 24 chromosomes in human (21 in mouse) with different fluorescent colors by combining five different fluorescent dyes and a fluorescence spectrum microscopy.

A number of FISH studies and other microscopic observations have revealed that in most mammalian cells, the interphase chromosomes are arranged in a nucleus so that the chromosomes occupy mutually exclusive volumes that are referred to as chromosome territories (for review see Cremer and Cremer 2001). The volume of each chromosome territory is roughly correlated with its DNA content but is affected by its overall transcription status (Croft et al. 1999). Chromosome territories do not show large-scale movement in an interphase nucleus and their overall positions within a nucleus are not affected by transcriptional status (Croft et al. 1999; Abney et al. 1997; Zink et al. 1998; Manders et al. 1999; Vazquez et al. 2001).

Recent studies have demonstrated that transcriptionally active sites are scattered throughout the territory and not localized to the surface of the territory (Abranches et al. 1998; Belmont et al. 1999; Verschure et al. 1999; Mahy et al. 2002). The chromosome territory seems to be a kind of sponge structure and many soluble proteins and protein complexes, such as transcriptional machineries, can diffuse into the territory. Occasionally, some gene-rich loci, in which overall transcriptional activity is high, extend away from the territory (Mahy et al. 2002). Thus, overall transcriptional status of a gene-rich locus, but not the activity of individual genes, may influence the organization of the chromosome territory.

Each chromosome territory has been known to have a preferential positioning within a nucleus. Two types of positioning have been found so far (Parada and Misteli 2002). The "relative positioning" is the relationship between the neighboring chromosomes. Chromosome territories occupy preferred positions relative to each other and a certain neighborhood environment is preferentially established (Parada et al. 2002). The "radial positioning" describes the distance between the territory and the center of the nucleus. Gene-rich chromosomes tend to localize in the interior of the nucleus and gene-poor chromosomes are located in the nuclear periphery. For example, chromosome 19 is one of the most gene-rich chromosomes in human cells and is preferentially positioned in the interior of the nucleus. On the other hand, chromosome 18, one of the gene-poor chromosomes, is preferentially found in the nuclear periphery (Croft et al. 1999; Sun et al. 2000; Boyle et al. 2001; Cremer et al. 2001). This tendency of chromosome positioning can also be seen in other human chromosomes (Boyle et al. 2001) and in the chromosomes of higher primates (Tanabe et al. 2002). Although the physiological significances of these positionings are still unclear, there might be a functional cross-talk between different chromosomes.

GFP-Coupled Chromosome-Labeling Techniques Reveal Overall Chromosome Dynamics Throughout the Cell Cycle

Real-time visualization of single chromosome or single chromosome locus can be achieved by labeling proteins that bind specifically to the locus of interest, although this application is limited to the specific regions of chromosome such as centromere, telomere, heterochromatin, and replication foci, due to the limited number of site-specific proteins. To label the entire chromosome, fluorescent protein-coupled histone is most commonly used and is now overwhelming the conventional DNA-specific fluorescent probes (see page 3) (see also Fig. 2). The exogenously introduced fluorescent core histone does not have a significant effect on the chromosome morphology or the cell viability in most cell lines.

Are the chromosome positionings in a nucleus inherited to the daughter cells after mitosis, and if so, how? Two recent papers addressed these questions by utilizing fluorescence-based chromosome observation techniques. Gerlich et al. utilized CFP- and YFP-coupled core histones and a patterned photobleaching technique combined with *in vivo* time-lapse microscopy and computer simulation. They found that the relative positioning of chromosome in mammalian cells was faithfully transmitted during mitosis (Gerlich et al. 2003). Namely, the global chromosome positioning seems to be heritable through the cell cycle. On the other hand, Walter et al. reported that chromosome positions significantly change during mitosis and early G1 phase, and the interphase pattern of chromosome territory was not perfectly maintained in daughter cells (Walter et al. 2003). These two groups observed the same chromosome event with different approaches. The apparent discrepancy between these two independent studies might result from how they define the "same chromosome positioning." Walter et al. utilized very stringent criterion for pattern maintenance and any deviations from the original pattern were considered as changes (Parada et al. 2003). On the other hand, the criterion used by Gerlich et al. was less stringent and asks whether the chromosomal patterns observed deviate from random arrangement of chromosome.

LacI-GFP/lacO System and Chromosome Mobility

In addition to the labeling of the entire chromosome by GFP-histone, the GFP-coupled labeling technique can also be applied to the labeling of a specific chromosomal locus. However, as mentioned above, this application largely depends on the existence of a site-specific protein and, thus, has been limited to several chromosome regions (centromere, telomere, etc.). A key technical breakthrough in the studies of the chromosome dynamics was introduced by the use of tandemly repeated arrays of a bacterial operon (lacO) in a cell expressing a fusion protein between the lac repressor and GFP (Robinett et al. 1996; Belmont and Straight 1998; Belmont 2001). In this system, tandem repeats of lacO are inserted into the desired position of the genome (Fig. 7). The live imaging of these cells provide information on the dynamic movement of specific chromosomal loci as well as other chromosomal regions such as telomere (Straight et al. 1997; Heun et al. 2001; Ancelin et al. 2002; Bressan et al. 2004), centromere (Straight et al. 1996; Nabeshima et al. 1998; Goshima and Yanagida 2000; Heun et al. 2001; Thrower and Bloom 2001; Yamamoto and Hiraoka 2003), and replication origins (Heun et al. 2001), as well as that of plasmids in bacterial cells (Pogliano et al. 2001). The biggest advantage of the technique is that one can visualize any regions of interest within a genome (of course, one has to spend some time to obtain the cell line carrying lacO repeat in a desired position in a genome) and high specificity of the lacI-GFP protein to the lacO locus.

The recent study using the lacI/lacO system also revealed that centromeric and telomeric regions of budding yeast chromosome remain at the periphery of a nucleus and do not show any large-scale movement during the interphase (Heun et al. 2001). Contrarily, noncentromeric and nontelomeric regions showed small and large movements in G1 phase (distance ≤ 0.2 μm and ≥ 0.5 μm). During S and G2 phases, only the small-scale movement was detected and the large-scale movement could not be seen, indicating that the dynamic rearrangement of the chromosome occurs mainly in G1 phase. Since the large chromosomal movement was abolished when F0F1-ATPase was suppressed, it involves an energy-dependent step.

Mammalian chromosomes, which show a nonpolar arrangement within a nucleus, do not undergo long-range movement in the interphase (Abney et al. 1997). However, detailed analysis of chromosomal loci by high-resolution fluorescence microscopy revealed that the chromosomal loci were moving in a manner of "constraint diffusion" (the diffusion coefficient is ~$5x10^{-12}$ cm^2/s and the area is limited within 1 μm) (Marshall et al. 1997). The diffusion coefficient and the size of the constraint largely varied according to each locus. For example, nucleolar-attached region and the nuclear periphery were less mobile than other loci (Chubb et al. 2002). Thus, in the living cell, chromatin fibers are constantly in a diffusion motion confined by some anchoring mechanisms, depending on the cell types. Both the range and the rate of chromatin mobility appear to be modulated by interactions between chromatin and internal nuclear structures, such as the nuclear envelope and the nucleolus.

1.4 Nanoscale Dynamics of the Nucleus

Nanoscale Visualization: Electron Microscopy Versus Scanning Probe Microscopy

Although the techniques in fluorescence microscopy have advanced re-markably, it has a physical limitation in the spatial resolution (several hundreds of nanometers). This is because, in all types of optical microscopes, the spatial resolution (ε_0) is limited by the wavelength of light utilized (l) and the numerical aperture of the objective (NA) (ε_0=0.61l/NA). Furthermore, the fluorescence signal in the image indicates the "location" of the molecule but not the real "shape" of the molecule, which is much smaller than the spatial resolution of fluorescence microscopy. To visualize the real "shape" of the molecule (~nm), scanning probe microscopy (SPM) or electron microscopy (EM) can be utilized. Both of these methods were originally developed to observe the surface of metals and polymers, but now are indispensable tools for the visualization of biological samples and are widely used in molecular and cellular biology. Due to the requirement of fixation and staining processes, EM has limitations in the observation of molecular dynamics. On the other hand, atomic force microscopy (AFM), one type of SPM, does not require such procedures, and is now the only device that can visualize nanometer structure in solution without fixation and staining of the sample (Fig. 8A).

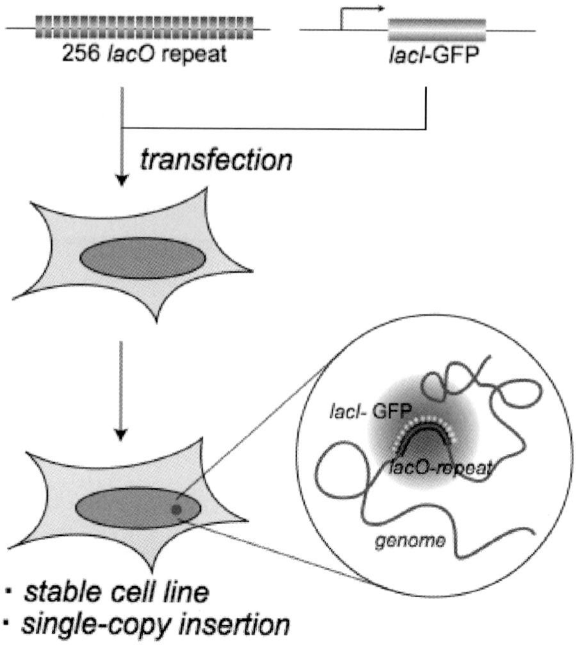

Fig. 7. Chromosome labeling (GFP–lacI system). The lacI–GFP/*lacO* system was developed by Belmont. The binding site for lacI, which is made up of 256 repeats of *lacO* DNA, should be stably incorporated into genomic DNA. When the plasmid encoding GFP-fused lacI is also introduced into the same cell line, the expressed GFP–lacI specifically binds to *lacO* repeats and, under the fluorescence microscope, indicates the position of *lacO* locus in the genome. Since the microscopic observation does not require cell fixation, this approach has a great advantage over the fluorescence in situ hybridization (FISH) procedure, which requires many steps of sample preparation (fixation, hybridization, etc.)

Nanoscale Architectures of DNA, Chromatin, Chromosome, and Inner Nuclear Compartment

The first application of AFM to biological samples dates back to the late 1980s, when several groups observed double-stranded DNA and DNA-protein complexes (Hansma et al. 1988). In the 1990s, the observation technique of AFM improved remarkably, together with a number of biochemical purification/reconstitution procedures. In the case of the observation of DNA-protein complex (for example, a transcription factor binding to a specific DNA sequence within a promoter region), the following information can be deduced from the analysis of the AFM image: (i) the protein binding site along several-kilobase DNA strand (with an accuracy of ~10 bp), (ii) stoichiometry of bound proteins (dimer, tetramer, or higher multimer) and (iii) structural change in a DNA strand caused by protein binding (bending, twisting, untwisting, wrapping, looping, etc). Our group previously reported the observation of stem-loop structure in a bacterial promoter region and specific binding of transcription factor to that structure (Ohta et al. 1996), dynamic relaxation of supercoiled DNA by replication factor (Yoshimura et al. 2000a), long-range DNA loop formation in β-globin enhancer region by transcription regulators (Yoshimura et al. 2000b), and DNA end-loop formation by telomere-specific proteins (Yoshimura et al. 2004).

In the 1970s, chromatin fibers were extensively studied by EM, and the higher-order arrangements of chromatin fibers such as 30-nm fibers and 300-nm loops were identified (Paulson and Laemmli 1977; Rattner and Hamkalo 1978a,b; Marsden and Laemmli 1979). The AFM was also utilized to observe chromatin fibers in the 1980s and 1990s. In-liquid and in-air imaging of extracted chromatin fiber has clearly revealed that the higher-order packing of chromatin fiber largely depends on the salt concentrations (Leuba et al. 1994; Zlatanova et al. 1994) and the presence of linker histones (Bustamante et al 1997; Leuba et al. 1998). In the late 1990s, when an *in vitro* chromatin reconstitution method was established, AFM was immediately applied to visualize those reconstituted chromatin fibers. The physical properties of the reconstituted chromatin fiber and the effect of linker histone H1 were investigated from AFM observation (Sato et al. 1999; Hizume et al. 2002, 2003) (see later section for details of the chromatin fiber observation).

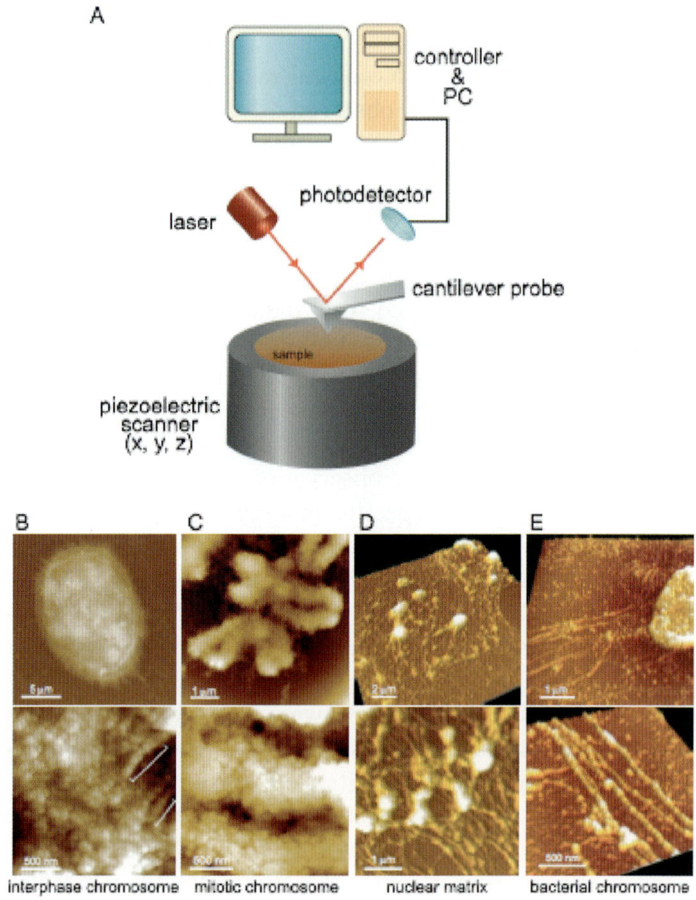

Fig. 8. Nanoscale visualization of nuclear components by atomic force microscopy (AFM). AFM generates three-dimensional surface profiles (topographical images) of molecules with a nanometer resolution. **A** A key element of AFM is its microscopic force sensor (cantilever), which is made of silicon or silicon nitride and measures 100–500 μm in length by 0.5–5 μm in width. Mounted on the end of the cantilever is a sharp tip that is used to sense the force between the sample and itself. By scanning a sample under the cantilever and recording the deflection of the cantilever, the local height of the sample is measured. Three-dimensional topographical maps of the surface are then constructed. **B–D** AFM images of human interphase chromosome (**B**), rye mitotic chromosome (**C**), human nuclear matrix (**D**), and bacterial chromosome in log-phase (**E**)

When nanoscale chromosomal architecture is argued based on the microscopic images, “how you purify the sample” and “how you prepare the specimen” are always important issues to be considered. Indeed, the EM and AFM observations of mitotic chromosome revealed that purified chromosomes might be covered by a significant amount of nonchromosomal proteins and RNA (Tamayo et al. 1999; Tamayo and Miles 2000). Our group recently developed an “on-substrate” specimen preparation procedure for AFM observation in order to exclude artifacts resulting from “bulk” purification procedures (Yoshimura et al. 2003). When this technique was applied to HeLa cells cultured on a cover glass, a distinct chromatin fiber of 80 nm width was observed in an interphase nucleus (Yoshimura et al. 2003) (Fig. 8) (see later section for details). Considering that the ~80 nm “fibrous” or “granular” structures were also identified in various chromosomes (Allen et al. 1993a,b ; Tamayo et al. 1999; Tamayo and Miles 2000), and that a fluorescence microscopy study also suggested the existence of 80-nm chromatin fibers (Belmont and Bruce 1994), this might be the one-step higher-order folding of the 30-nm chromatin fibers, although its molecular mechanism is not yet fully understood.

The nuclear envelope has also been a good target of AFM observation, not only because of its distinct structure and dimension but also because of its easiness in the specimen preparation on a flat surface. The groups of Oberleithner and others utilized AFM to observe the NPC in the nuclear envelope of *Xenopus* oocyte and cultured cells to demonstrate that the structure of the NPC changes depending on ATP (Rakowska et al. 1998), Ca^{2+} (Wang and Clapham 1999) and extracellular hormone stimulation (Folprecht et al. 1996) (see Chapter 2 by Oberleithner).

AFM Visualizes Nanoscale “Dynamics” of Nuclear Components

The biggest advantage of the AFM is its ability to image in liquid. This has two important meanings. First, the sample does not have to be dried, and thus, the damage of the sample can be minimized. Second, the enzymatic action and the conformational change of a protein (or a protein complex) can be monitored by time-lapse imaging, enabling the visualization of a single molecule in action.

The first visualization of a single molecule in motion was achieved in 1994 by Hansma’s group. They were successful in obtaining time-lapse images of DNA being digested by DNase I at 30-s intervals (Bezanilla et al. 1994). The Bustamante group successfully visualized the molecular events of RNA polymerase sliding along DNA. They observed RNA polymerase binding to the promoter and forming an initiation complex with DNA wrapping around the protein (Rivetti et al. 1999). They made a "stalled complex" by depriving cytidine triphosphate (CTP) from nucleotide triphosphate mix. When CTP was added to the stalled complex, the RNA polymerase started to slide along the DNA template at 0.5-2 bp/s (Kasas et al. 1997; Bustamante et al. 1999; Guthold et al. 1999). The interaction of a restriction enzyme and DNA (Ellis et al. 1999; Berge et al. 2000) was also analyzed by in-liquid time-lapse imaging, producing a clear image of how a site-specific protein finds the target sequence.

The time resolution (scanning speed) has been an important property of AFM. Thanks to intensive studies by many researchers, the scanning rate of AFM was improved with newly designed piezoelectric scanners and feedback systems. Now, ~30 frames/s (for tapping mode) (Ando et al. 2001) and ~100 frames/s (for contact mode) are already being achieved. The fast-scanning AFM revealed a rapid movement of plasmid DNA on the mica surface (Fig. 9) as well as dynamic movement of an enzyme on DNA. In the case of HindIII (type II restriction enzyme), a dimer first randomly binds to DNA and immediately starts to slide along the DNA fiber. The sliding is completely random and can go in both directions. Most of the enzymes stay on DNA for ~1.5 s and travel ~250 bp. When the enzyme encounters its recognition sequence during such a short binding, it digests DNA into two fragments and the dimer dissociates into two monomers. The digestion and dissociation into component monomers take place simultaneously and each DNA fragment retains one of the monomers. These monomers will dissociate from the DNA on the order of several seconds and then reform a dimmer in order to start the process over again.

The Combination of AFM and Fluorescence Microscopy (AFM/FM) Reveals the Nanoscale Structure of Eukaryotic Genome and Inner Nuclear Architectures

The fluorescence microscope localizes a specific molecule, but does not visualize the "real shape" of the molecule in nanoscale. On the other hand, AFM visualizes the nanoscale structure of the specimen but does not provide any information on the "identity" of the molecule. When you are using AFM to observe a large protein complexes or inner cellular architectures such as chromosomes, you will always encounter the problem of how to distinguish or identify a specific protein in the AFM image. This situation has led many AFM researchers to combine the AFM with other

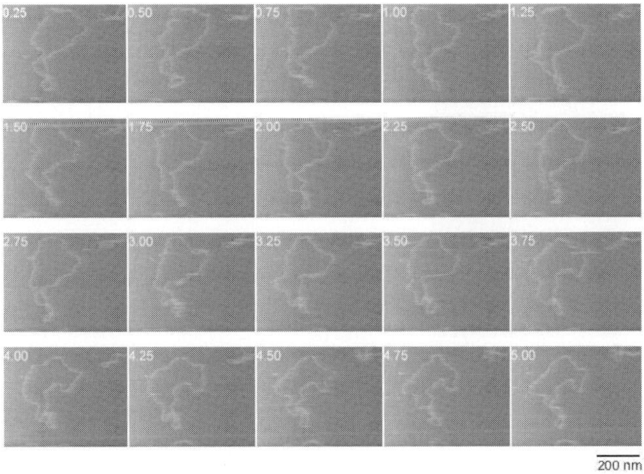

Fig. 9. Fast-scanning AFM. In-liquid fast-scanning AFM images of plasmid DNA (~2.9 kb) deposited on a mica surface. The resonance frequency of the cantilever is 500–1000 kHz in liquid. In the course of the experiment, images were captured at a scanning rate of 4 frames per second. The image analysis demonstrated that the movement is random and is not affected by the scanning direction

observation techniques in order to overlay the AFM image with an image that indicates the location of a specific protein.

The purification and isolation of the interphase chromosome is difficult because of its tight connection to inner-nuclear architectures such as nuclear lamina. Nanoscale analyses of interphase chromosome could be achieved by observing an ultrathin section of a nucleus or chemically dissected cultured cells. Recently, we applied a chemical dissection procedure of cultured cells for AFM and successfully observed the nanoscale structure of the interphase chromosome and the inner-nuclear architectures (Yoshimura et al. 2003). The interphase chromosome is composed of globular structural units of ~80 nm width. On the other hand, the inner nuclear architecture is composed of a network of thin filaments (10-60 nm width) and small globules (100-500 nm diameter) (Fig. 8B), which is similar to so-called nuclear matrix structures previously identified by EM (Fey et al. 1986; He et al. 1990; Nickerson et al. 1997; Nickerson 2001).

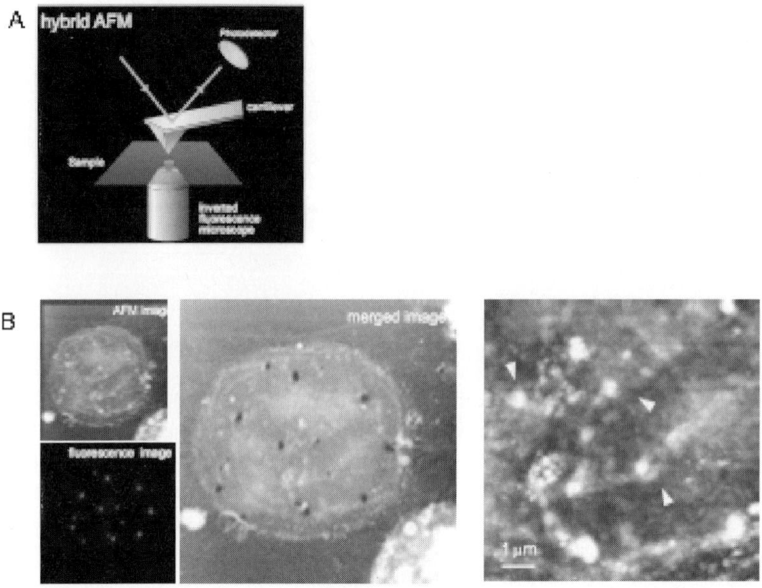

Fig. 10A,B. Identification of a specific protein on an AFM image by AFM/fluorescence microscopy (FM). **A** Schematic illustration of the AFM/FM system. The AFM scanner is mounted on the stage of an inverted microscope. **B** Hybrid AFM image of HeLa cell nuclear matrix. HeLa cells expressing GFP-fused PML protein were grown on a cover glass and the glass was then treated to remove the plasma membrane, cytosol, and chromosomes, leaving purified nuclear matrix on the glass. A Fluorescence signal was obtained with an inverted fluorescence microscope and an AFM image was also obtained from the same cell by the scanner mounted on the microscope stage

When we carefully looked at the image of inner nuclear fibers, the small globules were located at the "junction" of the thin filaments, suggesting that it connects filaments into a mesh-like network. To examine whether these globular structures are homogeneous in the protein composition, the AFM was combined with fluorescence microscopy (AFM/FM hybrid imaging) (Fig. 10). The cultured cells expressing GFP-labeled PML protein, one of the components of PML bodies, were first observed by FM and then by AFM. The alignment of the AFM image and fluorescence image of the nuclear matrix fraction purified on a cover glass revealed that not all globular structures contained PML protein (Fig. 10B). There is an increasing number of proteins found in the inner-nuclear components. The utilization of this combined AFM/FM will reveal the nanoscale structures of each component together with their protein compositions. However, since the AFM image and the fluorescence image must be overlaid manually, the accuracy of the image overlay is on the order of ~10-100 nm. To obtain the merged image at several-nanometer resolution, we need another technique.

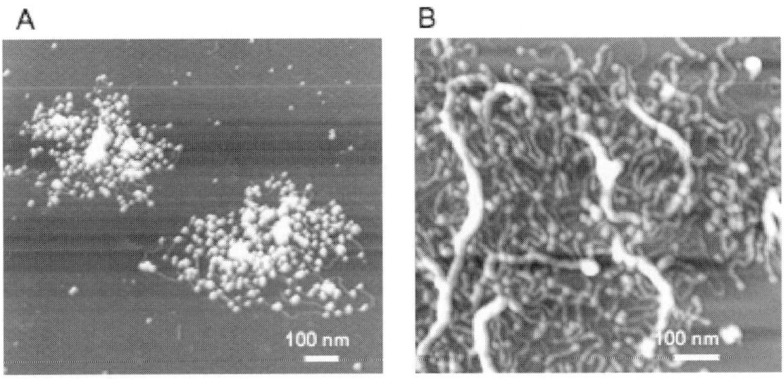

Fig. 11A,B. In vitro chromatin reconstitution. Nucleosomes are reconstituted on ~100 kb closed circular plasmid by salt-dialysis procedure and observed by AFM. The so-called beads-on-a-string structure of chromatin fiber can be seen (**A**). The addition of H1 induces the formation of thick chromatin fiber of ~30 nm width (**B**).

The Combination of Molecular Recognition (PicoTREC) and AFM, As Well as Biochemical Reconstitution Procedure, Reveals the Dynamics of Chromatin Fiber and Chromatin-Associated Proteins

The development of chromatin reconstitution technique has been playing critical roles in the recent progress in chromatin dynamics. The highest reconstitution efficiency with a proper nucleosome spacing (~200 bp) can be achieved by the utilization of the embryo extract from *Droshophila melanogaster* (Becker and Wu 1992). This chromatin template has been utilized for the analyses of transcriptional control by nucleosome. The salt-dialysis method (Ura et al. 1995; Li and Wrange 1997) constitutes another procedure, where purified DNA and core histone octamer are mixed in the presence of a high concentration of salt (2 M NaCl) and the sample is then dialyzed to the final salt concentration of 50 mM or below. Although the reconstitution efficiency is not as high as in the method using embryo extract, this method is still useful for the characterization of chromatin fiber itself because of its simple components (Allen et al. 1993a,b; Sato et al. 1999; Yodh et al. 1999).

Recently, we developed a salt-dialysis method to efficiently re-constitute chromatin fibers without using any nucleosome-positioning signals. By comparing the reconstitution efficiency of plasmid DNA ranging from 3 to 100 kb, we found that a 100-kb plasmid with negative supercoils had highest efficiency in the salt-dialysis reconstitution (one nucleosome per ~250 bp) (Hizume et al. 2002, 2003). The most striking property of this long chromatin fiber was observed when it was mixed with linker histone H1. An addition of histone H1 induced a further folding of the chromatin fiber into a thicker fiber of 20-30 nm width (Fig. 11), and the width of the fiber changed depending on the ion concentration. These results demonstrate that the higher-order folding of chromatin fiber largely depends both on the presence of linker histone and nucleosome-nucleosome interaction.

In the single molecule imaging of the chromatin fiber associated with various regulatory proteins, it is extremely difficult to identify a particular protein on the chromatin fiber, since the fluorescence microscope does not provide sufficient spatial resolution. The PicoTREC system was developed by Molecular Imaging (Tempe, AZ, USA) to recognize a specific protein by SPM probe, while obtaining a topographic image with the same probe (Stroh et al. 2004a,b) (Fig. 12). In this system, the antibody specific for the protein of interest was covalently attached on an AFM tip via a 10-20-nm polyethylene glycol (PEG) spacer. The modified AFM tip can collect normal topographical information as well as the attractive force exerted from antibody-antibody interaction. Since the PicoTREC system simultaneously obtains these two different signals and produces two different images (a topographical image and a recognition image), the overlay of these two different images can be achieved with an accuracy of several nanometers. So far, the core histones in nucleosome, as well as several proteins including lysozyme and BSA, have been successfully recognized by this system (Stroh et al. 2004a, b) (Fig. 12).

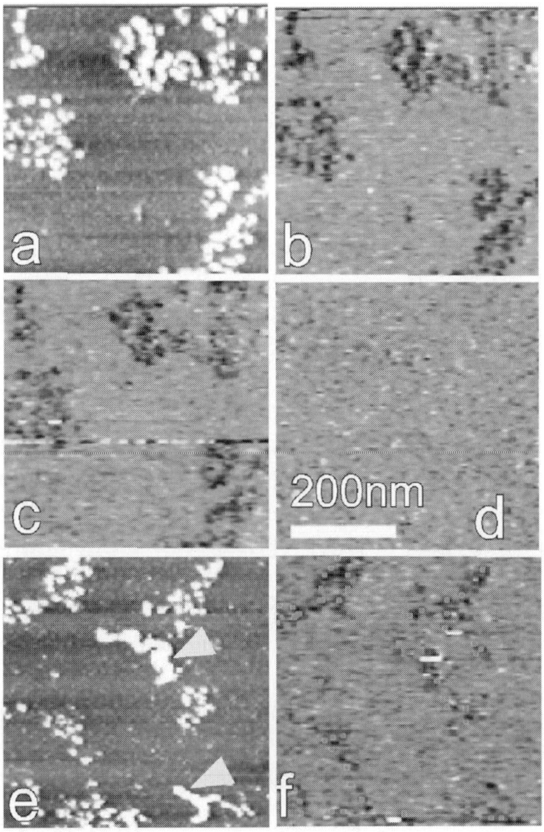

Fig. 12. Identification of a specific protein by PicoTREC AFM. PicoTREC is a new Topography and RECognition imaging solution that requires no fluorescence, radioactivity, enzyme-linked detection schemes, or other extraneous labels. PicoTREC can be used to map ligand–receptor interactions, study biological processes, or probe nanometer-scale molecular binding sites on a variety of surfaces.

Fig. 12(*continued*). A variety of molecules can be attached to AFM cantilevers, making them chemically or biologically selective sensors so that individual molecular interactions can be studied. It combines real-time detection of molecular recognition events and single-molecule sensitivity with the imaging capability of AFM. PicoTREC allows a researcher to quickly determine where specific, single molecule binding interactions occur on a sample and resolves them from surface topography information. For example, when an antibody is attached to an AFM tip, the forces required to break the bonds that are involved in binding events with target molecules are easily detected and resolved by PicoTREC. **a** A topographical image of chromatin fiber. **b** A recognition image with antihistone H3 antibody attached to the probe. **c,d** The recognition signal disappeared when the blocking peptide was added in the imaging chamber (**d**) but was not with bovine serum albumin (**c**). **e,f** The specificity of the recognition signal. The nonchromatin molecules in the topographical image (**e**, *arrowheads*) were not seen in the recognition image (**f**) (Images were provided by Dr. Lindsay, Arizona State University)

SNOM/AFM Simultaneously Provides AFM Image and Fluorescence Image with High Spatial Resolution

SNOM/AFM (scanning near-field optical microscope/AFM, also referred to as NSOM/AFM) is another instrument that allows one to simultaneously obtain AFM and fluorescence images. SNOM/AFM contains all of the equipment necessary for AFM but has a cantilever made of an optical fiber instead of silicon crystal. A sharp optical fiber probe with a tip radius of 50-100 nm is used not only to obtain the topographical image of the sample, but also to irradiate the excitation light to the sample. The excitation light, when coming out from a small aperture of the fiber tip, produces an evanescent field, which allows the excitation of a very limited area of the specimen. Thus, SNOM/AFM can obtain fluorescence images with a resolution much higher than the limit of the conventional fluorescence microscope. Furthermore, because of this unique probe, higher accuracy in image alignment can be achieved.

SNOM/AFM was first applied to the structural analysis of mitotic chromosomes. A specific locus in mitotic chromosome was labeled by the FISH technique. When this specimen was observed by SNOM/AFM, both the nanoscale topographical image and the fluorescence signal from the FISH probe were obtained (Yoshino et al. 2002; Fukushi et al. 2003; Oberringer et al. 2003). However, it was claimed that because of the harsh treatment in the FISH procedure (heat, denatured, and dry), the chromosomes could no longer maintain their native structure. To overcome this problem, low-temperature FISH was developed to reduce the sample damage (Winkler et al. 2003). SNOM/AFM was also applied to the protein on the cell surface. The location of the cell-surface receptor was also mapped on a cultured cell (Nagy et al. 1999). It was clearly observed on this cell line that the hormone stimulation induced clustering of the receptor on the cell surface (Nagy et al. 1999).

References

Abney JR, Cutler B, Fillbach ML, Axelrod D, Scalettar BA (1997) Chromatin dy-namics in interphase nuclei and its implications for nuclear structure. J Cell Biol 137:1459-1468

Abranches R, Beven AF, Aragon-Alcaide L, Shaw PJ (1998) Transcription sites are not correlated with chromosome territories in wheat nuclei. J Cell Biol 143:5-12

Allen MJ, Dong XF, O'Neill TE, Yau P, Kowalczykowski SC, Gatewood J, Bal-horn R, Bradbury EM (1993a) Atomic force microscope measurements of nu-cleosome cores assembled along defined DNA sequences. Biochemistry 32:8390-8396.

Allen MJ, Lee C, Lee JD, Pogany GC, Balooch M, Siekhaus WJ, Balhorn R (1993b) Atomic force microscopy of mammalian sperm chromatin. Chromo-soma 102:623-630

Ancelin K, Brunori M, Bauwens S, Koering CE, Brun C, Ricoul M, Pommier JP, Sabatier L, Gilson E (2002) Targeting assay to study the cis functions of hu-man telomeric proteins: evidence for inhibition of telomerase by TRF1 and for activation of telomere degradation by TRF2. Mol Cell Biol 22:3474-3487

Ando T, Kodera N, Takai E, Maruyama D, Saito K, Toda A (2001) A high-speed atomic force microscope for studying biological macromolecules. Proc Natl Acad Sci USA 98:12468-12472

Beaudouin J, Gerlich D, Daigle N, Eils R, Ellenberg J (2002) Nuclear envelope breakdown proceeds by microtubule-induced tearing of the lamina. Cell 108:83-96

Becker PB, Wu C (1992) Cell-free system for assembly of transcriptionally rep-ressed chromatin from Drosophila embryos. Mol Cell Biol 12:2241-2249

Belmont AS (2001) Visualizing chromosome dynamics with GFP. Trends Cell Biol 11:250-257

Belmont AS, Bruce K (1994) Visualization of G1 chromosomes: a folded, twisted, supercoiled chromonema model of interphase chromatid structure. J Cell Biol 127:287-302

Belmont AS, Straight AF (1998) In vivo visualization of chromosomes using lac operator-repressor binding. Trends Cell Biol 8:121-124

Belmont AS, Dietzel S, Nye AC, Strukov YG, Tumbar T (1999) Large-scale chromatin structure and function. Curr Opin Cell Biol 11:307-311

Berge T, Ellis DJ, Dryden DT, Edwardson JM, Henderson RM (2000) Trans-loca-tion-independent dimerization of the EcoKI endonuclease visualized by atomic force microscopy. Biophys J 79:479-484

Bezanilla M, Drake B, Nudler E, Kashlev M, Hansma PK, Hansma HG (1994) Motion and enzymatic degradation of DNA in the atomic force microscope. Biophys J 67:2454-2459

Boyle S, Gilchrist S, Bridger JM, Mahy NL, Ellis JA, Bickmore WA (2001) The spatial organization of human chromosomes within the nuclei of normal and emerin-mutant cells. Hum Mol Genet 10:211-219

Bressan DA, Vazquez J, Haber JE (2004) Mating type-dependent constraints on the mobility of the left arm of yeast chromosome III. J Cell Biol 164:361-371

Bretschneider T, Diez S, Anderson K, Heuser J, Clarke M, Muller-Taubenberger A, Kohler J, Gerisch G (2004) Dynamic actin patterns and Arp2/3 assembly at the substrate-attached surface of motile cells. Curr Biol 14:1-10

Burke B, Ellenberg J (2002) Remodelling the walls of the nucleus. Nat Rev Mol Cell Biol 3:487-497

Bustamante C, Zuccheri G, Leuba SH, Yang G, Samori B (1997) Visualization and analysis of chromatin by scanning force microscopy. Methods 12:73-83

Bustamante C, Guthold M, Zhu X, Yang G (1999) Facilitated target location on DNA by individual Escherichia coli RNA polymerase molecules observed with the scanning force microscope operating in liquid. J Biol Chem 274:16665-16668

Carmo-Fonseca M, Platani M, Swedlow JR (2002) Macromolecular mobility in-side the cell nucleus. Trends Cell Biol 12:491-495

Chen D, Huang S (2001) Nucleolar components involved in ribosome biogenesis cycle between the nucleolus and nucleoplasm in interphase cells. J Cell Biol 153:169-176

Chubb JR, Boyle S, Perry P, Bickmore WA (2002) Chromatin motion is con-strained by association with nuclear compartments in human cells. Curr Biol 12:439-445

Cremer T, Cremer C (2001) Chromosome territories, nuclear architecture and gene regulation in mammalian cells. Nat Rev Genet 2:292-301

Cremer M, von Hase J, Volm T, Brero A, Kreth G, Walter J, Fischer C, Solovei I, Cremer C, Cremer T (2001) Non-random radial higher-order chromatin ar-rangements in nuclei of diploid human cells. Chromosome Res 9:541-567

Croft JA, Bridger JM, Boyle S, Perry P, Teague P, Bickmore WA (1999) Dif-fer-ences in the localization and morphology of chromosomes in the human nu-cleus. J Cell Biol 145:1119-1131

Daigle N, Beaudouin J, Hartnell L, Imreh G, Hallberg E, Lippincott-Schwartz J, Ellenberg J (2001) Nuclear pore complexes form immobile networks and have a very low turnover in live mammalian cells. J Cell Biol 154:71-84

Damelin M, Silver PA (2000) Mapping interactions between nuclear transport factors in living cells reveals pathways through the nuclear pore complex. Mol Cell 5:133-140

Damelin M, Silver PA (2002) In situ analysis of spatial relationships between proteins of the nuclear pore complex. Biophys J 83:3626-3636

Day RN, Periasamy A, Schaufele F (2001) Fluorescence resonance energy transfer microscopy of localized protein interactions in the living cell nucleus. Meth-ods 25:4-18

Dundr M, Misteli T, Olson MO (2000) The dynamics of postmitotic reassembly of the nucleolus. J Cell Biol 150:433-446

Eils R, Gerlich D, Tvarusko W, Spector DL, Misteli T (2000) Quantitative imag-ing of pre-mRNA splicing factors in living cells. Mol Biol Cell 11:413-418

Ellis DJ, Dryden DT, Berge T, Edwardson JM, Henderson RM (1999) Direct ob-servation of DNA translocation and cleavage by the EcoKI endonuclease using atomic force microscopy. Nat Struct Biol 6:15-17

Fey EG, Krochmalnic G, Penman S (1986) The nonchromatin substructures of the nucleus: the ribonucleoprotein (RNP)-containing and RNP-depleted matrices analyzed by sequential fractionation and resinless section electron microscopy. J Cell Biol 102:1654-1665

Folprecht G, Schneider S, Oberleithner H (1996) Aldosterone activates the nuclear pore transporter in cultured kidney cells imaged with atomic force microscopy. Pflugers Arch 432:831-838

Fukushi D, Shichiri M, Sugiyama S, Yoshino T, Hagiwara S, Ohtani T (2003) Scanning near-field optical/atomic force microscopy detection of fluorescence in situ hybridization signals beyond the optical limit. Exp Cell Res 289:237-244

Gerlich D, Beaudouin J, Kalbfuss B, Daigle N, Eils R, Ellenberg J (2003) Global chromosome positions are transmitted through mitosis in mammalian cells. Cell 112:751-764

Goshima G, Yanagida M (2000) Establishing biorientation occurs with precocious separation of the sister kinetochores, but not the arms, in the early spindle of budding yeast. Cell 100:619-633

Greene EC, Mizuuchi K (2004) Visualizing the assembly and disassembly mecha-nisms of the MuB transposition targeting complex. J Biol Chem

Griffis ER, Altan N, Lippincott-Schwartz J, Powers MA (2002) Nup98 is a mobile nucleoporin with transcription-dependent dynamics. Mol Biol Cell 13:1282-1297

Guthold M, Zhu X, Rivetti C, Yang G, Thomson NH, Kasas S, Hansma HG, Smith B, Hansma PK, Bustamante C (1999) Direct observation of one-dimensional diffusion and transcription by Escherichia coli RNA polymerase. Biophys J 77:2284-2294

Handwerger KE, Murphy C, Gall JG (2003) Steady-state dynamics of Cajal body components in the Xenopus germinal vesicle. J Cell Biol 160:495-504

Hansma PK, Elings VB, Marti O, Bracker CE (1988) Scanning tunneling mi-cro-scopy and atomic force microscopy: application to biology and technology. Science 242:209-216

Harada Y, Funatsu T, Murakami K, Nonoyama Y, Ishihama A, Yanagida T (1999) Single-molecule imaging of RNA polymerase-DNA interactions in real time. Biophys J 76:709-715

Haraguchi T, Ding D, Yamamoto A, Kaneda T, Koujin T, Hiraoka Y (1999) Multi-ple-color fluorescence imaging of chromosomes and microtubules in living cells. Cell Struct. Funct. 24:291-298

He DC, Nickerson JA, Penman S (1990) Core filaments of the nuclear matrix. J Cell Biol 110:569-580

Heun P, Laroche T, Shimada K, Furrer P, Gasser SM (2001) Chromosome dy-namics in the yeast interphase nucleus. Science 294:2181-2186

Hizume K, Yoshimura SH, Maruyama H, Kim J, Wada H, Takeyasu K (2002) Chromatin reconstitution: development of a salt-dialysis method monitored by nano-technology. Arch Histol Cytol 65:405-413

Hizume K, Yoshimura SH, Wada H, Takeyasu K (2003) Atomic force microscopy demonstrates a critical role of DNA superhelicity in the nucleosome dynamics. Cell Biochem. Biophys 40:249-261

Huang S, Deerinck TJ, Ellisman MH, Spector DL (1998) The perinucleolar com-partment and transcription. J Cell Biol 143:35-47

Jones JT, Myers JW, Ferrell JE, Meyer T (2004) Probing the precision of the mi-totic clock with a live-cell fluorescent biosensor. Nat Biotechnol 22:306-312

Kalab P, Weis K, Heald R (2002) Visualization of a Ran-GTP gradient in inter-phase and mitotic Xenopus egg extracts. Science 295:2452-2456

Kasas S, Thomson NH, Smith BL, Hansma HG, Zhu X, Guthold M, Bustamante C, Kool ET, Kashlev M, Hansma PK (1997) Escherichia coli RNA polym-erase activity observed using atomic force microscopy. Biochemistry 36:461-468

Kruhlak MJ, Lever MA, Fischle W, Verdin E, Bazett-Jones DP, Hendzel MJ (2000) Reduced mobility of the alternate splicing factor (ASF) through the nucleoplasm and steady state speckle compartments. J Cell Biol 150:41-51

Leuba SH, Yang G, Robert C, Samori B, van Holde K, Zlatanova J, Bustamante C (1994) Three-dimensional structure of extended chromatin fibers as revealed by tapping-mode scanning force microscopy. Proc Natl Acad Sci USA 91:11621-11625

Leuba SH, Bustamante C, van Holde K, Zlatanova J (1998) Linker histone tails and N-tails of histone H3 are redundant: scanning force microscopy studies of reconstituted fibers. Biophys J 74:2830-2839

Lever MA, Th'ng JP, Sun X, Hendzel MJ (2000) Rapid exchange of histone H1.1 on chromatin in living human cells. Nature 408:873-876

Li Q, Wrange O (1997) Assays for transcription factors access to nucleosomal DNA. Methods 12:96-104

Macara IG (2002) Why FRET about Ran? Dev Cell 2:379-380

Mahy NL, Perry PE, Bickmore WA (2002) Gene density and transcription in-flu-ence the localization of chromatin outside of chromosome territories de-tectable by FISH. J Cell Biol 159:753-763

Mahy NL, Perry PE, Gilchrist S, Baldock RA, Bickmore WA (2002) Spatial or-ganization of active and inactive genes and noncoding DNA within chromo-some territories. J Cell Biol 157:579-589

Manders EM, Kimura H, Cook PR (1999) Direct imaging of DNA in living cells reveals the dynamics of chromosome formation. J Cell Biol 144:813-821

Marcello A, Cinelli RA, Ferrari A, Signorelli A, Tyagi M, Pellegrini V, Beltram F, Giacca M (2001) Visualization of in vivo direct interaction between HIV-1 TAT and human cyclin T1 in specific subcellular compartments by fluo-res-cence resonance energy transfer. J Biol Chem 276:39220-39225

Marsden MP, Laemmli UK (1979) Metaphase chromosome structure: evidence for a radial loop model. Cell 17:849-858

Marshall WF, Straight A, Marko JF, Swedlow J, Dernburg A, Belmont A, Murray AW, Agard DA, Sedat JW (1997) Interphase chromosomes undergo con-strained diffusional motion in living cells. Curr Biol 7:930-939

Mergny JL, Maurizot JC (2001) Fluorescence resonance energy transfer as a probe for G-quartet formation by a telomeric repeat. Chembiochem 2:124-132

Misteli T (2001) Protein dynamics: implications for nuclear architecture and gene expression. Science 291:843-847

Misteli T, Gunjan A, Hock R, Bustin M, Brown DT (2000) Dynamic binding of histone H1 to chromatin in living cells. Nature 408:877-881

Moir RD, Yoon M, Khuon S, Goldman RD (2000) Nuclear lamins A and B1: dif-ferent pathways of assembly during nuclear envelope formation in living cells. J Cell Biol 151:1155-1168

Muratani M, Gerlich D, Janicki SM, Gebhard M, Eils R, Spector DL (2002) Metabolic-energy-dependent movement of PML bodies within the mam-malian cell nucleus. Nat Cell Biol 4:106-110

Nabeshima K, Nakagawa T, Straight AF, Murray A, Chikashige Y, Yamashita YM, Hiraoka Y, Yanagida M (1998) Dynamics of centromeres during meta-phase-anaphase transition in fission yeast: Dis1 is implicated in force balance in metaphase bipolar spindle. Mol Biol Cell 9:3211-3225

Nagy P, Jenei A, Kirsch AK, Szollosi J, Damjanovich S, Jovin TM (1999) Ac-tiva-tion-dependent clustering of the erbB2 receptor tyrosine kinase detected by scanning near-field optical microscopy. J Cell Sci 112(Pt II):1733-1741

Nickerson J (2001) Experimental observations of a nuclear matrix. J Cell Sci 114:463-474

Nickerson JA, Krockmalnic G, Wan KM, Penman S (1997) The nuclear matrix re-vealed by eluting chromatin from a cross-linked nucleus. Proc Natl Acad Sci USA 94:4446-4450

Oberringer M, Englisch A, Heinz B, Gao H, Martin T, Hartmann U (2003) Atomic force microscopy and scanning near-field optical microscopy studies on the characterization of human metaphase chromosomes. Eur Biophys J 32:620-627

Ohta T, Nettikadan S, Tokumasu F, Ideno H, Abe Y, Kuroda M, Hayashi H, Take-yasu K (1996) Atomic force microscopy proposes a novel model for stem-loop structure that binds a heat shock protein in the Staphylococcus aureus HSP70 operon. Biochem Biophys Res Commun 226:730-734

Parada L, Misteli T (2002) Chromosome positioning in the interphase nucleus. Trends Cell Biol 12:425-432

Parada LA, McQueen PG, Munson PJ, Misteli T (2002) Conservation of relative chromosome positioning in normal and cancer cells. Curr Biol 12:1692-1697

Parada LA, Roix JJ, Misteli T (2003) An uncertainty principle in chromosome po-sitioning. Trends Cell Biol 13:393-396

Paulson JR, Laemmli UK (1977) The structure of histone-depleted metaphase chromosomes. Cell 12:817-828

Pederson T (2000) Diffusional protein transport within the nucleus: a message in the medium. Nat Cell Biol 2: E73-74

Pederson T (2001) Protein mobility within the nucleus-what are the right moves? Cell 104:635-638

Phair RD, Misteli T (2000) High mobility of proteins in the mammalian cell nu-cleus. Nature 404:604-609

Plafker K, Macara IG (2002) Fluorescence resonance energy transfer biosensors that detect Ran conformational changes and a Ran x GDP-importin-beta-RanBP1 complex in vitro and in intact cells. J Biol Chem 277:30121-30127

Platani M, Goldberg I, Swedlow JR, Lamond AI (2000) In vivo analysis of Cajal body movement, separation, and joining in live human cells. J Cell Biol 151:1561-1574

Platani M, Goldberg I, Lamond AI, Swedlow JR (2002) Cajal body dynamics and association with chromatin are ATP-dependent. Nat Cell Biol 4:502-508

Pogliano J, Ho TQ, Zhong Z, Helinski DR (2001) Multicopy plasmids are clus-tered and localized in Escherichia coli. Proc Natl Acad Sci USA 98:4486-4491

Rakowska A, Danker T, Schneider SW, Oberleithner H (1998) ATP-Induced shape change of nuclear pores visualized with the atomic force microscope. J Membr Biol 163:129-136

Rappoport JZ, Taha BW, Simon SM (2003) Movement of plasma-membrane-associated clathrin spots along the microtubule cytoskele-ton. Traffic 4:460-467

Rattner JB, Hamkalo BA (1978a) Higher order structure in metaphase chromo-somes. I. The 250 Å fiber. Chromosoma 69:363-372

Rattner JB, Hamkalo BA (1978b) Higher order structure in metaphase chromo-somes. II. The relationship between the 250 Å fiber, superbeads and beads-on-a-string. Chromosoma 69:373-379

Rivetti C, Guthold M, Bustamante C (1999) Wrapping of DNA around the E. coli RNA polymerase open promoter complex. EMBO J 18:4464-4475

Robinett CC, Straight A, Li G, Willhelm C, Sudlow G, Murray A, Belmont AS (1996) In vivo localization of DNA sequences and visualization of large-scale

chromatin organization using lac operator/repressor recognition. J Cell Biol 135:1685-1700
Salina D, Bodoor K, Eckley DM, Schroer TA, Rattner JB, Burke B (2002) Cyto-plasmic dynein as a facilitator of nuclear envelope breakdown. Cell 108:97-107
Sato MH, Ura K, Hohmura KI, Tokumasu F, Yoshimura SH, Hanaoka F, Take-yasu K (1999) Atomic force microscopy sees nucleosome positioning and his-tone H1- induced compaction in reconstituted chromatin. FEBS Lett 452:267-271
Seksek O, Biwersi J, Verkman AS (1997) Translational diffusion of macro-mole-cule-sized solutes in cytoplasm and nucleus. J Cell Biol 138:131-142
Shimi T, Koujin T, Segura-Totten M, Wilson KL, Haraguchi T, Hiraoka Y (2004) Dynamic interaction between BAF and emerin revealed by FRAP, FLIP, and FRET analyses in living HeLa cells. J Struct Biol 147:31-41
Shopland LS, Lawrence JB (2000) Seeking common ground in nuclear com-plex-ity. J Cell Biol 150: F1-4
Sleeman JE, Trinkle-Mulcahy L, Prescott AR, Ogg SC, Lamond AI (2003) Cajal body proteins SMN and Coilin show differential dynamic behaviour in vivo. J Cell Sci 116:2039-2050
Snaar S, Wiesmeijer K, Jochemsen AG, Tanke HJ, Dirks RW (2000) Mutational analysis of fibrillarin and its mobility in living human cells. J Cell Biol 151:653-662
Straight AF, Belmont AS, Robinett CC, Murray AW (1996) GFP tagging of bud-ding yeast chromosomes reveals that protein-protein interactions can me-diate sister chromatid cohesion. Curr Biol 6:1599-1608
Straight AF, Marshall WF, Sedat JW, Murray AW (1997) Mitosis in living bud-ding yeast: anaphase A but no metaphase plate. Science 277:574-578
Stroh CM, Ebner A, Geretschlager M, Freudenthaler G, Kienberger F, Kam-ruzza-han AS, Smith-Gill SJ, Gruber HJ, Hinterdorfer P (2004a) Simultane-ous to-pography and recognition imaging using force microscopy. Biophys J 87:1981-1990
Stroh C, Wang H, Bash R, Ashcroft B, Nelson J, Gruber H, Lohr D, Lindsay SM, Hinterdorfer P (2004b) Single-molecule recognition imaging microscopy. Proc Natl Acad Sci USA 101:12503-12507
Sun HB, Shen J, Yokota H (2000) Size-dependent positioning of human chromo-somes in interphase nuclei. Biophys J 79:184-190
Suntharalingam M, Wente SR (2003) Peering through the pore: nuclear pore com-plex structure, assembly, and function. Dev Cell 4:775-789
Tamayo J, Miles M (2000) Human chromosome structure studied by scanning force microscopy after an enzymatic digestion of the covering cell material. Ultramicroscopy 82:245-251
Tamayo J, Miles M, Thein A, Soothill P (1999) Selective cleaning of the cell de-bris in human chromosome preparations studied by scanning force mi-cros-copy. J Struct Biol 128:200-210
Tanabe H, Muller S, Neusser M, von Hase J, Calcagno E, Cremer M, Solovei I, Cremer C, Cremer T (2002) Evolutionary conservation of chromosome

terri-tory arrangements in cell nuclei from higher primates. Proc Natl Acad Sci USA 99:4424-4429

Thrower DA, Bloom K (2001) Dicentric chromosome stretching during anaphase reveals roles of Sir2/Ku in chromatin compaction in budding yeast. Mol Biol Cell 12:2800-2812

Tokunaga M, Kitamura K, Saito K, Iwane AH, Yanagida T (1997) Single molecule imaging of fluorophores and enzymatic reactions achieved by objective-type total internal reflection fluorescence microscopy. Biochem Biophys Res Commun 235:47-53

Ura K, Hayes JJ, Wolffe AP (1995) A positive role for nucleosome mobility in the transcriptional activity of chromatin templates: restriction by linker histones. EMBO J 14:3752-3765

Vasu SK, Forbes DJ (2001) Nuclear pores and nuclear assembly. Curr Opin Cell Biol 13:363-375

Vazquez J, Belmont AS, Sedat JW (2001) Multiple regimes of constrained chro-mosome motion are regulated in the interphase Drosophila nucleus. Curr Biol 11:1227-1239

Verschure PJ, van Der Kraan I, Manders EM, van Driel R (1999) Spatial rela-tion-ship between transcription sites and chromosome territories. J Cell Biol 147:13-24

Walter J, Schermelleh L, Cremer M, Tashiro S, Cremer T (2003) Chromosome or-der in HeLa cells changes during mitosis and early G1, but is stably main-tained during subsequent interphase stages. J Cell Biol 160:685-697

Wang H, Clapham DE (1999) Conformational changes of the in situ nuclear pore complex. Biophys J 77:241-247

Winkler R, Perner B, Rapp A, Durm M, Cremer C, Greulich KO, Hausmann M (2003) Labeling quality and chromosome morphology after low temperature FISH analysed by scanning far-field and near-field optical microscopy. J Mi-crosc 209:23-33

Yamamoto A, Hiraoka Y (2003) Monopolar spindle attachment of sister chro-mat-ids is ensured by two distinct mechanisms at the first meiotic division in fis-sion yeast. EMBO J 22:2284-2296

Yodh JG, Lyubchenko YL, Shlyakhtenko LS, Woodbury N, Lohr D (1999) Evi-dence for nonrandom behavior in 208-12 subsaturated nucleosomal array populations analyzed by AFM. Biochemistry 38:15756-15763

Yoshimura SH, Ohniwa RL, Sato MH, Matsunaga F, Kobayashi G, Uga H, Wada C, Takeyasu K (2000a) DNA phase transition promoted by replication ini-tia-tor. Biochemistry 39:9139-9145

Yoshimura SH, Yoshida C, Igarashi K, Takeyasu K (2000b) Atomic force mi-cro-scopy proposes a "kiss and pull" mechanism for enhancer function. J Electron Microsc (Tokyo) 49:407-413

Yoshimura SH, Kim J, Takeyasu K (2003) On-substrate lysis treatment combined with scanning probe microscopy revealed chromosome structures in eu-karyo-tes and prokaryotes. J Electron Microsc (Tokyo) 52:415-423

Yoshimura SH, Maruyama H, Ishikawa F, Ohki R, Takeyasu K (2004) Molecular mechanisms of DNA end-loop formation by TRF2. Genes Cells 9:205-218

Yoshino T, Sugiyama S, Hagiwara S, Ushiki T, Ohtani T (2002) Simultaneous collection of topographic and fluorescent images of barley chromosomes by scanning near-field optical/atomic force microscopy. J Electron Microsc (To-kyo) 51:199-203

Zink D, Cremer T, Saffrich R, Fischer R, Trendelenburg MF, Ansorge W, Stelzer EH (1998) Structure and dynamics of human interphase chromosome ter-rito-ries in vivo. Hum Genet 102:241-251

Zlatanova J, Leuba SH, Yang G, Bustamante C, van Holde K (1994) Linker DNA accessibility in chromatin fibers of different conformations: a reevaluation. Proc Natl Acad Sci USA 91:5277-5280

2 Nuclear Envelope: Nanoarray Responsive to Aldosterone

Hans Oberleithner

Institute of Physiology II, Nanolab, University Münster, Robert-Koch-Strasse. 27a, 48149 Münster, Germany

2.1 Introduction

In 1994 a paper was published that contained a rather unusual observation made with a rather unusual technique (Oberleithner et al. 1994). The unusual observation was the increase in number of nuclear pore complexes (NPCs) in nuclear envelopes of kidney cells in response to aldosterone and the unusual technique applied in this study was atomic force microscopy (AFM). At that time, aldosterone had been considered as a hormone that controlled fluid and electrolyte balance in kidney through regulation of plasma membrane ion channels and transporters but virtually nothing was known about its interaction with the nuclear barrier. Possibly, those AFM experiments were born of the desperate desire of a few renal physiologists who wanted to apply a new nanotechnique, originally developed by physicists working in the material sciences (Binnig and Quate 1986), on a biological membrane with some relevance for kidney function. In the meantime 10 years have passed. Aldosterone underwent a dazzling metamorphosis in terms of site and mode of action (Oberleithner 2004). Atomic force microscopy developed into a useful tool in the biological sciences (Roco 2003). Finally, the nuclear envelope advanced to an extensively explored membrane system that selectively passes signals from outside into the nucleus (Gerasimenko et al. 2003). In this short review chapter I will focus on recent developments in this field

2.2 Atomic Force Microscopy

The AF microscope works by moving a probe back and forth across a surface and recording features as the probe encounters them. Atomic force microscopy produces images that are not compromised by the limitations of the wavelengths of the various types of electromagnetic radiation. This means that very high resolution (below 1 nm) can be obtained. Biological work using the AFM has shown considerable growth since its invention in 1986 (Binnig and Quate 1986). The AFM probe (i.e., the AFM tip which scans the sample) is typically made from a pyramidal crystal of silicon nitride deposited onto a gold-coated, flexible cantilever. A sample (e.g. the nuclear envelope) is prepared on a flat substrate (i.e., glass or mica) and moved so that it makes contact with the probe. The sample is then moved in a raster pattern and the cantilever carrying the probe is deflected vertically as features in the sample move under it. The movement is controlled by a series of piezoelectric drivers and the control is such that the probe can be positioned in either horizontal "x" or "y" dimensions or in the vertical "z" dimension very accurately. A low-powered laser is focused onto the cantilever and is reflected onto a series of photomultiplier detector elements. As a result, when the probe scans across a surface and meets some sort of obstacle, the cantilever is deflected, changing the reflected angle of the laser, and therefore affecting the signal detected by the photomultipliers. The signal is fed into a computer which then constructs a three-dimensional image from the information received. The principle of the AFM is therefore quite simple, and its implementation is dependent on the availability of suitable sensitive photomultipliers and piezo control mechanisms. More about the different modes of recording high resolution AFM images can be found in a recent review (Henderson and Oberleithner 2000).

2.3 Aldosterone

The traditional concept of aldosterone action is the binding of the ligand to classic mineralocorticoid receptors (MRs) in target cells of kidney, colon, and salivary and sweat glands. It is followed by a genomic response and an appropriate change in physiological function of the target cell. Over the past years, the endocrine properties of aldosterone have taken on a broader perspective involving nonclassic actions in nonepithelial cells found in nonclassic target tissues (Epstein 2001). At least four new perspectives shape up, related to aldosterone action. The fact that the MRs play a crucial role in the function of a broad variety of cell types should not lead to an underestimation of the potential role of rapid (apparently nongenomic) effects elicited by this hormone.

1. Many more different cell types than previously thought express the MR. Besides the above mentioned classic aldosterone-responsive tissues MR are found also in endothelial cells, cardiac myocytes, fibroblasts, vascu-

lar smooth muscle cells and many others. In parallel, evidence for distinct changes in cellular functions elicited by aldosterone and blocked by specific inhibitors is accumulating in the literature (Delyani 2000).

2. Aldosterone exerts rapid effects, nongenomic in nature (Falkenstein et al. 2000). These effects have a rapid time course (seconds to minutes) and are not inhibited by actinomycin D. They have been considered to reflect aldosterone acting via a high-affinity membrane receptor, distinct from the classic intracellular MR on two reasons. One is that the rapid effects have not been inhibited by spironolactone and another is that glucocorticoids do not mimic the effects of aldosterone at nanomolar concentrations.
3. Rapid, nongenomic effects of aldosterone may also reflect actions via the classic intracellular MR. Such a scenario can be derived from experiments in vascular smooth muscle cells (Alzamora et al. 2000), cardiac myocytes (Mihailidou et al. 2000) and vascular endothelial cells (Golestaneh et al. 2001) discussed in more detail in a recent review (Funder 2001).

Table 1. Changes in nuclear envelope structure and function in response to aldosterone

Parameter	0–2 min after aldosterone	8–15 min after aldosterone	18–24 min after aldosterone
Nuclear envelope electrical resistance	Increased	Decreased	Increased
Nuclear pore configuration	Flags on pores	Pores free	Plugs in pores
Nuclear envelope dextran permeability	Unchanged	Strongly increased	Unchanged
Interpretation (Schafer et al. 2002; Oberleithner et al. 2003)	Mineralocorticoid receptor docking on nuclear pore surface blocking small peripheral channels	Transcription processes going on in the cell nucleus paralleled by nuclear swelling	Export of early transcripts (mRNA) through nuclear pore central channels

4. Aldosterone action may not even depend upon aldosterone secreted classically from the suprarenal glands but may be triggered by aldosterone released directly at the site of action. Indeed, recent experiments in endothelial cells and cardiomyocytes show that the complete cascade, the renin angiotensin aldosterone system (RAAS) , including MR, can be localized within the heart (Hatakeyama et al. 1994; Silvestre et al. 1998). This suggests that observations like mineralocorticoid-induced vascular remodeling and hypertension may stem from the paracrine and/or autocrine action of locally produced aldosterone released in response to angiotensin II derived from the vascular wall. In other words, the vascular bed is endowed with all of the cellular effectors that had been believed to exist exclusively in selected organs such as the kidney.

2.4 *Xenopus laevis* Oocyte

Insight into aldosterone action at subcellular level was recently gained by investigating a rather nonclassic target cell, namely *Xenopus laevis* oocyte, with AFM and electrical methods. The biological model, the technical approaches and the major experimental results have been published in detail previously (Danker et al. 1999; Shahin et al. 2001; Schafer et al. 2002). The reason why oocytes advanced to a suitable model for studying aldosterone action at molecular level was the following: Stage VI *Xenopus laevis* oocytes are inactive in terms of transcription but can be activated by exogenous stimuli (Golden et al. 1980; Leonard and La Marca 1975). They respond to aldosterone by triggering a fast signal cascade involving the cell nucleus. Some molecules of this signal cascade interact with the nuclear barrier and thus can be studied at single molecule level.

2.5 Nuclear Pores

In eukaryotic cells two concentric membranes in parallel separate the nucleus from the cytoplasm. This so-called nuclear envelope is punctured by nuclear pore complexes (NPCs) that serve as regulated pathways for macromolecules entering and exiting the nuclear compartment. Transport across NPCs (each NPC has a molecular mass of 120 MDa) occurs through central channels. Such import and export of macromolecules through individual NPCs could be elicited in *Xenopus laevis* oocyte by injection of aldosterone and visualized with AFM.

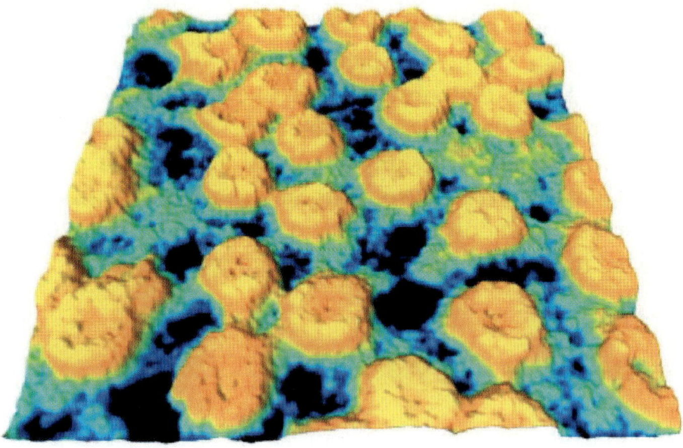

Fig. 1. Nuclear envelope of *Xenopus laevis* oocyte, a "biological nanoarray," imaged with atomic force microscopy (AFM). The area is about 600×600 nm. The image height (*z-axis*) is 15 nm. Nuclear pore complexes are the *yellow structures*. The lipid bilayer membrane visible between the pores is more or less *dark blue*

2.6 Flags on Pores

Mineralocorticoid receptors, usually located in the cytosol in an inactive state, move rapidly toward the cell nucleus upon activation. Two minutes after aldosterone injection into a *Xenopus laevis* oocyte a homogeneous population of macromolecules was found attached to the cytoplasmic surface of NPCs. These macromolecules were termed "flags" since they appeared "shaky" when scanned. Such a "shaky" appearance indicated a rather loose interaction of the macromolecule with the NPC. A few minutes later, flags had disappeared (Schafer et al. 2002). Flags could resemble the aldosterone receptors. From the volume of the individual macromolecules, measured by AFM (Henderson et al. 1996; Schneider et al. 1998; Pietrasanta et al. 1999; Berge et al. 2000), a molecular mass of about 100 kDa per molecule could be estimated. This is a fair match of the molecular mass of a monomeric MR (Arriza et al. 1987). Nevertheless, AFM can

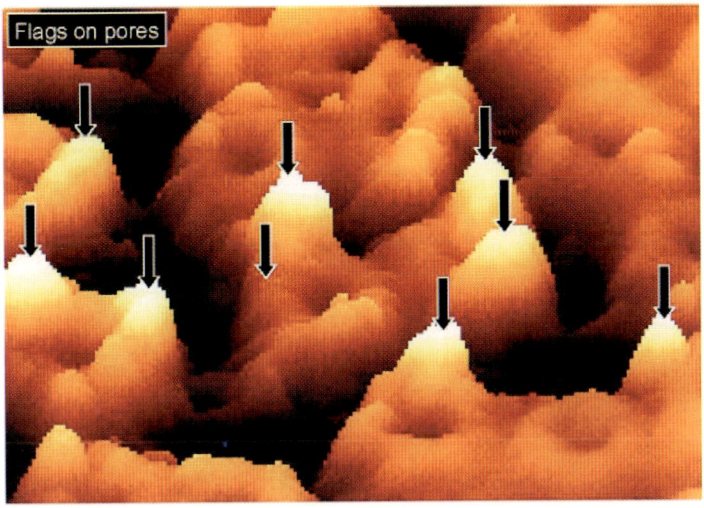

Fig. 2. Atomic force microscopy imaging of "flags" on nuclear pores. Putative mineralocorticoid receptors ("flags" indicated by *arrows*) attached to the rims of nuclear pores (ring-like structures with central channels) 2 min after aldosterone stimulation. Individual pores are about 80–100 nm. The image height (*z-axis*) is 10 nm. Flags (about 8 nm in height) disappear a minute later probably due to nuclear import

only estimate the volume with considerable scatter and there is yet no proof for any specificity. The fact that the putative receptors disappeared a few minutes after aldosterone injection was probably caused by transport through the NPC central channels into the nucleoplasm.

2.7 Plugs in Pores

Twenty minutes after hormone injection large plugs (≈1 MDa) appeared in the central NPC channels (Schafer et al. 2002). They could be ribonucleoproteins (RNPs) exiting the nucleus. If plugs were located close to the surface of individual NPC they could be detected by AFM. They had large dimensions (about 40 nm in diameter and 8 nm in height) and could be even harvested by the AFM stylus. Indications that plugs could be the first products of the genomic response came from the molecular mass that fairly matched the mass of rRNPs (Sperling et al. 1997). Further support for this theory could be derived from the delayed response, i.e., plug formation 20 min after hormone injection.

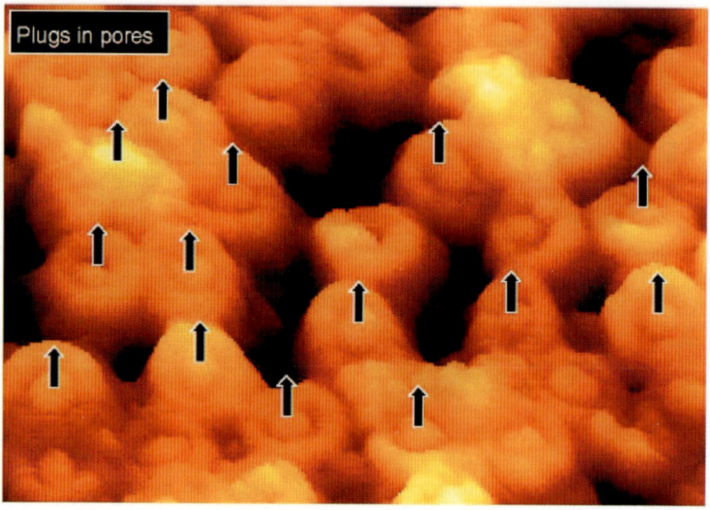

Fig. 3. Atomic force microscopy imaging of "plugs" in nuclear pores. Putative ribonucleoproteins ("plugs" indicated by *arrows* that point in export direction) located in the central channels of nuclear pores 20 min after aldosterone stimulation. Individual pores are dilated to outer diameters of about 120–140 nm. The image height (*z-axis*) is 10 nm. Plugs disappear a few minutes later probably due to export into the cytosol

2.8 Nuclear Pore Channels

Aldosterone transiently changed the electrical properties of the nuclear envelope (Schafer et al. 2002). The observed changes correlated with the morphological changes of the NPCs observed with AFM. Electrical resistance measurements in isolated nuclei revealed transient electrical nuclear envelope resistance peaks following aldosterone treatment, an early (2 min) peak and a late (20 min) peak. Electrical peaks reflected macromolecule interaction with NPCs. As worked out by elegant studies of Bustamante et al (Bustamante et al. 1995), macromolecules can serve as dielectric insulators when plugging the central channels. Thus, the early peak in nuclear envelope electrical resistance induced by aldosterone was not due to plugging of the central channel with putative MR but rather due to blockage of the so-called peripheral channels of NPCs. Evidence for such small peripheral channels located in the NPC rings surrounding the large central channel, came from electron microscopy (Hinshaw et al.

1992). We recently confirmed the existence of peripheral channels with AFM (Shahin et al. 2001). The fact that the 100 kDa molecules (flags) decorated the cytoplasmic rings explained the early peak in electrical resistance. The flags (presumably aldosterone receptors) physically blocked the small peripheral channels and prevented ion fluxes that finally determined nuclear envelope electrical resistance. The late peak was probably caused by the plugs that exited the central channels. Based on these experiments, a model was proposed in which the central channel (60 nm in length) is more or less always congested with macromolecules that prevent ion flux. The "real" diffusional ion permeability, however, is based on the function of small peripheral channels that can be in the open configuration (resting pores that allow ion equilibrium between cytosol and nucleoplasm), in the "blocked" configuration (macromolecules on pore surface occupying the entrances of the small peripheral channels and causing a transient ion disequilibrium between cytosol and nucleoplasm) or in the "squeezed" configuration (large RNPs exiting the central channels causing a transient ion disequilibrium between cytosol and nucleoplasm similar as mentioned above (for details see Mazzanti et al. 2001).

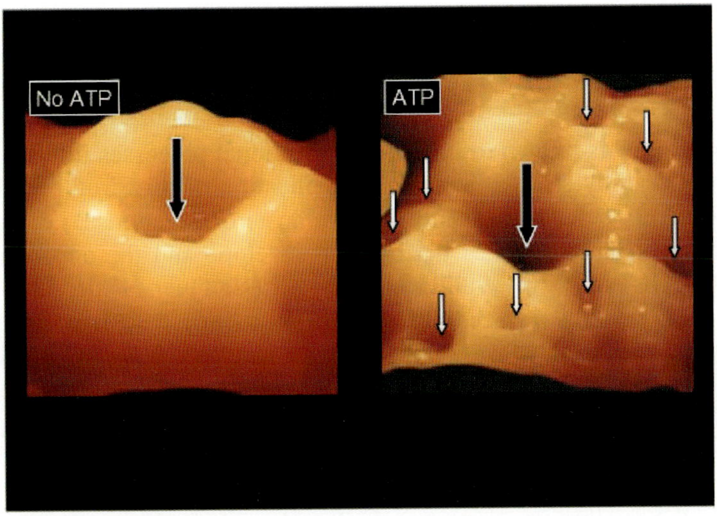

Fig. 4. Effect of adenosine triphosphate (*ATP*) (and calcium) on the conformation of individual NPCs of *Xenopus laevis* oocyte. ATP (plus calcium) induces the formation of so-called peripheral channels (*small arrows*) circularly arranged around the central channel (*large arrow*). Each AFM image is about 100 nm in width. The image height (*z-axis*) is 10 nm

2.9 Spironolactone

Spironolactone is an aldosterone antagonist that competes for the same binding site on the MR. Upon spironolactone binding the release of heat shock protein hsp 90 is facilitated leaving behind a destabilized form of MR (Couette et al. 1996). Surprisingly, structural as well as functional studies revealed that spironolactone interfered already with MR vectorial transport, directed from cytosol to nucleus. With AFM putative MRs (flags) could not be detected on the NPC surface if spironolactone was coinjected with aldosterone (Schafer et al. 2002). At the same time, the early electrical peak was missing. Both AFM and electrical data suggest that the putative MRs do not move towards the nuclear envelope if bound to spironolactone. Data were in agreement with studies using confocal fluorescence microscopy to trace the translocation of MRs from the cytosol into the cell nucleus (Fejes-Toth et al. 1998). The researchers convincingly showed that spironolactone strongly attenuated intranuclear MR accumulation. It was not surprising that spironolactone also inhibited the late response (i.e., the late electrical resistance peak and the plug appearance in the NPC central channels). This was the logic consequence, since the blocker already interfered with the pregenomic signal cascade.

2.10 Actinomycin D

Actinomycin D is known to inhibit transcriptional processes in the cell nucleus triggered by steroid hormones. Application of this drug in aldosterone-injected oocytes prevented the late aldosterone response, i.e., the late electrical resistance peak of the nuclear envelope and the appearance of plugs (rRNPs) in NPC central channels (Oberleithner et al. 2000). This observation strongly indicated that NPC plugs originate from transcriptional processes elicited by aldosterone. In contrast, Actinomycin D was ineffective at blocking the early response, which confirms that the early electrical peak usually observable after aldosterone injection was pregenomic in nature. The aldosterone-induced transcriptional process in the oocyte nucleus was found to have an electrical correlate. Electrical resistance of the nuclear envelope transiently decreased within a narrow time segment (Schafer et al. 2002). This electrical leak of the nuclear envelope was accompanied by a passive permeability leak for small macromolecules (Buchholz et al. 2004). The leak pathway induced by aldosterone lasted only minutes. Evidence for the transient formation of so-called megapores could be derived from recent AFM experiments on nuclear envelopes

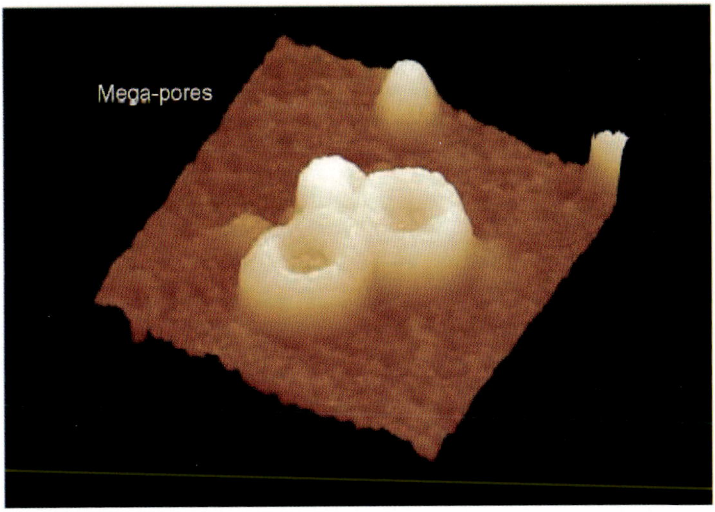

Fig. 5. Atomic force microscopy images of so-called megapores found in the nuclear envelope of *Xenopus laevis* oocyte 10 min after aldosterone injection. The megapores have outer diameters in the range of 1000 nm and a height of about 150 nm. They are composed of close to 100 individual nuclear pores (visible in the flat part of this image). The estimated lifetime of these large pores is in the range of a few minutes. They are candidates that mediate a large transient macromolecule permeability of the nuclear envelope

studied right at the time when aldosterone-induced transcription processes occurred in the nucleus (unpublished observation in the author's laboratory). Megapores resembled suprastructures in the nuclear envelope made up by a large number of individual NPCs arranged around a "megachannel."

2.11 Intracellular Calcium

A rise in intracellular free calcium is the typical early response of a target cell when exposed to aldosterone (Gekle et al. 1996; Schneider et al. 1997a; Harvey and Higgins 2000). It occurs few seconds after hormone application and is usually transient in nature. Beyond doubt, it is a pregenomic cellular event, that still lacks an explanation concerning its physiological relevance. Recent data indicate a potentially important role for intracellular free calcium (Schafer et al. 2003). When EGTA was coinjected with aldosterone, the early and the late electrical peaks were missing. Obviously, free calcium is necessary for an appropriate aldosterone response. Since the early electrical resistance increase reflects MR docking to NPCs it was assumed that MR docking is missing under low calcium conditions. The mechanism by which calcium ions could interfere with the pregenomic signaling pathway remains obscure. Calcium could be necessary for hormone–receptor interaction, for vectorial movement of activated receptors towards the nuclear envelope and for docking of HR to NPCs. Mineralocorticoid receptors are known to be associated to cytoskeletal structures (Jalaguier et al. 1996; Golestaneh et al. 2001). According to the literature aldosterone abolishes such interactions and obviously allows MRs to move on. This could be a crucial step of MR translocation into the cell nucleus. Calcium ions are supposed to play a role in this scenario. It is concluded that the intracellular calcium increase observed in response to aldosterone enables MRs to travel to NPCs. Then, MRs are translocated into the nucleoplasm. Only in presence of appropriate concentrations of ionized calcium can a cell properly respond to the steroid.

2.12 Endothelium

Some years ago it was observed in living aortic endothelial cells that aldosterone transiently increased cell volume (Schneider et al. 1997b). The aldosterone response occurred in minutes and could be inhibited by amiloride. Since AFM was applied in this previous study it was possible to analyze the three-dimensional morphology of the adherent endothelial cells together with cell volume. Although AFM images of living cells have only poor resolution it became obvious in a later analysis (Oberleithner et al. 2000) that the volume change mainly occurred in the cell nucleus. Based on these data it was postulated that volume "cycles" between intracellular compartments induced by aldosterone.

2.13 Nuclear Volume Cycling

To shift volume from the cytosol to the nucleoplasm, osmotic driving forces are necessary. Macromolecule movements occur between cytosol and nucleoplasm in response to steroid hormones. Most likely, changes in water balance are mediated by receptor import into the cell nucleus and export of transcribed mRNA into the cytosol. Such macromolecule shifts together with condensation/decondensation processes of the chromatin induced by steroid interaction with DNA response elements could explain changes in volume.

The changes of cell volume in response to aldosterone could be divided into two phases, a pregenomic and a genomic response (Oberleithner et al. 2003). The onset was immediate (within 2 min) and independent of the classical receptors since spironolactone was ineffective. This was typical for a pregenomic response. It smoothly intercalated with the genomic response indicated by sensitivity to spironolactone 5 min after hormone exposure. Noteworthy was the transient nature of the volume change and the biphasic response observed in the nuclear compartment. The sharp volume increase of the nuclear compartment at the onset of the hormone response and the sharp volume decrease about 15–20 min later strongly indicated that the volume change occurred in the nucleus. It matched the previous observations of receptor import into the nucleus within 2 min and mRNA export into the opposite direction after 20 min. We proposed a model where the aldosterone-induced initial nuclear swelling was indicative for receptor import while the late nuclear shrinkage indicated nuclear mRNA export (Oberleithner et al. 2003).

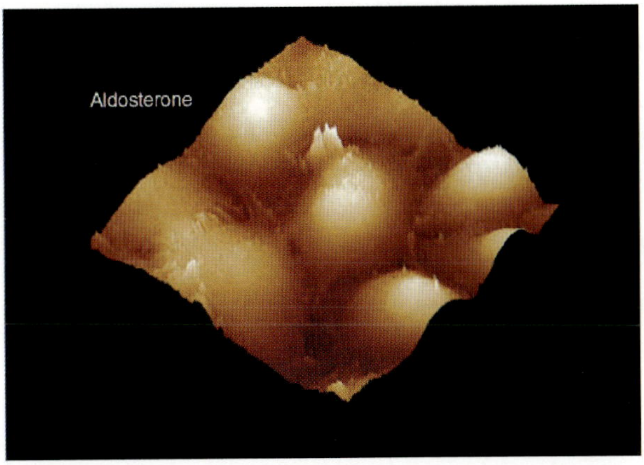

Fig. 6. Atomic force microscopy imaging of human umbilical vein endothelial cells (HUVEC). Shown is the apical endothelial surface. An individual HUVEC cell, exposed for 3 days to aldosterone, has a volume of about 2000 femtoliters. Clearly visible are the swollen nuclei (light upper parts of cells). Cell heights are in the range of 5 μm

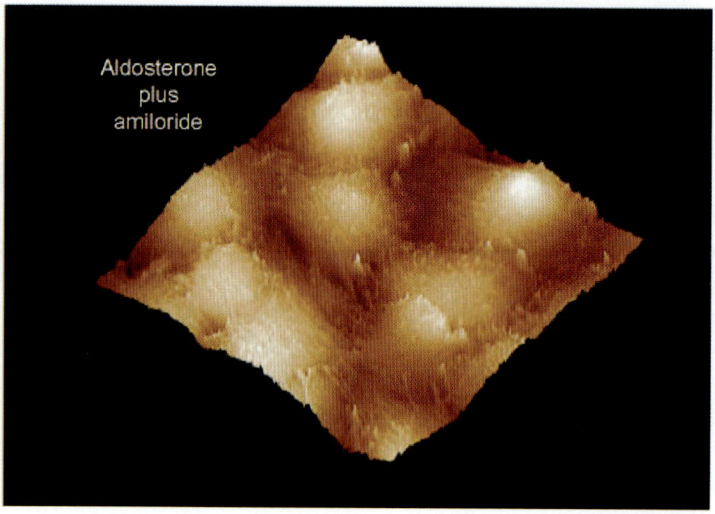

Fig. 7. Atomic force microscopy imaging of HUVEC. Shown is the apical endothelial surface. An individual HUVEC cell, exposed for 3 days to aldosterone and then treated with a plasma membrane sodium channel blocker (amiloride) for 1 h, has a volume of about 1400 femtoliters. Cells including nuclei appear shrunken as compared to cells in absence of the blocker (compare with Fig. 6). Cell heights are in the range of 3–4 µm

2.14 Amiloride

To our surprise, cariporide, a potent inhibitor of plasma membrane Na^+/H^+ exchange (Symons and Schaefer 2001), did not block the volume changes. Usually, the antiporter mediates sodium uptake whenever a cell needs to gain volume (Hayashi et al. 2002). This situation should happen at the onset of aldosterone action when receptors, accompanied by water, move into the nucleus and thus shrink the cytosolic compartment (while the nucleus swells). In contrast to cariporide, amiloride was effective. A low dose of amiloride wiped out the aldosterone-induced volume increase indicating that epithelial Na^+ channels mediated the volume changes across plasma membrane. Indeed, epithelial Na^+ channels have been shown to exist in vascular endothelial cells (Golestaneh et al. 2001). They are regulated by aldosterone and require an intact cytoskeleton.

2.15 Conclusion and Outlook

The nuclear envelope is a smart barrier. Transport of molecules across this barrier is mediated by perfectly tuned NPCs. Aldosterone and probably other steroid hormones regulate the nuclear pores and make them more or less permeable depending on the metabolic state of the cell. From the physiological point of view this knowledge could be useful in understanding the signal cascade underlying steroid hormone action. From the medical point of view a clearly defined, hormonally induced transient permeability change of the nuclear barrier could help to specifically deliver material from the cytosol into the nucleus. Such treatments could be eventually applicable also in the human organism.

References

Alzamora R, Michea L, Marusic ET (2000) Role of 11beta-hydroxysteroid dehydrogenase in nongenomic aldosterone effects in human arteries. Hypertension 35:1099–1104

Arriza JL, Weinberger C, Cerelli G, Glaser TM, Handelin BL, Housman, DE, Evans RM (1987) Cloning of human mineralocorticoid receptor complementary DNA: structural and functional kinship with the glucocorticoid receptor. Science 237:268–275

Berge T, Ellis DJ, Dryden DT, Edwardson JM, Henderson RM (2000) Translocation-independent dimerization of the EcoKI endonuclease visualized by atomic force microscopy. Biophys J 79:479–484

Binnig G, Quate CF (1986) Atomic force microscope. Phys Rev Lett 56(9):930–934

Buchholz I, Enss K, Schafer C, Shahin V, Albermann L, Schlune A, Oberleithner H (2004) Transient permeability leak of nuclear envelope induced by aldosterone. J Membr Biol. 199:135-141

Bustamante JO, Hanover JA, Liepins A (1995) The ion channel behavior of the nuclear pore complex. J Membr Biol 146:239–251

Couette B, Fagart J, Jalaguier S, Lombes M, Souque A, Rafestin-Oblin ME (1996) Ligand-induced conformational change in the human mineralocorticoid receptor occurs within its hetero-oligomeric structure. Biochem J 315 (Pt 2):421–427

Danker T, Schillers H, Storck J, Shahin V, Krämer B, Wilhelmi M, Oberleithner H (1999) Nuclear hourglass technique: novel approach detects electrically open pores in *Xenopus laevis* oocyte. Proc Natl Acad Sci USA 96:13531–13535

Delyani JA (2000) Mineralocorticoid receptor antagonists: the evolution of utility and pharmacology. Kidney Int 57:1408–1411

Epstein M (2001) Aldosterone and the hypertensive kidney: its emerging role as a mediator of progressive renal dysfunction: a paradigm shift. J Hypertens 19:829–842

Falkenstein E, Christ M, Feuring M, Wehling M (2000) Specific nongenomic actions of aldosterone. Kidney Int 57:1390–1394

Fejes-Toth G, Pearce D, Naray-Fejes-Toth A (1998) Subcellular localization of mineralocorticoid receptors in living cells: effects of receptor agonists and antagonists. Proc Natl Acad Sci USA 95:2973–2978

Funder JW (2001) Non-genomic actions of aldosterone: role in hypertension. Curr Opin Nephrol Hypertens 10:227–230

Gekle M, Golenhofen N, Oberleithner H, Silbernagl S (1996) Rapid activation of Na^+/H^+ exchange by aldosterone in renal epithelial cells requires Ca^{2+} and stimulation of a plasma membrane proton conductance. Proc Natl Acad Sci USA 93:10500–10504

Gerasimenko J, Maruyama Y, Tepikin A, Petersen OH, Gerasimenko O (2003) Calcium signalling in and around the nuclear envelope. Biochem Soc Trans 31:76–78

Golden L, Schafer U, Rosbash M (1980) Accumulation of individual pA+ RNAs during oogenesis of *Xenopus laevis*. Cell 22:835–844

Golestaneh N, Klein C, Valamanesh F, Suarez G, Agarwal MK, Mirshahi M (2001) Mineralocorticoid receptor-mediated signaling regulates the ion gated sodium channel in vascular endothelial cells and requires an intact cytoskeleton. Biochem Biophys Res Commun 280:1300–1306

Harvey BJ, Higgins M (2000) Nongenomic effects of aldosterone on Ca^{2+} in M-1 cortical collecting duct cells. Kidney Int 57:1395–1403

Hatakeyama H, Miyamori I, Fujita T, Takeda Y, Takeda R, Yamamoto H (1994) Vascular aldosterone. Biosynthesis and a link to angiotensin II-induced hypertrophy of vascular smooth muscle cells. J Biol Chem 269:24316–24320

Hayashi H, Szaszi K, Grinstein S (2002) Multiple modes of regulation of Na^+/H^+ exchangers. Ann N Y Acad Sci 976:248–258

Henderson RM, Oberleithner H (2000) Pushing, pulling, dragging, and vibrating renal epithelia by using atomic force microscopy. Am J Physiol Renal Physiol 278:F689–F701

Henderson RM, Schneider S, Li Q, Hornby D, White SJ, Oberleithner H (1996) Imaging ROMK1 inwardly rectifying ATP-sensitive K+ channel protein using atomic force microscopy. Proc Natl Acad Sci USA 93:8756–8760

Hinshaw JE, Carragher BO, Milligan RA (1992) Architecture and design of the nuclear pore complex. Cell 69:1133–1141

Jalaguier S, Mornet D, Mesnier D, Leger JJ, Auzou G (1996) Human mineralocorticoid receptor interacts with actin under mineralocorticoid ligand modulation. FEBS Lett 384:112–116

Leonard DA, La Marca MJ (1975) In vivo synthesis and turnover of cytoplasmic ribosomal RNA by stage 6 oocytes of *Xenopus laevis*. Dev Biol 45:199–202

Mazzanti M, Bustamante JO, Oberleithner H (2001) Electrical dimension of the nuclear envelope. Physiol Rev 81:1–19

Mihailidou AS, Bundgaard H, Mardini M, Hansen PS, Kjeldsen K, Rasmussen HH (2000) Hyperaldosteronemia in rabbits inhibits the cardiac sarcolemmal Na(+)-K(+) pump. Circ Res 86:37–42

Oberleithner H (2004) Unorthodox sites and modes of aldosterone action. News Physiol Sci 19:51-54

Oberleithner H, Brinckmann E, Schwab A, Krohne G (1994) Imaging nuclear pores of aldosterone sensitive kidney cells by atomic force microscopy. Proc Natl Acad Sci USA 91:9784–9788

Oberleithner H, Reinhardt J, Schillers H, Pagel P, Schneider SW (2000) Aldosterone and nuclear volume cycling. Cell Physiol Biochem 10:429–434

Oberleithner H, Schneider SW, Albermann L, Hillebrand U, Ludwig T, Riethmuller C, Shahin V, Schafer C, Schillers H (2003) Endothelial cell swelling by aldosterone. J Membr Biol 196:163–172

Oberleithner H, Ludwig T, Riethmüller C, Hillebrand U, Albermann L, Schafer C, Shahin V, Schillers H (2004) Human endothelium: target for aldosterone. Hypertension 43:952-956

Pietrasanta LI, Thrower D, Hsieh W, Rao S, Stemmann O, Lechner J, Carbon J, Hansma H (1999) Probing the *Saccharomyces cerevisiae* centromeric DNA (CEN DNA)-binding factor 3 (CBF3) kinetochore complex by using atomic force microscopy. Proc Natl Acad Sci USA 96:3757–3762

Roco MC (2003) Nanotechnology: convergence with modern biology and medicine. Curr Opin Biotechnol 14:337–346

Schafer C, Shahin V, Albermann L, Hug MJ, Reinhardt J, Schillers H, Schneider SW, Oberleithner H (2002) Aldosterone signaling pathway across the nuclear envelope. Proc Natl Acad Sci USA 99:7154–7159

Schafer C, Shahin V, Albermann L, Schillers H, Hug MJ, Oberleithner H (2003) Intracellular calcium: a prerequisite for aldosterone action. J Membr Biol 196:157–162

Schneider M, Ulsenheimer A, Christ M, Wehling M (1997a) Nongenomic effects of aldosterone on intracellular calcium in porcine endothelial cells. Am J Physiol 272:E616–E620

Schneider SW, Yano Y, Sumpio BE, Jena BP, Geibel JP, Gekle M, Oberleithner H (1997b) Rapid aldosterone-induced cell volume increase of endothelial cells measured by the atomic force microscope. Cell Biol Int 21:759–768

Schneider SW, Larmer J, Henderson RM, Oberleithner H (1998) Molecular weights of individual proteins correlate with molecular volumes measured by atomic force microscopy [In Process Citation]. Pflügers Arch 435:362–367

Shahin V, Danker T, Enss K, Ossig R, Oberleithner H (2001) Evidence for Ca^{2+}- and ATP-sensitive peripheral channels in nuclear pore complexes. FASEB J 15:1895–1901

Silvestre JS, Robert V, Heymes C, Aupetit-Faisant B, Mouas C, Moalic JM, Swynghedauw B, Delcayre C (1998) Myocardial production of aldosterone and corticosterone in the rat. Physiological regulation. J Biol Chem 273:4883–4891

Sperling R, Koster AJ, Melamed-Bessudo C, Rubinstein A, Angenitzki M, Berkovitch-Yellin Z, Sperling J (1997) Three-dimensional image reconstruction of large nuclear RNP (lnRNP) particles by automated electron tomography. J Mol Biol 267:570–583

Symons JD, Schaefer S (2001) Na(+)/H(+) exchange subtype 1 inhibition reduces endothelial dysfunction in vessels from stunned myocardium. Am J Physiol Heart Circ Physiol 281:H1575-H1582

3. Mitotic Chromosome Segregation Control

Yu Xue[1], Chuanhai Fu[1], Yong Miao[1], Jianhui Yao[1], Zhen Dou[1], Jie Zhang[1], Larry Brako[2], and Xuebiao Yao[1,2]

[1]Laboratory of Cell Dynamics, School of Life Sciences, University of Science & Technology of China, Hefei 230027, China

[2]Department of Physiology, Morehouse School of Medicine, Atlanta, GA 30310, USA

3.1 Introduction

The somatic division, called mitosis, is characterized by equal distribution of parental genome into two daughter cells. Mitosis involves a dramatic reorganization of both nucleus and cytoplasm driven by protein kinase cascades including master controller Cdk1-cyclin B. Mitosis is an ancient eukaryotic event, and some divergence emerged during evolution. Many single cell eukaryotes, including yeast and slime molds, undergo a closed mitosis, in which mitotic spindle formation and chromosome segregation occur within an intact nuclear envelope. However, higher eukaryotes such as animal and plant cells use open mitosis, in which nuclear envelope disassembles before the chromosomes segregate. This review primarily focuses on mitotic chromosome segregation in animal cells and refers to other organisms when regulation is mechanistically conserved. For convenience of discussion, mitotic chromosome dynamics are subdivided into six phases: prophase, prometaphase, metaphase, anaphase, telophase and cytokinesis.

This chapter highlights the research progress made over the past 10 years, which has sought to identify and illustrate specific roles for proteins involved in kinetochore dynamics, kinetochore–spindle interaction, and mitotic checkpoint, which underlie mitotic chromosome segregation control. Table 1 provides a summary of the various proteins that have been implicated in the regulation of chromosome segregation. In the course of this review, relevant earlier studies are briefly discussed and references given to related review articles. We discuss how centromeres and their kinetochores assemble, how they power mitotic chromosome movements, how, as signaling elements of the mitotic checkpoint, they control cell cycle advance during cell division.

Table 1. Summary of parietal cell proteins implicated in the cell activation process

Name	Yeast homologue	Cellular location	Presumed function
INNER KINETOCHORE			
CENP-A	CSE4	Hetero-chromatin	Centromere specification
CENP-B	PDC2	Inner kinetochore	α-Satellite DNA binding
CENP-C	CBF1	Inner kinetochore	Centromere biogenesis
DAXX	BIG1	Inner kinetochore	Binding to CENP-C in interphase
PASS1	NDC10	Inner kinetochore	Centromere specification
CENTRAL ZONE			
CENP-I	CFF3	Interzone kinetochore	Kinetochore assembly? Responsible for CENP-C targeting
HsMIS12	MTW1	Interzone kinetochore	Kinetochore assembly?
CENP-H	CBF1	Inner kinetochore	Centromere biogenesis
HEC1	TID3	Inner kinetochore	Responsible for TTK, Mad1 localization
NUF1	SPEC	Inner kinetochore	Centromere specification
HsSPC24	Spc24	Inner kinetochore	Centromere maturation
HsSPC25	Spec25	Inner kinetochore	Centromere maturation

Name	Yeast homologue	Cellular location	Presumed function
OUTER ZONE			
CENP-E	CIN8 (?)	Kinetochore corona	Chromosome segregation
			Spindle checkpoint signaling
MCAK	KIP3	Interzonal kinetochore	Spindle microtubule depolymerase
KIFC1	KAR3	Kinetochore corona (?)	Chromosome congression
DYNEIN complex	DYN1	Kinetochore corona	Chromosome segregation
Dynamitin	JNM1	Kinetochore corona?	Dynein complex assembly?
P150glued	NIP100	Kinetochore corona?	Dynein complex assembly?, MAPs
CH-TOG	STU2	Kinetochore corona?	Kinetochore–spindle association?
CLIP-170	BIK1	Kinetochore corona?	Kinetochore–spindle association?
CLASP1	STU1	Kinetochore corona?	Kinetochore–spindle association?
DKC1	CBF5	Kinetochore corona?	Kinetochore–spindle association?
PRC1?	ASE	Kinetochore corona?	Kinetochore–spindle association?
LIS1	LIS1		
Other outer KT proteins			
CENP-F	SLK19(?)	Kinetochore corona?	Kinetochore–spindle association?
53BP1	MLP1(?)	Kinetochore corona?	Kinetochore–spindle association?
HKLP2	RAD9	Kinetochore corona?	Kinetochore–spindle association?
REGULATORY ELEMENTS			
Aurora B	Ipl1	Kinetochore inner	Kinetochore assembly and protein modification
NCENP	SLI15	Kinetochore inner	Kinetochore protein active regulation
Survivin	BIR	Kinetochore inner?	

Name	Yeast homologue	Cellular location	Presumed function
SPINDLE CHECKPOINT			
HsBUB1	BUB1	Kinetochore	Kinetochore assembly and protein modification
HsBUB3	BUB3	Kinetochore	Kinetochore protein activity regulation
HsMAD1	MAD1	Kinetochore	Kinetochore protein activity regulation
HsMAD2	MAD2	Kinetochore	Kinetochore protein activity regulation
HsMAD3	MAD3	Kinetochore	Kinetochore assembly and protein modification
TTK	MPS1	Kinetochore	Kinetochore assembly and protein modification
HsCDC20	CDC20	Kinetochore?	APC/C activation
ZW10-ROD	BUB1	Kinetochore	Kinetochore assembly and protein modification
PP2A dephosphorylation	PPH22	Kinetochore	Kinetochore assembly and protein
SGO1	SGO1	Kinetochore	Kinetochore assembly and protein modification

3.2 Mitotic Spindle: An Elaborate Structure for Chromosome Segregation

Chromosome condensation, the landmark event at the onset of prophase, often begins at the nuclear periphery with simultaneous disassembly of nucleolus. In the cytoplasm, the interphase microtubule arrays emanating from a single centrosome is converted into two sets of astral arrays surrounding the duplicated centrosomes. These two asters then separate and migrate over the surface of the nucleus. Later, chromosomes condense into distinct paired threads, termed sister chromatids, which are closely paired along their entire length. Although chromosome condensation was first observed more than a century ago (e.g., Sutton 1903), the molecular mechanism is just beginning to be uncovered.

Mitosis involves a series of complex chromosome movements coordinated by the mitotic spindle. During prometaphase, spindle microtubules nucleated at centrosomes grow and shrink rapidly until they encounter and bind to a kinetochore via a "search and capture" process. If the sister kinetochore captures a spindle microtubule from the same pole, a syntelic attachment is created; if the captured microtubule is from the opposite pole, the attachment is bipolar. In metaphase, sister chromatids with bipolar attachments align in the middle of the spindle to form the metaphase plate. The spindle assembly checkpoint senses the state of chromosome-spindle attachment, delaying cell cycle progression until all pairs of sister chromatids have formed bipolar attachments. It is not yet known how monopolar, syntelic, and bipolar attachments are distinguished, but only bipolar attachments are stable and give rise to tension across paired sister kinetochores. When all pairs of sister chromatids have made stable bipolar attachments to microtubules, the mitotic checkpoint is silenced, the anaphase-promoting complex is then activated to degrade the cohesin. Chromatids then begin to move toward the opposite poles while maintaining end-on microtubule attachment. Chromosome-to-pole movements during anaphase A and subsequent separation of the poles during anaphase B create two equal and separated sets of sister chromatids. The rate at which this process goes wrong in yeast is on the order of 1×10^{-5} errors per chromosome per cell division, demonstrating the high fidelity of the mechanical and regulatory mechanisms that orchestrate chromosome segregation.

3.3 Kinetochore Dynamics in Mitosis

Chromosome movements on the spindle fibers during mitosis is powered and regulated by the kinetochore. The kinetochore is the site for spindle microtubule–centromere association. (In general, centromere refers to the highly repetitive DNA segment that confers centromere function while kinetochore often refers to the proteinaceous structure assembled onto the surface of the centromere.) Structurally, it is composed of four layers: an innermost plate that apparently consists of a specialized layer of chromatin, an interzone, an outer plate that has been argued to consist of tightly packed fibers (e.g., Rattner 1986), and an outermost fuzzy, fibrous corona that is most clearly seen after microtubule disassembly (e.g., Yao et al. 1997). Under electron microscopy, the kinetochore appears as a narrow band of dense chromatin just at the surface of the primary constriction, the inner kinetochore, and a laminar outer kinetochore domain that contains many of the microtubule binding and signal transduction molecules discussed below. The cohesin protein complex linking two sister kinetochores will be addressed as well.

Although the molecular composition and structure differ markedly respective to the kinetochore among various organisms, biochemical and genetic studies have revealed some common aspects of the organization, which provides a sketch of a simple "core centromere architecture." Chromatin is the key feature and the centromere domain is built on a unique class of nucleosome found only on centromere, in which atypical histone H3, referred to as CENP-A, governs the characteristics of the nucleosome (Smith 2002). CENP-A is a variant of histone H3 with more than 60% sequence identity at the C-terminal histone fold domain. Using CENP-A purified from HeLa cells, Takeyasu and colleagues employed atomic force microscopy to evaluate whether CENP-A can replace histone H3 in an in vitro nucleosome reconstitution assay (Yoda et al. 2000). They found that typical "beads on a string" images obtained from histone H3-organized nucleosomes were similar to those obtained with CENP-A-mediated assemblies. In fact, mononucleosomes isolated by glycerol gradient sedimentation had a relative molecular mass of approximately 200 kDa and were composed of 120–150 bp of DNA and equimolar amounts of CENP-A, and histones H4, H2A, and H2B. Thus, CENP-A forms an octameric complex with histones H4, H2A, and H2B in the presence of DNA, indicating the exchangeability between CENP-A and histone H3.

Great progress has been made toward a better understanding of the molecular composition of the kinetochore protein complex mediating the attachment of spindle microtubules to mitotic chromosomes in budding yeast (e.g., He et al. 2000). He and his colleagues show that four major protein complexes comprised of >20 components are central players in outer kinetochore assembly and microtubule binding. The Ctf19 complex and the Ndc80 complexes appear to represent "adaptors" that interact with both core centromere and distal kinetochore or spindle components (Ortiz et al. 1999; Wigge and Kilmartin 2001; Janke et al. 2001). The Dam1p complex (Cheeseman et al. 2001; Janke et al. 2002) (also known as DASH [Li et al. 2002]) may be the central component of microtubule binding activity, with its activity regulated by the yeast Aurora B kinase (Ipl1) (Biggins et al. 1999; Tanaka et al. 2002). Interestingly, Ipl1-dependent phosphorylation of the three Dam1p components is essential for microtubule capture of yeast kinetochore (Cheeseman et al. 2002a), suggesting a critical role of Ipl1 in kinetochore dynamics. While the yeast homologues of INCENP (Sli15) and survivin (Bir1) are highly conserved in eukaryotic cells (Bolton et al. 2002), it would be interesting to see whether animal cells use similar components to govern chromosome movements.

Mammalian cell centromere organization is in much more complex than that in the yeasts. For example, human centromeres contain extensive (1500 to >30 000 copies), tandemly repeated arrays of a 171 bp sequence element called α satellite. Centromere function has been mapped to α-satellite arrays by centromeric deletions, either naturally occurring on the X chromosome (Schueler et al. 2001) or those induced by telomere insertion into the Y (Brown et al. 1994). Centromeric satellite DNAs are not conserved in sequence among metazoans, but share: (1) their presence in very large tandem arrays, and (2) unit repeat lengths that tend toward multiples of the nucleosome repeat length (Henikoff et al. 2001).

CENP-A nucleosomes are bound to α-satellite DNA in human centromeres (Vafa and Sullivan 1997), but these are not uniformly distributed. Stretched chromatin fibers reveal interspersed CENP-A- and histone H3-containing nucleosomes (Blower et al. 2002). These foci may represent kinetochore "subunits" that assemble together to form multiple binding sites for the multiple microtubules that attach to mammalian centromeres (Zinkowski et al. 1991). In fact, our own attempt to dissection this complex structure using proteomic approach combined with in vitro reconstitution supports the notion that kinetochore is assembled by several protein subcomplexes (see below).

The fundamental link of centromere to a distinct histone also supports a role for chromatin structure in centromere determination. CENP-A appears to be at the foundation of the kinetochore assembly process and is required for assembly of most distal kinetochore components examined (Howman et al. 2000; Oegema et al. 2001). CENP-A assembled chromatin does not form a specialized domain of DNA replication (Shelby et al. 2000; Sullivan and Karpen 2001), and its loading is uncoupled from that of the conventional histones (Shelby et al. 1997; Takahashi et al. 2000).

3.4 Determinants of De Novo Centromere Formation

What determines the assembly of centromere? This has been answered, at least in part, by engineering artificial mammalian chromosomes. Minichromosomes were initially formed de novo by a mixture of a large synthetic chromosome 7 α-satellite array, telomeric sequences, a selectable marker, and genomic DNA fragments (Harrington et al. 1997). This produced stably transmitted microchromosomes with functional centromeres containing chromosome 7 α-satellite sequences and typical kinetochore components, but only in extremely rare frequency. Subsequently, a yeast artificial chromosome (YAC) carrying chromosome 21 α-satellite arrays was retrofitted with telomeres by recombination and introduced into human cells, which yields a much enhanced frequency of microchromosomes containing newly assembled functional centromeres, but only when the YAC contained α-I type satellite arrays, a frequent site marked by the CENP-B, a centromeric DNA-binding protein (Ikeno et al. 1998). This suggested a role for CENP-B in centromere biogenesis, an idea previously dismissed by demonstration that CENP-B null mice are viable and fertile with only mild phenotypic effects on gonad size (Hudson et al. 1998). By engineering two α-I satellite arrays such that they differed only by the presence of functional CENP-B boxes, CENP-B was shown to be necessary for de novo centromere formation, but it functions efficiently only in the context of α-satellite DNA. Since heterochromatin formation is a highly cooperative process driven by a series of mutually reinforcing reactions (Richards and Elgin 2002), it is likely that CENP-B functions in mediating centromeric chromatin modification, as demonstrated by the three CENP-B homologues in *Schizosaccharomyces pombe* (Nakagawa et al. 2002). Given the biochemical feature and complexity of kinetochore in mammalian cells and the availability of genomic database, proteomic analysis will shed light on molecular composition of kinetochore in mammalian cells. Precise delineation of centromere assembly machinery must await molecular dissection of kinetochore composition and reconstitution in vitro using sequentially added purified components.

Condensin, a complex of five proteins, is a major constituent of mitotic chromosome, and it appears to play an essential role for chromosome condensation (Hirano 2000). Phosphorylation of two condensin subunits by Cdk1-cyclin B stimulates their entry into the nucleus and association with chromatin as cells enter prophase. Condensin was discovered in a search for proteins that associated with mitotic chromosome using *Xenopus* egg extracts. Depletion of condensin caused defects in mitotic chromosome assembly, suggesting the importance of condensing in centromere formation and/or maturation. While the molecular function of condensin remains to be established, it seems that condensin induces a superhelical twist into DNA molecules (Hirano 2000). Recent studies reveal that the onset of condensation correlates to the phosphorylation of histone H1 by Cdk1-cyclin B and H3 by aurora-B. It has been hypothesized that local chromatin unfolding due to histone phosphorylation allows binding of other factors such as condensing proteins, which condense the chromosome.

To identify the molecular machinery underlying centromeric cohesion, Watanabe and his colleagues using fission yeast meiosis as a model system (Kitajima et al. 2004). Since meiosis produces haploid germ cells after a pair of specialized nuclear divisions, those authors reasoned that sister chromatids must segregate together during the first meiotic division (meiosis I), which requires that sister chromatid cohesion persists at centromeres. Using genetic assay, they indeed identified Sgo1 (shugoshin), a protector of the centromeric cohesin Rec8 in fission yeast. Moreover, they identified Sgo2, a paralogue of shugoshin in fission yeast, which is required for faithful mitotic chromosome segregation. Both Sgo1 and Sgo2 are well conserved in eukaryotic cells. Interestingly, localization of Sgo1 and Sgo2 at centromeres requires the kinase Bub1, identifying shugoshin as a crucial target for the kinetochore function of Bub1, which validates the role of Bub1 in maintaining chromosome stability during mitosis as mutation of Bub1 cause chromosome instability in human colorectal cell lines (Cahill et al. 1998; see below). These findings provide insights into the evolution of meiosis and kinetochore regulation during mitosis and meiosis. It would be interesting to delineate whether and how Bub1 kinase regulates Sgo1 and Sgo2 in metaphase I and II, respectively, and the precise function of Sgo2 in mitosis.

3.5 Kinetochores Power Chromosome Movements in Mitosis

It has been a century-long challenge and pursuit to understand how cells divide and faithfully transmit chromosomes at each cell division. In a typical somatic cell cycle, chromosomes in prophase initiate condensation, then, upon disassembly of the nuclear envelope and the interphase microtubule array, the fully compacted chromosomes spill into what was the cytoplasm to produce prometaphase. As a nascent mitotic spindle assembles, a dynamic process of repetitive search by unstable microtubules ensues for capture of chromosomes at their kinetochores (Fig. 1; modified). Initial capture is frequently by binding of one kinetochore of a duplicated chromosome pair along the side of a spindle microtubule (e.g., Yao et al. 1997), allowing rapid (up to 1 μm/s) poleward translocation along that microtubule powered by a minus-end directed, kinetochore bound microtubule motor, almost certainly cytoplasmic dynein (Rieder and Alexander 1990). This is followed by attachment of additional microtubules (up to 50 in humans [Rieder 1981] or 7 at mouse kinetochores [Putkey et al. 2002]), motor action at attachment sites, and oscillatory movements coupled to continued growth and shrinkage of those kinetochore bound microtubules.

Subsequent capture by the unattached kinetochore of a microtubule from the opposite spindle pole produces bi-orientation and congression to the cell center in a rapid, discontinuous series of movements, again mediated by kinetochore motors (e.g., Yao et al. 1997). The presence at kinetochores of active plus and minus end motors has been demonstrated, with net direction of movement redirectable in vitro by mitotic phosphorylation (Hyman and Mitchison 1991). In mammals, known functioning kinetochore motors include the kinesin family member CENP-E (e.g., Yao et al. 2000) and cytoplasmic dynein.

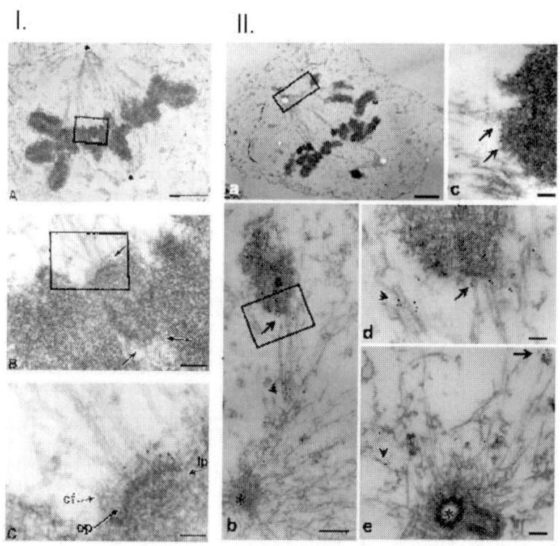

Fig. 1. Ultrastructural analyses of kinetochore-associated motor CENP-E and its binding protein TTK, a spindle checkpoint kinase. **I** CENP-E is a major component of corona fiber of the kinetochore. At metaphase CENP-E extends from the kinetochore outer plate at least 50 nm along spindle microtubules. **A** Low-magnification view of a metaphase HeLa cell with chromosomes aligned at the equator between the spindle poles (*asterisks*). **B** Magnified view of one metaphase chromosome showing that spindle microtubules indeed associate with a kinetochore with a trilaminar structure. Five 10-nm gold particles are located to each sister kinetochore (*arrows*). Five additional gold particles just to the right of the boxed area represent CENP-E associated with the kinetochore of another chromosome (more clearly seen in adjacent sections). **C** Higher magnification view shows that CENP-E is located to the corona fibers of the kinetochore. *op*, outer plate; *ip*, inner plate; *cf*, corona fibers. *Bars*: **A** 2 μm; **B** 170 nm; **C** 70 nm

Fig. 1. ***(continued)*** **II** TTK is localized at the developing kinetochore and spindle poles of mitotic HeLa cells. Visualization of TTK is achieved by 10-nm gold conjugated goat antirabbit IgG. **a** Low magnification of a prometaphase HeLa cells. *Asterisks* mark the two spindle poles of the bipolar spindle. An apparent mono-oriented chromosome is *boxed*, and higher magnification is shown in **b**. **b** Magnified view of shows that 10 nm particles represent the TTK localization at the kinetochore (*arrow*) and spindle microtubules adjacent to centrosome (*arrowhead*). **c** Enlarged view of **b** shows a bioriented chromosome with 10 nm gold particles deposit onto the kinetochore (*arrows*). **d** Higher magnification view of boxed area of **b** shows that 10 nm gold particles decorate the outer surface of the kinetochore interfaced with spindle microtubule (*arrow*). In addition, six 10-nm gold particles also mark an apparent protein complex associated with spindle microtubules (*arrowhead*). **e** Magnified view of a centrosome shows that a microtubule-associated protein complex decorated by eight 10-nm gold particles (*arrow*) traffics toward the pole. Several 10-nm gold particles (arrowhead) are also deposited to the pole 300–400 nm away from the centrioles. *Bars*: **a** 2 μm; **b** 500 nm; **c–e** 150 nm

3.6 Motors in Chromosome Congression

A long perplexing question is how chromosome congression is precisely achieved. Laser ablation to disconnect the two chromatids or destroy either kinetochore argues that the major force behind the chromosome motility is generated by the leading kinetochore (Khodjakov and Rieder 1996), i.e., the one whose bound microtubules are shortening and whose motors are moving toward the microtubule minus ends. Microinjection of p50 dynamitin and a dynein antibody, two soluble dynein inhibitors, disrupts the alignment of kinetochores at metaphase (Sharp et al. 2000). Mutations in the tethers (Rough deal [Rod] and Zeste white 10 [ZW10]) that link dynein to kinetochores attenuate the rate of poleward chromosome movement (Savoian et al. 2000), implicating dynein as a likely primary motor for congression. Depletion of CENP-E using antisense oligonucleotide effected a block of chromosomal alignment at the metaphase plate, indicating the essential role of CENP-E in chromosome congression and/or metaphase alignment (e.g., Yao et al. 2000). Indeed, chromosome alignment is precluded by disruption of CENP-E function in vitro using Xenopus egg extracts (Wood et al. 1997) and antibody injection in mammalian cells (Chan et al. 1997). A final class of kinetochore motor-like component, exemplified by the Kin I subgroup of the kinesin family (MCAK in mammals [Wordeman and Mitchison 1995] and XKCM1 in *Xenopus* [Walczak et al., 1996]), is a microtubule depolymerase [Desai et al. 1999]. Two most recent studies demonstrate that Aurora B phosphorylates and regulates MCAK activity both in vitro and in vivo (Andrews et al. 2004; Lan et al. 2004). Interestingly, Aurora B kinase activity was required for localization of MCAK to kinetochore, but not to spindle poles. Protein phosphorylation of serine 196 by Aurora B in the neck region of MCAK inhibited its microtubule depolymerization activity, perhaps due to confirmational changes. Using phospho-S196 specific antibody, this phosphorylation was shown at centromeres and anaphase spindle midzones in vivo. Addition of phospho-S196 antibodies to cultured cells or in vitro assembled spindle caused defects in chromosome positioning and/or segregation. It remains interesting to see whether Aurora B participates mitotic checkpoint by directing MCAK to depolymerize incorrectly oriented kinetochore microtubules, which allows removing mitotic brake.

Chromosomes also experience forces exerted along the chromosomes arms (polar wind) as a result of spindle microtubule interaction with plus-end directed microtubule motors bound to chromatin (chromokinesins). The first identified, *Drosophila* Nod, is required for proper alignment of meiotic chromosomes that have not undergone recombination (Afshar et al. 1995). Immunodepleting another (Kid) prevents normal chromosome alignment (Antonio et al. 2000; Funabiki and Murray, 2000), while antibody-induced inhibition of human Kid blocks chromosome oscillations, with chromosome arms atypically extending toward spindle poles during congression (Levesque and Compton 2001). It is generally believed that Kids provide an additional layer of surveillance for chromosome segregation.

Despite greater progress made toward molecular composition of kinetochore, the precise mechanics responsible for chromosome segregation remains elusive. Early biophysical and microscopic studies pointed to the role of dynein in chromosome congression. The ~50 pN force generated during chromosome movement (Alexander and Rieder 1991; Nicklas 1983, 1988) is equivalent to several dynein molecules per kinetochore (~6 pN per motor molecule [Shingyoji et al. 1998]). Since kinetochore is a multimotor protein complex, it is difficult to assign an individual motor for the complex chromosome movement based on evidence available. However, it would be possible and exciting to ascertain individual motor's function in different aspects of chromosome movements using inducible dominant negative mutation and real-time microscopic image.

3.7 Mitotic Complexities: Protein Subcomplexes and Circuitry

In close mitosis system, for example budding yeast, where the intranuclear spindle forms prior to centromere duplication during S phase, nonmotor microtubule-associated proteins appear to provide chief force for microtubule capture, while dynein plays a role in spindle positioning and perhaps maintaining, but not chromosome movement (Yeh et al. 1995). The Dam1p complex interacts physically with central kinetochore proteins of both the Ctf3 while Ndc80 complexes and binds to microtubules directly in vitro (Cheeseman et al. 2001), consistent with a direct role in mediating kinetochore–microtubule attachments. These authors further pursue the molecular regulation of kinetochore–microtubule by identifying Ipl1p targets using a combination of tandem affinity chromatography and mass spectrometry (Cheeseman et al. 2002b). Among 28 proteins recovered by this assay, ten of these phosphorylation proteins are targeted directly by Ipl1p. Their systematic mutational analysis of the Ipl1p phosphorylation sites demonstrated that the essential microtubule binding protein Dam1p is a key Ipl1p target for regulating kinetochore–microtubule attachments in vivo.

A group of microtubule plus-end binding proteins (e.g., the EB1 protein family and CLIP-170 in mammals) might also be involved in mediating interactions between microtubules and kinetochores. The yeast EB1 homologue BIM1 localizes to the plus ends of cytoplasmic microtubules and increases dynamic instability (Tirnauer et al. 1999) while removal of EB1 nails down the spindle microtubule stability (Rogers et al. 2002). During congression, EB1 is found at the ends of kinetochore of a chromatid pair which are bound to microtubules that are growing, but not at those that are shrinking (Tirnauer et al. 2002). EB1 interacts (Su et al. 1995) with the human adenomatous polyposis coli (APC) tumor suppressor protein. Adenomatous polyposis coli also binds to and stabilizes microtubules (Zumbrunn et al. 2001), localizes to the ends of microtubules embedded in kinetochores, forms a complex with mitotic checkpoint proteins Bub1 and Bub3, and is a substrate for both of the Bub1 and BubR1 kinases in vitro (Kaplan et al. 2001). As truncations of the *APC* gene are found in most colorectal tumors, Fodd and colleagues (2001) reasoned that mutations in *APC* might be responsible for chromosomal instability and examined mouse ES cells homozygous for *Min* (multiple intestinal neoplasia) or *Apc*1638T alleles. They show that *Apc* mutant ES cells display extensive chromosome and spindle aberrations, providing genetic evidence for a role of APC in chromosome segregation. Consistent with this, APC accumulates at the kinetochore during mitosis. *Apc* mutant cells form mitotic spindles with an abundance of microtubules that inefficiently connect with kinetochores. This phenotype is recapitulated by the induced expression of a 253-amino-acid carboxy-terminal fragment of APC in microsatellite unstable colorectal cancer cells. One possible explanation for these observations is that EB1/APC complex may be one of the nonmotor linker(s) that connect microtubule attachment and the spindle checkpoint signaling machinery on the kinetochore.

Although there has been a recent explosion in the identification of budding yeast kinetochore components, the physical interactions that underlie kinetochore function remain obscure, in particular to kinetochore of mammals, where open mitosis governs cell duplication. To better understand how mammalian cell kinetochores attach to microtubules and how this attachment is regulated, we sought to characterize the kinetochore composition of human cells using a combination of affinity chromatography and mass spectrometric analyses. Potential protein–protein interactions among kinetochore proteins were assessed using yeast hybrid screen and epitope tagging. Our current analyses provide a draft map for kinetochore protein subcomplexes at the kinetochore of human cells (Fig. 2). Further examination of these interactions in living cells will delineate protein–protein interaction circuitry at the kinetochore and consolidate these interactions into mitotic regulation, and elucidate how aberrant protein–protein interactions cause chromosome instability phenotype.

3.8 Anaphase Movements Driven by Motors and Flux

During anaphase identical sister chromatids separate and move towards opposite poles of the mitotic spindle. In the mitotic spindle, kinetochore microtubules have their plus ends embedded in the kinetochore and their minus ends at the spindle pole, revealed by photobleaching fluorescent microtubules (Gorbsky et al. 1987) and confirmed using fluorescence photoactivation of tubulin assembled into kinetochore bound microtubules (Mitchison and Salmon 1992), flux represents continuous addition of tubulin subunits at kinetochores, coupled to disassembly at the poles driven by plus-end directed, pole bound microtubule motors presumably pulling on kinetochore microtubules and sliding them poleward. The major mechanism for chromosome movement in anaphase in vertebrate somatic cells is motor-powered kinetochore movement coupled to microtubule disassembly at the kinetochore (e.g., Mitchison and Salmon 1992; Walters et al. 1996). Early studies show that poleward flux makes a relatively minor

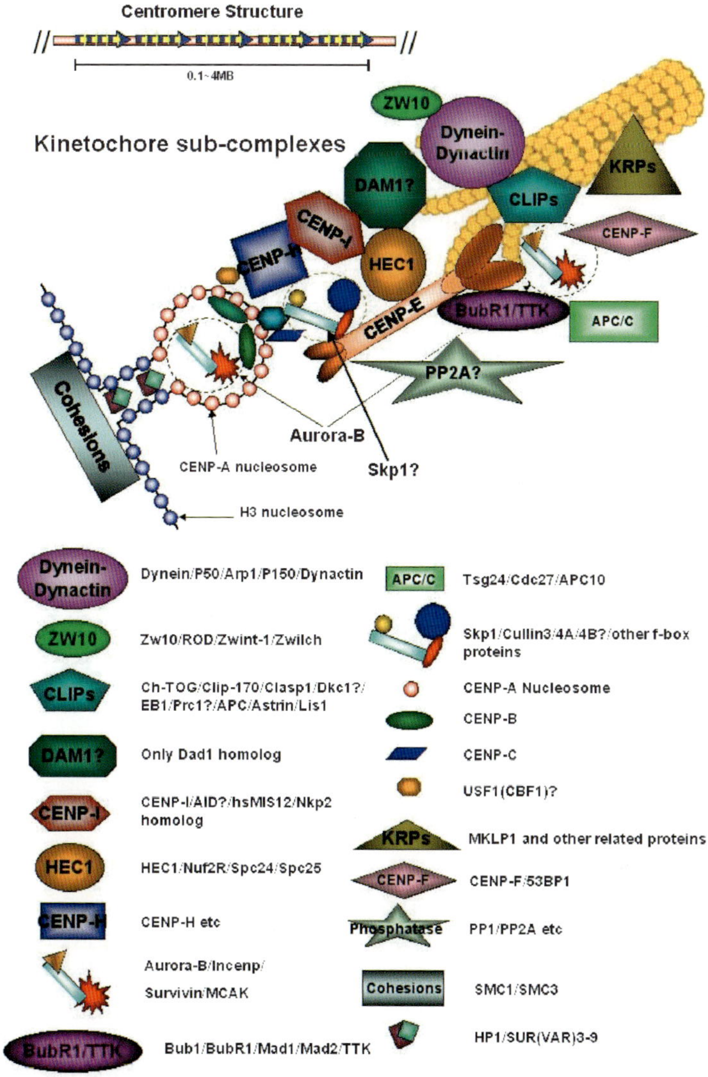

Fig. 2. Schematic drawing of hypothetical composition of kinetochore subcomplexes. Information presented here is derived from published studies and our own unpublished observation (pull-down assay, yeast genetic screen, and computational analyses)

contribution with chromosome-to-pole movement three to eight times faster than flux. Yeast appears to lack microtubule depolymerization at poles and poleward microtubule flux during anaphase (Mallavarapu et al.

1999). In *Xenopus* egg extracts, however, anaphase A movement occurs at rates similar to poleward spindle microtubule flux (Desai et al. 1998), consistent with flux as the predominant mechanism. Elsewhere, the situation is controversial: fluorescent speckle microscopy has been used to claim a dominant role for flux (Maddox et al. 2002) in syncytial *Drosophila* embryos, while other efforts have found that dynein inhibitors disrupt chromatid-to-pole movement during anaphase A (Savoian et al. 2000; Sharp et al. 2000).

Rogers and his colleagues have recently shown (2004) that two functionally distinct microtubule-destabilizing KinI kinesin enzymes are responsible for anaphase chromosome motion in *Drosophila*. One of them, KLP59C, is required to depolymerize kinetochore microtubules at their kinetochore-associated plus ends, thereby contributing to chromatid motility through a Pac-Man-based mechanism. The other, KLP10A, is required to depolymerize microtubules at their pole-associated minus ends, thereby moving chromatids by means of poleward flux. One question remain unanswered is how microtubules maintain their attachment to kinetochores and spindle poles while undergoing polymerization and depolymerization. Also, do other motile motors at the kinetochore (e.g., dynein and CENP-E) contribute to the process?

3.9 Signaling Cascade for Spatial-temporal Control of Chromosome Segregation

To assure accurate segregation, the mitotic checkpoint (also known as the spindle assembly checkpoint) acts to block entry into anaphase until both kinetochores of every duplicated chromatid pair have attached correctly to spindle microtubules. It has been proposed that unattached kinetochores and/or those not under microtubule-exerted tension are the central signaling elements that produce a "wait anaphase" signal. By filming mitoses, it was initially found that anaphase ensues about 20 min after the last kinetochore attaches to the spindle (Rieder et al. 1994) and that by repeated detachment of a meiotic chromosome from a spindle by manipulation with a microneedle delayed anaphase indefinitely (Li and Nicklas 1995). That it was an unattached kinetochore that was responsible came from the seminal demonstration that laser ablation of the last unattached kinetochore produces anaphase onset within about 15 min (Rieder et al. 1995). A kinetochore-dependent wait anaphase signal is also suggested in budding yeast: blocking centromere assembly (by destruction of the Cbf3 component

Ndc10) eliminates mitotic delay in the presence of microtubule assembly inhibitors (Gardner et al. 2001).

Yeast genetic screen initially identified seven components of the mitotic checkpoint, Mad1–Mad3 (Mitotic arrest defective) (Li and Murray 1991), Bub1–Bub3 (Budding uninhibited by benomyl) (Hoyt et al. 1991), and Mps1, a kinase that is also essential for spindle pole body duplication (Weiss and Winey 1996). There are vertebrate homologues of all of these except Bub2. As initially demonstrated for Mad2 (Chen et al. 1996; Li and Benezra 1996), other mitotic checkpoint proteins have now been identified to bind to and act at unattached kinetochores including Mad1 (Chen et al. 1998), Bub1 (Taylor and McKeon 1997), Bub3 (Taylor et al. 1998), BubR1 (the mammalian Mad3) (Taylor et al. 1998), and Mps1 (Abrieu et al. 2001). In mammals, there are several additional components involving mitotic checkpoint. For example, loss of the kinetochore-associated microtubule motor protein CENP-E, a binding partner of BubR1, the checkpoint cannot be established or maintained *Xenopus* egg extracts (Abrieu et al. 2000), HeLa cells (Yao et al. 2000), in mice (Putkey et al. 2002). Most recently, Cleveland and his colleagues show that single unattached kinetochores due to depletion of CENP-E cannot block entry into anaphase but result in aneuploidy in 25% of divisions in primary mouse fibroblasts in vitro and in 95% of regenerating hepatocytes of knock-out mice. Significantly, they further demonstrate that CENP-E binds to and directly stimulates the kinase activity of purified BubR1 in vitro. Thus, CENP-E is required for enhancing recruitment of its binding partner BubR1 to each unattached kinetochore and for stimulating BubR1 kinase activity, implicating it as an essential amplifier of a basal mitotic checkpoint signal. It would be interesting to further establish the inter-relationship of BubR1–CENP-E and provide structural view as how CENP-E association enhances BubR1 activity in vitro and whether this holds in living cells.

Although the nature of the direct molecular interaction(s) between checkpoint proteins and kinetochores and the inter-relationship among checkpoint proteins has not been determined, the basic scheme of the signaling cascade has been established. Mad2 is recruited to unattached kinetochores in a complex with Mad1 (Chen et al. 1998). BubR1 and Bub1, both kinases, are required for generation and then rapid release from kinetochores of one or more inhibitors of Cdc20 (Fizzy in flies [Dawson et al. 1993], p55 in mammals [Kallio et al. 1998] or Slp1p in fission yeast [Kim et al. 1998]). Interestingly, recent studies show that TTK (human homologue of yeast MPS1) interacts with CENP-E, a mitotic kinesin located to corona fiber of kinetochore (Zhang et al. 2002). To elucidate the molecular function of TTK1, Dou and his colleagues (2003) conducted untrastructural studies and revealed its dynamic distribution profile. TTK is present at the nuclear pore adjacent complex of interphase HeLa cells. Upon nuclear envelope fragmentation, TTK targets to the outermost region of the developing kinetochores of mono-orient chromosome as well as to spindle poles. After stable attachment, throughout chromosome congression. TTK is a constituent of the corona fibers, extending up to 90 nm away from the kinetochore outer plate. Upon metaphase alignment, TTK departs from the kinetochore and migrates toward the centrosomes. Taken together, this evidence strongly supports a model in which TTK functions in spindle checkpoint signaling cascades at both kinetochore and centrosome, which were supported by several independent studies (Liu et al. 2003; Fisk et al. 2003).

In an attempt to elucidate the signaling cascade of Mad1, Lou and his colleagues (2004) carried out yeast genetic screen to identify Nek2A as a potential binding partner. Chromosome segregation in mitosis is orchestrated by protein kinase signaling cascades. Although it was established that Nek2A is a centrosome-associated protein kinase. Like Mad1, Nek2A is localized to HeLa cell kinetochore of mitotic cells. Significantly, elimination of Nek2A by siRNA does not arrest cells in mitosis but causes aberrant premature chromosome segregation. Nek2A is required for Mad2 but not Mad1, Bub1 and Hec1 to associate with kinetochores. Moreover, loss of Nek2A impairs mitotic checkpoint signaling in response to spindle damage by nocodazole, which effected mitotic escape and led to generation of cells with multiple nuclei. These studies demonstrate that Nek2A is a kinetochore-associated protein kinase essential for faithful chromosome segregation. The dynamic distribution of Nek2A at kinetochore and centrosome provide another example of checkpoint molecular dynamics linking between kinetochore and centrosome.

The recruitment of Mad2 at kinetochores has often been taken as a measure of ongoing checkpoint signal generation (and release). However, inhibition of ZW10/Rod yields an inactive checkpoint despite prominent Mad2 binding at kinetochores (Chan et al. 2000). Diminution of Hec1, the homologue of yeast Ndc80, on the other hand, yields a chronically activated checkpoint with no Mad2 bound to kinetochores (Martin-Lluesma et al. 2002). Thus, it is now abundantly clear that the steady-state level of Mad2 bound to kinetochores is not a faithful reporter for checkpoint activation or inactivation.

Another candidate for spindle checkpoint reporters is BubR1, which directly binds Cdc20 and APC/C elements Cdc16, Cdc27, and APC7 (Chan et al. 1999; Wu et al. 2000) and by doing so can block Cdc20 activation of APC/C for mitotic substrates. The inhibitory activity has been argued to be BubR1 alone without a contribution of Mad2 (Tang et al.

2001) or in a complex (named MCC) that apparently contains stoichiometric amounts of BubR1, Bub3, Mad2, and Cdc20 (Sudakin et al. 2001). An equivalent quaternary complex containing Mad3 (the yeast BubR1), Bub3, Mad2, and Cdc20 has also been observed in budding yeast (Fraschini et al. 2001; Hardwick et al. 2000). Further, the much higher APC/C inhibitory activity in vitro of MCC (>3000-fold more potent than tetrameric Mad2) (Sudakin et al. 2001) would make this complex an attractive candidate for a diffusible inhibitory signal were it not that it (and its yeast counterpart) is formed in a kinetochore-independent manner (Sudakin et al. 2001; Fraschini et al. 2001). This has led to the suggestion that the kinetochore contribution may modify APC/C itself to increase its affinity for MCC (Sudakin et al. 2001). However, equally plausible are models with spatial signal amplification, for example, MCC gradient near kinetochores, thereby producing a high concentration of an inhibitor to saturate available Cdc20. In fact, such intracellular gradients can be evaluated using a FRET assay.

3.10 Silencing the Checkpoint Signal: Attachment and Tension

There is a continuing debate as to whether the mitotic checkpoint is silenced by microtubule attachment (Rieder et al. 1995) or by the tension exerted between bioriented kinetochore pairs after attachment (Li and Nicklas 1995) and whether activities of subsets of the known components are selectively silenced by one or the other (Skoufias et al. 2001; Zhou et al. 2002).

McIntosh initially proposed the tension model under which the mechanical tension generated by poleward-directed forces acting on both kinetochores of a bioriented chromatid pair turns off checkpoint signaling (McIntosh 1991). Compelling evidence for a tension requirement initially emerged using praying mantis spermatocytes in meiosis I, which have a Y chromosome and two genetically different X chromosomes. Although the Y is supposed to pair with both X chromosomes, occasionally it pairs only with one, thus yielding a mono-oriented X. When this occurs, anaphase onset is delayed by up to 9 hr but can be triggered to initiate almost immediately by application of mechanical tension applied across the mono-oriented kinetochore by use of a force-calibrated microneedle (Li and Nicklas 1995).

This is a general finding in meiosis: eliminating tension between homologous chromosomes by preventing recombination delayed anaphase onset in yeast, presumably from checkpoint activation. This delay was eliminated by genetically allowing sister kinetochores to inappropriately separate during meiosis I, thereby allowing each homologue to biorient in the absence of recombination and restore tension across the sisters (Shonn et al. 2000). Similarly, in cells that enter mitosis without a prior round of DNA replication, the unreplicated chromatids attach to spindle microtubules, but no tension can be developed and the mitotic checkpoint is chronically activated (Stern and Murray 2001).

However, in the initial demonstration that kinetochores are the signaling elements for the checkpoint, destruction of the last unattached kinetochore eliminated checkpoint signaling despite lack of tension on the kinetochore of the sister chromatid (Rieder et al. 1995). Similarly, in maize, satisfying the checkpoint requires tension in meiosis, but in mitosis, attachment is sufficient (Yu et al. 1999). Furthermore, in PtK1 cells, loss of Mad2 recruitment to kinetochores depends on microtubule attachment, not tension (Waters et al. 1998). Similarly, very low doses of the microtubule inhibitor vinblastine produce loss of tension and kinetochore bound Mad2 and a 'checkpoint' arrest that, unlike the case after inhibition of spindle assembly (Gorbsky et al. 1998), is insensitive to inactivation with Mad2 antibody injection (Skoufias et al. 2001).

This has led to the proposal that there are two branches of checkpoint signaling and silencing. One, in which the signal released is an activated, inhibitory form of Mad2, is silenced by microtubule attachment, while the other, presumably involving conversion of BubR1 into a Cdc20 inhibitor, is silenced by tension. The challenge ahead is to distinguish these two possibilities and illustrate the checkpoint signaling regulation.

3.11 Mitotic Checkpoint Defects Promote Aneuploidy and Tumorigenesis

The mitotic checkpoint in budding yeast, where the intranuclear spindle forms quickly after centromeres are duplicated in S, is a real checkpoint activated only to arrest mitosis in the relatively rare instances when attachment is delayed. However, in most other higher organisms such as mammals, it becomes an essential cell cycle control pathway activated at every mitosis/meiosis immediately upon nuclear envelope disassembly. Loss of Mad2, Bub3, or Rae1 in mice is lethal early, with cells accumulating mitotic errors and undergoing apoptosis by embryonic day 5 or 6 (Dobles et al. 2000; Kalitsis et al. 2000; Babu et al. 2003). Similarly, in *Drosophila*, loss of Bub1 causes chromosome mis-segregation and lethality (Basu et al. 1999). Microinjection of antibodies to Mad2 yields premature anaphase onset and chromosome mis-segregation (Gorbsky et al. 1998). Haplo-insufficiency in Mad2 provokes late onset, self-limiting lung tumors (Michel et al. 2001). Elimination of Nek2A effects insufficient targeting and/or retention of Mad2 to the kinetochore and causes premature chromosome segregation in cultured cells, which displays chromosome instability phenotype (Lou et al. 2004). Reduction in Bub3 or Rae1 generates aneuploidy in vitro and a sharply increased susceptibility to chemically induced tumorigenesis (Babu et al. 2003), suggesting the checkpoint control is essential for mitotic regulation. While the primary mission of the checkpoint is to prevent such errors in chromosome segregation, a hallmark of human tumor progression (Hartwell and Kastan 1994), it would be of great interests to illustrate how perturbation of kinetochore protein–protein interaction effects chromosome instability.

3.12 The Epigenetic Regulation of Centromere Structure and Function

Human centromere formation involves the assembly of the mitotic kinetochore onto chromosomal locations that contain the interphase prekinetochore. In budding yeast, centromere DNA alone can nucleate centromere formation de novo, centromeres in *S. pombe* depend strongly, and those of metazoan cells primarily, on epigenetic factors rather than DNA sequence for activity. Three lines of evidence support this statement. First, centromere sequences are evolving at an unusually high rate, coevolving with their essential partner CENP-A, and show no obvious sequence conservation that links divergent species or even different chromosomes in the case of *Drosophila* (Henikoff et al. 2001). Second, centromere DNA sequences by themselves are unable to specify centromere function: stable dicentric chromosomes have been found in which one centromere has been silenced with no obvious rearrangement of centromere DNA (Sullivan and Willard 1998). Third, acquisition of centromere function has been found on certain rearranged chromosomes lacking an endogenous centromere.

Syndromes of disordered "chromatin remodeling" are unique in medicine because they arise from a general deregulation of DNA transcription caused by mutations in genes encoding enzymes which mediate changes in chromatin structure. Chromatin is the packaged form of DNA in the eukaryotic cell. It consists almost entirely of repeating units, called nucleosomes, in which short segments of DNA are wrapped tightly around a disk-like structure comprising two subunits of each of the histone proteins H2A, H2B, H3 and H4. Histone proteins are covalently modified by a number of different adducts (i.e., acetylation and phosphorylation) that regulate the tightness of the DNA–histone interactions. Mutations in genes encoding enzymes that mediate chromatin structure can result in a loss of proper regulation of chromatin structure, which in turn can result in deregulation of gene transcription and inappropriate protein expression. Significantly, there are several diseases whose defects in chromatin remodeling are tied to chromosome instability (e.g., Rett syndrome (RS); immunodeficiency-centromeric instability–facial anomalies syndrome). Molecular elucidation of these regulatory elements will not only provide insights into our understanding of centromere assembly but shed light on pathogenesis of the aforementioned diseases.

The tumor suppressor gene RASSF1A is frequently silenced in lung cancer and other sporadic tumors as a result of hypermethylation of a CpG island in its promoter. However, the precise mechanism by which RASSF1A functions in cell cycle regulation and tumor suppression has remained unknown. Song and his colleague show that RASSF1A regulates the stability of mitotic cyclins and the timing of mitotic progression by interacting with Cdc20, an activator of the APC/C, resulting in the inhibition of APC/C activity. Although RASSF1A does not contribute to either the Mad2-dependent spindle assembly checkpoint or the function of Emi1, depletion of RASSF1A by RNA interference accelerated the mitotic cyclin degradation and mitotic progression as a result of premature APC activation. It also caused a cell division defect characterized by centrosome abnormalities and multipolar spindles. Thus, these findings link epigenetic regulation of spindle protein activity to mitotic progression.

3.13 Conclusions

It becomes increasingly clear that the centromere and its associated kinetochore are much more than simple attachment sites for spindle microtubules. Mammalian centromeres are much more complex than initially imagined, representing repeated assemblies of the simple, one-nucleosome centromeres in budding yeast. Central to genetic inheritance, in almost all examples known they are also epigenetically determined and regulated. In addition, kinetochore regulators include active components in microtubule capture, stabilization, and in powering chromosome movements essential for faithful segregation. More than that, they are the signaling elements for controlling cell cycle advance through mitosis by as yet identified protein–protein interaction circuitry.

3.14 Future Directions

In the next few years, we can expect proteomic and biochemical analyses to generate a complete list of kinetochore components and subcomplexes, encompassing perhaps 150–180 proteins, and interaction circuitry about the modes and roles of these proteins in orchestrating sophistic chromosome dynamics in dividing cells. Real-time studies will also provide a framework of kinetochore architecture and dynamics at various stages of cell cycle. Given the spatial-temporal trait of protein–protein interactions, we believe that functional studies will require the development of high-throughput image robotics for studying real-time protein dynamics in live-cell chromosome movements in wild-type and "mutant" cells with high precision. It would be exciting and challenging task to consolidate the protein–protein interaction informatics from different genetic and biochemical background into a model for kinetochore dynamics in mitotic chromosome segregation. It will be also necessary to develop new in vitro assays to reconstitute and evaluate the interactions of kinetochore complexes to mimic kinetochore dynamics of mitosis. A combination of proteomic, genetic, biochemical and biophotonic analyses with computational modeling will enable us to consolidate the mechanistic view of heredity envisioned more than 100 years ago.

3.15 Acknowledgments

We thank members of our groups for insightful discussion during the course of this study and apologize to the many colleagues whose work we were unable to cite due to the space limit. This work was supported by grants from Chinese Natural Science Foundation (39925018) and the American Cancer Society (RPG-56292) to X.Y.

References

Abrieu A, Kahana JA, Wood KW, Cleveland DW (2000) CENP-E as an essential component of the mitotic checkpoint in vitro. Cell 102:817–826

Abrieu A, Magnaghi-Jaulin L, Kahana JA, Peter M, Castro A, Vigneron S, Lorca T, Cleveland DW, Labbe J (2001) Mps1 is a kinetochore-associated kinase essential for the vertebrate mitotic checkpoint. Cell 106:83–93

Andrews PD, Ovechkina Y, Morrice N, Wagenbach M, Duncan K, Wordeman L, Swedlow JR. (2004) Aurora B regulates MCAK at the mitotic centromere. Dev Cell 6:253–68

Afshar K, Barton NR, Hawley RS, Goldstein LS (1995) DNA binding and meiotic chromosomal localization of the Drosophila nod kinesin-like protein. Cell 81:129–138

Ando S, Yang H, Nozaki N, Okazaki T, Yoda K (2002) CENP-A, -B, and -C chromatin complex that contains the I-type alpha-satellite array constitutes the prekinetochore in HeLa cells. Mol Cell Biol 22:2229–2241

Antonio C, Ferby I, Wilhelm H, Jones M, Karsenti E, Nebreda AR, Vernos I (2000) Xkid, a chromokinesin required for chromosome alignment on the metaphase plate. Cell 102:425–435

Ault JG, Nicklas RB (1989) Tension, microtubule rearrangements, and the proper distribution of chromosomes in mitosis. Chromosoma 98:33–39

Babu JR, Jeganathan KB, Baker DJ, Wu X, Kang-Decker N, van Deursen JM (2003) Rae1 is an essential mitotic checkpoint regulator that cooperates with Bub3 to prevent chromosome missegregation. J Cell Biol 160:341–353

Basu J, Bousbaa H, Logarinho E, Li Z, Williams BC, Lopes C, Sunkel CE, Goldberg ML (1999) Mutations in the essential spindle checkpoint bub1 cause chromosome missegregation and fail to block apoptosis in Drosophila. J Cell Biol 46:13–28

Barry AE, Howman EV, Cancilla MR, Saffery R, Choo KH (1999) Sequence analysis of an 80 kb human neocentromere. Hum Mol Genet 8:217–227

Basto R, Gomes R, Karess R (2000) Rough Deal and ZW10 are required for the Metaphase checkpoint in Drosophila. Nat Cell Biol 2:939–943

Bernard P, Maure JF, Partridge JF, Genier S, Javerzat JP, Allshire RC (2001) Requirement of heterochromatin for cohesion at centromeres. Science 294:2539–2542

Biggins S, Severin FF, Bhalla N, Sassoon I, Hyman AA, Murray AW (1999) The conserved protein kinase Ipl1 regulates microtubule binding to kinetochores in budding yeast. Genes Dev 13:532–544

Blower MD, Sullivan BA, Karpen GH (2002) Conserved organization of centromeric chromatin in flies and humans. Dev Cell 2:319–330

Bolton MA, Lan W, Powers SE, McCleland ML, Kuang J, Stukenberg PT (2002) Aurora B kinase exists in a complex with survivin and INCENP and its kinase activity is stimulated by survivin binding and phosphorylation. Mol Biol Cell 13:3064–3077

Cahill DP, Lengauer C, Yu J, Riggins GJ, Willson JK, Markowitz SD, Kinzler KW, Vogelstein B (1998) Mutations of mitotic checkpoint genes in human cancers. Nature 392:300–303

Chan GK, Jablonski SA, Sudakin V, Hittle JC, Yen TJ (1999) Human BUBR1 is a mitotic checkpoint kinase that monitors CENP-E functions at kinetochores and binds the cyclosome/APC. J Cell Biol 146:941–954

Chan GK, Jablonski SA, Starr DA, Goldberg ML, Yen TJ (2000) Human Zw10 and ROD are mitotic checkpoint proteins that bind to kinetochores. Nat Cell Biol 2:944–947

Cheeseman IM, Brew C, Wolyniak M, Desai A, Anderson S, Muster N, Yates JR, Huffaker TC, Drubin DG, Barnes G (2001) Implication of a novel multiprotein Dam1p complex in outer kinetochore function. J Cell Biol 155:1137–1145

Cheeseman IM, Anderson S, Jwa M, Green EM, Kang J, Yates JR, Chan CS, Drubin DG, Barnes G (2002a) Phospho-regulation of kinetochore-microtubule attachments by the aurora kinase ipl1p. Cell 111:163–172

Cheeseman IM, Drubin DG, Barnes G (2002b) Simple centromere, complex kinetochore: linking spindle microtubules and centromeric DNA in budding yeast. J Cell Biol 157:199–203

Chen RH, Waters JC, Salmon ED, Murray AW (1996) Association of spindle assembly checkpoint component XMAD2 with unattached kinetochores. Science 274:242–246

Chen RH, Shevchenko A, Mann M, Murray AW (1998) Spindle checkpoint protein Xmad1 recruits Xmad2 to unattached kinetochores. J Cell Biol 143:283–295

Clarke L (1998) Centromeres: proteins, protein complexes, and repeated domains at centromeres of simple eukaryotes. Curr Opin Genet Dev 8:212–218

Cohen-Fix O, Peters JM, Kirschner MW, Koshland D (1996) Anaphase initiation in Saccharomyces cerevisiae is controlled by the APC-dependent degradation of the anaphase inhibitor Pds1p. Genes Dev 10:3081–3093

Dawson IA, Roth S, Akam M, Artavanis-Tsakonas S (1993) Mutations of the fizzy locus cause metaphase arrest in *Drosophila melanogaster* embryos. Development 117:359–376

Desai A, Maddox PS, Mitchison TJ, Salmon ED (1998) Anaphase A chromosome movement and poleward spindle microtubule flux occur at similar rates in Xenopus extract spindles. J Cell Biol 141:703–713

Desai A, Verma S, Mitchison TJ, Walczak CE (1999) Kin I kinesins are microtubule-destabilizing enzymes. Cell 96:69–78

Dobles M, Liberal V, Scott ML, Benezra R, Sorger PK (2000) Chromosome missegregation and apoptosis in mice lacking the mitotic checkpoint Mad2. Cell 101:635–645

Dou Z, Sawagechi A, Zhang J, Luo H, Brako L, Yao XB (2003) Dynamic distribution of TTK in HeLa cells: insights from an ultrastructural study. Cell Res 13:443–449

Ekwall K, Olsson T, Turner BM, Cranston G, Allshire RC (1997) Transient inhibition of histone deacetylation alters the structural and functional imprint at fission yeast centromeres. Cell 91:1021–1032

Fang G, Yu H, Kirschner MW (1998) The checkpoint protein MAD2 and the mitotic regulator CDC20 form a ternary complex with the anaphase-promoting complex to control anaphase initiation. Genes Dev 12:1871–1883

Fisk HA, Mattison CP, Winey M (2003) Human Mps1 protein kinase is required for centrosome duplication and normal mitotic progression. Proc Natl Acad Sci USA 100:14875–14880

Fraschini R, Beretta A, Sironi L, Musacchio A, Lucchini G, Piatti S (2001) Bub3 interaction with Mad2, Mad3 and Cdc20 is mediated by WD40 repeats and does not require intact kinetochores. EMBO J 20:6648–6659

Funabiki H, Murray AW (2000) The Xenopus chromokinesin Xkid is essential for metaphase chromosome alignment and must be degraded to allow anaphase chromosome movement. Cell 102:411–424

Gardner GD, Poddar A, Yellman C, Tavormina PA, Monteagudo MC, Burke DJ (2001) The spindle checkpoint of the yeast Saccharomyces cerevisiae requires kinetochore function and maps to the CBF3 domain. Genetics 157:1493–1502

Gorbsky GJ, Sammak PJ, Borisy GG (1987) Chromosomes move poleward in anaphase along stationary microtubules that coordinately disassemble from their kinetochore ends. J Cell Biol 104:9–18

Gorbsky GJ, Chen RH, Murray AW (1998) Microinjection of antibody to Mad2 protein into mammalian cells in mitosis induces premature anaphase. J Cell Biol 141:1193–1205

Goshima G, Saitoh S, Yanagida M (1999) Proper metaphase spindle length is determined by centromere proteins Mis12 and Mis6 required for faithful chromosome segregation. Genes Dev 13:1664–1677

Hardwick KG, Johnston RC, Smith DL, Murray AW (2000) MAD3 encodes a novel component of the spindle checkpoint which interacts with Bub3p, Cdc20p, and Mad2p. J Cell Biol 148:871–882

Harrington JA, Van Bokkelen G, Mays RW, Gustashaw K, Willard HF (1997) Formation of de novo centromeres and construction of first-generation human artificial microchromosomes. Nat Genet 15:345–355

Hartwell LH, Kastan MB (1994) Cell cycle control and cancer. Science 266:1821–1828

Hays TS, Wise D, Salmon ED (1982) Traction force on a kinetochore at metaphase acts as a linear function of kinetochore fiber length. J Cell Biol 93:374–389

He D, Zeng C, Woods K, Zhong L, Turner D, Busch RK, Brinkley BR, Busch H (1998) Cenp-g: a new centromeric protein that is associated with the alpha-1 satellite DNA subfamily. Chromosoma 107:189–197

He X, Asthana S, Sorger PK (2000) Transient sister chromatid separation and elastic deformation of chromosomes during mitosis in budding yeast. Cell 101:763–775

Hemmerich P, Stoyan T, Wieland G, Koch M, Lechner J, Diekmann S (2000) Interaction of yeast kinetochore proteins with centromere-protein/transcription factor Cbf1. Proc Natl Acad Sci USA 97:12583–12588

Henikoff S, Ahmad K, Malik HS (2001) The centromere paradox: stable inheritance with rapidly evolving DNA. Science 293:1098–1102

Hiruno T (2000) Chromosome cohesion, condensation, and separation. Ann Rev Biochem 69:115–144

Howell BJ, Hoffman DB, Fang G, Murray AW, Salmon ED (2000) Visualization of mad2 dynamics at kinetochores, along spindles fibers, and at spindle poles in living cells. J Cell Biol 150:1233–1250

Howell BJ, McEwen BF, Canman JC, Hoffman DB, Farrar EM, Rieder CL, Salmon ED (2001) Cytoplasmic dynein/dynactin drives kinetochore protein transport to the spindle poles and has a role in mitotic spindle checkpoint inactivation. J Cell Biol 155:1159–1172

Howman EV, Fowler KJ, Newson AJ, Redward S, MacDonald AC, Kalitsis P, Choo KH (2000) Early disruption of centromeric chromatin organization in centromere protein A (Cenpa) null mice. Proc Natl Acad Sci USA 97:1148–1153

Hoyt MA, Totis L, Roberts BT (1991) S. cerevisiae genes required for cell cycle arrest in response to loss of microtubule function. Cell 66:507–517

Hudson DF, Fowler KJ, Earle E, Saffery R, Kalitsis P, Trowell H, Hill J, Wreford NG, de Kretser DM, Cancilla MR et al. (1998) Centromere protein B null mice are mitotically and meiotically normal but have lower body and testis weights. J Cell Biol 141:309–319

Hyman AA, Mitchison TJ (1991) Two different microtubule-based motor activities with opposite polarities in kinetochores. Nature 351:206–211

Ikeno M, Grimes B, Okazaki T, Nakano M, Saitoh K, Hoshino H, McGill NI, Cooke H, Masumoto H (1998) Construction of YAC-based mammalian artificial chromosomes. Nat Biotechnol 16:431–439

Janke C, Ortiz J, Lechner J, Shevchenko A, Shevchenko A, Magiera MM, Schramm C, Schiebel E (2001) The budding yeast proteins Spc24p and Spc25p interact with Ndc80p and Nuf2p at the kinetochore and are important for kinetochore clustering and checkpoint control. EMBO J 20:777–791

Janke C, Ortiz J, Tanaka TU, Lechner J, Schiebel E (2002) Four new subunits of the Dam1-Duo1 complex reveal novel functions in sister kinetochore biorientation. EMBO J 21:181–193

Kalitsis P, Earle E, Fowler KJ, Choo KH (2000) Bub3 gene disruption in mice reveals essential mitotic spindle checkpoint function during early embryogenesis. Genes Dev 14:2277–2282

Kallio M, Weinstein J, Daum JR, Burke DJ, Gorbsky GJ (1998) Mammalian $p55^{CDC}$ mediates association of the spindle checkpoint protein Mad2 with the cyclosome/anaphase-promoting complex, and is involved in regulating anaphase onset and late mitotic events. J Cell Biol 141:1393–1406

Kallio MJ, Beardmore VA, Weinstein J, Gorbsky GJ (2002) Rapid microtubule-independent dynamics of Cdc20 at kinetochores and centrosomes in mammalian cells. J Cell Biol 158:841–847

Kaplan KB, Burds AA, Swedlow JR, Bekir SS, Sorger PK, Nathke IS (2001) A role for the Adenomatous Polyposis Coli protein in chromosome segregation. Nat Cell Biol 3:429–432

Ke YW, Dou Z, Zhang J, Yao XB (2003) Function and regulation of Aurora/Ipl1p kinase family in cell division. Cell Res 13:69–81

Keith KC, Fitzgerald-Hayes M (2000) CSE4 genetically interacts with the Saccharomyces cerevisiae centromere DNA elements CDE I and CDE II but not CDE III. Implications for the path of the centromere DNA around a cse4p variant nucleosome. Genetics 156:973–981

Khodjakov A, Rieder CL (1996) Kinetochores moving away from their associated pole do not exert a significant pushing force on the chromosome. J Cell Biol 135:315–327

Khodjakov A, Cole RW, McEwen BF, Buttle KF, Rieder CL (1997) Chromosome fragments possessing only one kinetochore can congress to the spindle equator. J Cell Biol 136:229–240

Kim SH, Lin DP, Matsumoto S, Kitazono A, Matsumoto T (1998) Fission yeast Slp1: an effector of the Mad2-dependent spindle checkpoint. Science 279:1045–1047

Kitajima TS, Kawashima SA, Watanabe Y (2004) The conserved kinetochore protein shugoshin protects centromeric cohesion during meiosis. Nature 427:510–7

Lachner M, O'Carroll D, Rea S, Mechtler K, Jenuwein T (2001) Methylation of histone H3 lysine 9 creates a binding site for HP1 proteins. Nature 410:116–120

Laloraya S, Guacci V, Koshland D (2000) Chromosomal addresses of the cohesin component Mcd1p. J Cell Biol 151:1047–1056

Lan W, Zhang X, Kline-Smith SL, Rosasco SE, Barrett-Wilt GA, Shabanowitz J, Hunt DF, Walczak CE, Stukenberg PT (2004) Aurora B phosphorylates centromeric MCAK and regulates its localization and microtubule depolymerization activity. Curr Biol 14:273–86

Levesque AA, Compton DA (2001) The chromokinesin Kid is necessary for chromosome arm orientation and oscillation, but not congression, on mitotic spindles. J Cell Biol 154:1135–1146

Li Y, Benezra R (1996) Identification of a human mitotic checkpoint gene: hsMAD2. Science 274:246–248

Li R, Murray AW (1991) Feedback control of mitosis in budding yeast. Cell 66:519–531

Li X, Nicklas RB (1995) Mitotic forces control a cell-cycle checkpoint. Nature 373:630–632

Li Y, Bachant J, Alcasabas AA, Wang Y, Qin J, Elledge SJ (2002) The mitotic spindle is required for loading of the DASH complex onto the kinetochore. Genes Dev 16:183–197

Liu ST, Chan GK, Hittle JC, Fujii G, Lees E, Yen TJ (2003) Human MPS1 kinase is required for mitotic arrest induced by the loss of CENP-E from kinetochores. Mol Biol Cell 14:1638–1651

Lou Y, Yao J, Zenreski A, Ahmed K, Wang H, Hu J, Wang Y, Yao X (2004) Nek2A interacts with Mad1 and possibly functions as a novel integrator of the spindle checkpoint signaling. J Biol Chem 279: 20049-20057

Maddox P, Desai A, Oegema K, Mitchison TJ, Salmon ED (2002) Poleward microtubule flux is a major component of spindle dynamics and anaphase a in mitotic Drosophila embryos. Curr Biol 12:1670–1674

Marston AL, Tham WH, Shah H, Amon A (2004) A genome-wide screen identifies genes required for centromeric cohesion. Science 303:1367–1370

Martin-Lluesma S, Stucke VM, Nigg EA (2002) Role of hec1 in spindle checkpoint signaling and kinetochore recruitment of mad1/mad2. Science 297:2267–2270

McEwen BF, Chan GKT, Zubrowski B, Savoian MS, Sauer MT, Yen TJ (2001) CENP-E is essential for reliable bioriented spindle attachment, but chromosome alignment can be achieved via redundant mechanisms in mammalian cells. Mol Biol Cell 12:2776–2789

McIntosh JR (1991) Structural and mechanical control of mitotic progression. Cold Spring Harb Symp Quant Biol 56:613–619

Megee PC, Mistrot C, Guacci V, Koshland D (1999) The centromeric sister chromatid cohesion site directs Mcd1p binding to adjacent sequences. Mol Cell 4:445–450

Meluh PB, Koshland D (1995) Evidence that the MIF2 gene of Saccharomyces cerevisiae encodes a centromere protein with homology to the mammalian centromere protein CENP-C. Mol Biol Cell 6:793–807

Meluh PB, Yang P, Glowczewski L, Koshland D, Smith MM (1998) Cse4p is a component of the core centromere of Saccharomyces cerevisiae. Cell 94:607–613

Michel LS, Liberal V, Chatterjee A, Kirchwegger R, Pasche B, Gerald W, Dobles M, Sorger PK, Murty VV, Benezra R (2001) Mad2 haplo-insufficiency causes premature anaphase and chromosome instability in mammalian cells. Nature 409:355–359

Mitchison TJ, Salmon ED (1992) Poleward kinetochore fiber movement occurs during both metaphase and anaphase-A in newt lung cell mitosis. J Cell Biol 119:569–582

Musacchio A, Hardwick KG (2002) The spindle checkpoint: structural insights into dynamic signalling. Nat Rev Mol Cell Biol 3:731–741

Nicklas RB (1983) Measurements of the force produced by the mitotic spindle in anaphase. J Cell Biol 97:542–548

Nicklas RB (1988) The forces that move chromosomes in mitosis. Annu Rev Biophys Biophys Chem 17:431–449

Oegema K, Desai A, Rybina S, Kirkham M, Hyman AA (2001) Functional analysis of kinetochore assembly in *Caenorhabditis elegans*. J Cell Biol 153:1209–1226

Ohzeki J, Nakano M, Okada T, Masumoto H (2002) CENP-B box is required for de novo centromere chromatin assembly on human alphoid DNA. J Cell Biol 159:765–775

Ortiz J, Stemmann O, Rank S, Lechner J (1999) A putative protein complex consisting of Ctf19, Mcm21, and Okp1 represents a missing link in the budding yeast kinetochore. Genes Dev 13:1140–1155

Ostergren G (1951) The mechanism of co-orientation in bivalents and multivalents. Hereditas 37:85–156

Partridge JF, Borgstrom B, Allshire RC (2000) Distinct protein interaction domains and protein spreading in a complex centromere. Genes Dev 14:783–791

Partridge J, Scott K, Bannister A, Kouzarides T, Allshire R (2002) cis-acting DNA from fission yeast centromeres mediates histone H3 methylation and recruitment of silencing factors and cohesin to an ectopic site. Curr Biol 12:1652–1660

Putkey FR, Cramer T, Morphew MK, Silk AD, Johnson RS, McIntosh JR, Cleveland DW (2002) Unstable kinetochore-microtubule capture and chromosomal instability following deletion of CENP-E. Dev Cell 3:351–365

Raff JW, Jeffers K, Huang JY (2002) The roles of Fzy/Cdc20 and Fzr/Cdh1 in regulating the destruction of cyclin B in space and time. J Cell Biol 157:1139–1149

Richards EJ, Elgin SC (2002) Epigenetic codes for heterochromatin formation and silencing: rounding up the usual suspects. Cell 108:489–500

Rieder CL (1981) The structure of the cold-stable kinetochore fiber in metaphase PtK1 cells. Chromosoma 84:145–158

Rieder CL, Alexander SP (1990) Kinetochores are transported poleward along a single astral microtubule during chromosome attachment to the spindle in newt lung cells. J Cell Biol 110:81–95

Rieder CL, Cole RW, Khodjakov A, Sluder G (1995) The checkpoint delaying anaphase in response to chromosome monoorientation is mediated by an inhibitory signal produced by unattached kinetochores. J Cell Biol 130:941–948

Rieder CL, Khodjakov A, Paliulis LV, Fortier TM, Cole RW, Sluder G (1997) Mitosis in vertebrate somatic cells with two spindles: implications for the metaphase/anaphase transition checkpoint and cleavage. Proc Natl Acad Sci USA 94:5107–5112

Roberts BT, Farr KA, Hoyt MA (1994) The Saccharomyces cerevisiae checkpoint gene BUB1 encodes a novel protein kinase. Mol Cell Biol 14:8282–8291

Rogers SL, Rogers GC, Sharp DJ, Vale RD (2002) Drosophila EB1 is important for proper assembly, dynamics, and positioning of the mitotic spindle. J Cell Biol 158:873–884

Rogers GC, Rogers SL, Schwimmer TA, Ems-McClung SC, Walczak CE, Vale RD, Scholey JM, Sharp DJ (2004) Two mitotic kinesins cooperate to drive sister chromatid separation during anaphase. Nature 427:364–370

Russell ID, Grancell AS, Sorger PK (1999) The unstable F-box protein p58-Ctf13 forms the structural core of the CBF3 kinetochore complex. J Cell Biol 145:933–950

Savoian MS, Goldberg ML, Rieder CL (2000) The rate of chromosome poleward motion is attenuated in Drosophila zw10 and rod mutants. Nat Cell Biol 2:948–952

Scholey JM, Brust-Mascher I, Mogilner A (2003) Cell division. Nature 422:746–752

Schueler MG, Higgins AW, Rudd MK, Gustashaw K, Willard HF (2001) Genomic and genetic definition of a functional human centromere. Science 294:109–115

Sharp DJ, Rogers GC, Scholey JM (2000) Cytoplasmic dynein is required for poleward chromosome movement during mitosis in *Drosophila* embryos. Nat Cell Biol 2:922–930

Shelby RD, Vafa O, Sullivan KF (1997) Assembly of CENP-A into centromeric chromatin requires a cooperative array of nucleosomal DNA contact sites. J Cell Biol 136:501–513

Shelby RD, Monier K, Sullivan KF (2000) Chromatin assembly at kinetochores is uncoupled from DNA replication. J Cell Biol 151:1113–1118

Shingyoji C, Higuchi H, Yoshimura M, Katayama E, Yanagida T (1998) Dynein arms are oscillating force generators. Nature 393:711–714

Skibbens RV, Skeen VP, Salmon ED (1993) Directional instability of kinetochore motility during chromosome congression and segregation in mitotic newt lung cells: a push-pull mechanism. J Cell Biol 122:859–875
Skoufias DA, Andreassen PR, Lacroix FB, Wilson L, Margolis RL (2001) Mammalian mad2 and bub1/bubR1 recognize distinct spindle-attachment and kinetochore-tension checkpoints. Proc Natl Acad Sci USA 98:4492–4497
Smith MM (2002) Centromeres and variant histones: what, where, when and why? Curr Opin Cell Biol 14:279–285
Stoler S, Keith KC, Curnick KE, Fitzgerald-Hayes M (1995) A mutation in CSE4, an essential gene encoding a novel chromatin-associated protein in yeast, causes chromosome nondisjunction and cell cycle arrest at mitosis. Genes Dev 9:573–586
Su LK, Burrell M, Hill DE, Gyuris J, Brent R, Wiltshire R, Trent J, Vogelstein B, Kinzler KW (1995) APC binds to the novel protein EB1. Cancer Res 55:2972–2977
Sudakin V, Chan GK, Yen TJ (2001) Checkpoint inhibition of the APC/C in HeLa cells is mediated by a complex of BUBR1, BUB3, CDC20, and MAD2. J Cell Biol 154:925–936
Sullivan B, Karpen G (2001) Centromere identity in Drosophila is not determined in vivo by replication timing. J Cell Biol 154:683–690
Sullivan BA, Willard HF (1998) Stable dicentric X chromosomes with two functional centromeres. Nat Genet 20:227–228
Sullivan BA, Blower MD, Karpen GH (2001) Determining centromere identity: cyclical stories and forking paths. Nat Rev Genet 2:584–596
Sun X, Wahlstrom J, Karpen GH (1997) Molecular structure of a functional Drosophila centromere. Cell 91:1007–1019
Takahashi K, Chen ES, Yanagida M (2000) Requirement of Mis6 centromere connector for localizing a CENP-A-like protein in fission yeast. Science 288:2215–2219
Takenaka K, Moriguchi T, Nishida E (1998) Activation of the protein kinase p38 in the spindle assembly checkpoint and mitotic arrest. Science 280:599–602
Tanaka T, Cosma MP, Wirth K, Nasmyth K (1999) Identification of cohesin association sites at centromeres and along chromosome arms. Cell 98:847–858
Tanaka TU, Rachidi N, Janke C, Pereira G, Galova M, Schiebel E, Stark MJ, Nasmyth K (2002) Evidence that the Ipl1-Sli15 (Aurora kinase-INCENP) complex promotes chromosome bi-orientation by altering kinetochore–spindle pole connections. Cell 108:317–329
Tang Z, Bharadwaj R, Li B, Yu H (2001) Mad2-independent inhibition of APC^{cdc20} by the mitotic checkpoint protein Bubr1. Dev Cell 1:227–237
Taylor SS, McKeon F (1997) Kinetochore localization of murine Bub1 is required for normal mitotic timing and checkpoint response to spindle damage. Cell 89:727–735
Taylor SS, Ha E, McKeon F (1998) The human homologue of Bub3 is required for kinetochore localization of Bub1 and a Mad3/Bub1-related protein kinase. J Cell Biol 142:1–11

Tirnauer JS, Canman JC, Salmon ED, Mitchison TJ (2002) EB1 targets to kinetochores with attached, polymerizing microtubules. Mol Biol Cell 13:4308–4316

Tugendreich S, Tomkiel J, Earnshaw W, Hieter P (1995) CDC27Hs colocalizes with CDC16Hs to the centrosome and mitotic spindle and is essential for the metaphase to anaphase transition. Cell 81:261–268

Tyler-Smith C, Oakey RJ, Larin Z, Fisher RB, Crocker M, Affara NA, Ferguson-Smith MA, Muenke M, Zuffardi O, Jobling MA (1993) Localization of DNA sequences required for human centromere function through an analysis of rearranged Y chromosomes. Nat Genet 5:368–375

Tyler-Smith C, Gimelli G, Giglio S, Floridia G, Pandya A, Terzoli G, Warburton PE, Earnshaw WC, Zuffardi O (1999) Transmission of a fully functional human neocentromere through three generations. Am J Hum Genet 64:1440–1444

Vafa O, Sullivan KF (1997) Chromatin containing CENP-A and alpha-satellite DNA is a major component of the inner kinetochore plate. Curr Biol 7:897–900

Volpe TA, Kidner C, Hall IM, Teng G, Grewal SI, Martienssen RA (2002) Regulation of heterochromatic silencing and histone H3 lysine-9 methylation by RNAi. Science 297:1833–1837

Walczak CE, Mitchison TJ, Desai A (1996) XKCM1: a xenopus kinesin-related protein that regulates microtubule dynamics during mitotic spindle assembly. Cell 84:37–47

Walters JC, Mitchison TJ, Rieder CL, Salmon ED (1996) The kinetochore microtubule minus-end disassembly associated with poleward flux produces a force that can do work. Mol Biol Cell 7:1547–1558

Waters JC, Chen RH, Murray AW, Salmon ED (1998) Localization of Mad2 to kinetochores depends on microtubule attachment, not tension. J Cell Biol 141:1181–1191

Weaver BA, Bonday ZQ, Putkey FR, Kops GJ, Silk AD, Cleveland DW (2003) Centromere-associated protein-E is essential for the mammalian mitotic checkpoint to prevent aneuploidy due to single chromosome loss. J Cell Biol 162:551–563

Weiss E, Winey M (1996) The *Saccharomyces cerevisiae* spindle pole body duplication gene MPS1 is part of a mitotic checkpoint. J Cell Biol 132:111–123

Wigge PA, Kilmartin JV (2001) The Ndc80p complex from Saccharomyces cerevisiae contains conserved centromere components and has a function in chromosome segregation. J Cell Biol 152:349–360

Williams BC, Gatti M, Goldberg ML (1996) Bipolar spindle attachments affect redistributions of ZW10, a *Drosophila* centromere/kinetochore component required for accurate chromosome segregation. J Cell Biol134:1127–1140

Wordeman, L and Mitchison, TJ (1995) Identification and partial characterization of mitotic centromere-associated kinesin, a kinesin-related protein that associates with centromeres during mitosis. J Cell Biol 128:95–104

Yao X, Anderson K, Cleveland DW (1997) The microtubule-dependent motor centromere-associated protein E (CENP-E) is an integral component of kine-

tochore corona fibers that link centromeres to spindle microtubules. J Cell Biol 139:435–447
Yao X, Abrieu A, Zheng Y, Sullivan KF, Cleveland DW (2000) CENP-E forms a link between attachment of spindle microtubules to kinetochores and the mitotic checkpoint. Nat Cell Biol 2:484–491
Yeh E, Skibbens RV, Cheng JW, Salmon ED, Bloom K (1995) Spindle dynamics and cell cycle regulation of dynein in the budding yeast, *Saccharomyces cerevisiae*. J Cell Biol 130:687–700
Yoda K, Ando S, Morishita S, Houmura K, Hashimoto K, Takeyasu K, Okazaki T (2000) Human centromere protein A (CENP-A) can replace histone H3 in nucleosome reconstitution in vitro. Proc Natl Acad Sci USA 97: 7266–7271
Yu HG, Muszynski MG, Kelly Dawe R (1999) The maize homologue of the cell cycle checkpoint protein MAD2 reveals kinetochore substructure and contrasting mitotic and meiotic localization patterns. J Cell Biol 145:425–435
Zou H, McGarry TJ, Bernal T, Kirschner MW (1999) Identification of a vertebrate sister-chromatid separation inhibitor involved in transformation and tumorigenesis. Science 285:418–422
Zhou J, Panda D, Landen JW, Wilson L, Joshi HC (2002) Minor alteration of microtubule dynamics causes loss of tension across kinetochore pairs and activates the spindle checkpoint. J Biol Chem 277:17200–17208
Zinkowski RP, Meyne J, Brinkley BR (1991) The centromere-kinetochore complex: a repeat subunit model. J Cell Biol 113:1091–1110
Zumbrunn J, Inoshita K, Hyman AA, Nathke IS (2001) Binding of the Adenomatous Polyposis Coli protein to microtubules increases microtubules stability and is regulated by GSK3b phosphorylation. Curr Biol 11:44–49
Zur A, Brandeis M (2001) Securin degradation is mediated by fzy and fzr, and is required for complete chromatid separation but not for cytokinesis. EMBO J 20:792–801

4. Breakdown and Reformation of the Nuclear Envelope

Tokuko Haraguchi and Yasushi Hiraoka

Cell Biology Group, Kansai Advanced Research Center, National Institute of Information and Communications Technology, 588-2 Iwaoka, Iwaoka-cho, Nishi-ku, Kobe 651-2492, Japan

4.1 Introduction

In eukaryotes, the nuclear envelope encapsulates chromosomes and provides a physical framework for their organization; it also acts as a nucleo-cytoplasmic boundary for intracellular components providing a regulated chemical environment within the nucleus. The genetic activities of chromosomes are modulated within this distinct physicochemical domain. The nuclear envelope is an apparently stable structure during interphase in the cell cycle, but is dynamic during mitosis, proceeding through disassembly and reassembly in a short period of time. These processes must be precisely regulated to ensure proper progression of the cell cycle, and defects in such processes often cause cell death or disease. Owing to the advancement of imaging technologies, the dynamic behavior of the nuclear envelope during the cell cycle is now being studied in detail in the living cells of many organisms. In this chapter we describe the dynamic processes of disassembly and reassembly of the nuclear envelope as revealed by fluorescence microscopy.

4.2 Structure of the Nuclear Envelope

The nuclear envelope is composed of three structural components: a double membrane, the nuclear lamina, and the nuclear pore complex (NPC) (Fig. 1). The double membrane is composed of two lipid bilayers (an inner nuclear membrane and an outer nuclear membrane). The outer nuclear membrane is contiguous with the endoplasmic reticulum (ER) membrane. In contrast, the inner nuclear membrane contains integral membrane

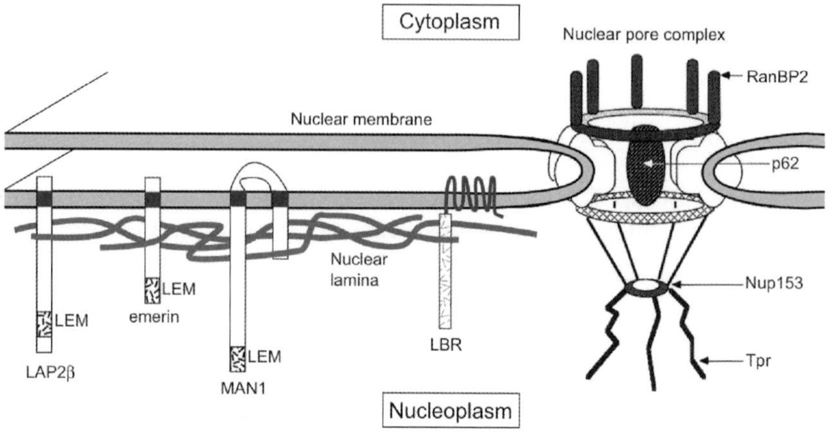

Fig. 1. A schematic diagram of the nuclear envelope in animal cells. LAP2β, emerin, and MAN1 belong to the LEM-domain proteins, which share the ~43-residue LEM domain. *LBR*, lamin B receptor

proteins specific to the nucleus, such as the lamin B receptor (LBR), lamina associated polypeptide 1 (LAP1), LAP2β, emerin, MAN1, otefin, nurim, nepsrin, Unc84, young arrest (YA), and LUMA (reviewed in Wilson 2000; Holaska et al. 2002; Rose et al. 2004). More recently, proteomic analysis has revealed about 80 protein components in total (Schirmer et al. 2003). Most of these inner nuclear membrane proteins bind either chromatin or the nuclear lamina, or both. These binding interactions are thought to influence gene expression, although details of the mechanism are largely unknown.

The NPC is embedded in the double membrane, and provides a molecular gate for nuclear-cytoplasmic transport across the nuclear envelope. The NPC is a gigantic protein complex with a molecular mass of about 125 MDa, which forms an 8-fold symmetrical structure. Proteomic analysis shows that in mammals and yeast, the NPCs are composed of about 30 proteins (Cronshaw et al. 2002; Rout et al. 2000).

The nuclear lamina is a mesh-like structure beneath the inner nuclear membrane composed mainly of A-type and B-type lamins. Although A-type and B-type lamins are similar in molecular structure, the biological features of the two lamins are different. B-type lamin is essential for cellular functions and is ubiquitously expressed in most tissues during all developmental stages. One the other hand, A-type lamin is non-essential for cell growth, and its expression is regulated during development.

Recently, several human disorders known as "nuclear envelopathy" or "nuclear laminopathy" have been reported (reviewed in Nagano and Arahata 2000; Wilson et al. 2001; Worman and Courvalin 2004). These disorders involve mutations in nuclear envelope components. Mutations in lamin A cause eight human diseases: the autosomal dominant form of Emery–Dreifuss muscular dystrophy (Bonne et al. 1999; Clements et al. 2000; Lee et al. 2001), limb-girdle muscular dystrophy type 1B, dilated cardiomyopathy type 1A, Dunnigan-type familial partial lipodystrophy, an axonal neuropathy known as Charcot–Marie–Tooth disorder type 2B1 (De Sandre-Giovannoli et al. 2002), a bone development disorder named mandibuloacral dysplasia (Novelli et al. 2002), Hutchinson–Gilford progeria syndrome (Eriksson et al. 2003; Mounkes et al. 2003), and atypical Werner syndrome (Chen et al. 2003). Emery–Dreifuss muscular dystrophy is also caused by mutations or loss of emerin (Bione et al. 1994). As defects in the nuclear envelope can cause many different diseases, the nuclear envelope must affect a wide variety of cellular functions. These diseases provide an insight into functions of the nuclear envelope, giving rise to the question of how the nuclear envelope is involved in cell aging or tissue-specific defects.

4.3 Breakdown of the Nuclear Envelope

Schemes of the Nuclear Envelope Breakdown

In the classical textbook description (e.g. Alberts et al. 2002), there are two main types of mitosis; these are defined by the behavior of the nuclear envelope. In most fungi the nuclear envelope remains intact throughout the cell cycle; the spindle-pole body (SPB), which is embedded in the nuclear envelope, acts as a spindle pole; and the mitotic spindle is formed within the intact nucleus (closed mitosis). In contrast, the nuclear envelope in animals and plants disassembles completely at prometaphase prior to mitotic chromosome segregation, and the mitotic spindle is formed between pairs of centrosomes (open mitosis). However, several observations now show that there are some variations of the typical closed or open mitoses, such as, partial breakdown of the nuclear envelope in closed mitosis or a persistent nuclear envelope remaining after spindle formation in open mitosis.

Partial breakdown of the nuclear envelope has been reported in several fungi that had previously been believed to undergo closed mitosis. Typical closed mitosis is observed in the budding yeast *Saccharomyces cerevisiae*; in this organism the SPB is embedded in an intact nuclear envelope throughout the cell cycle. In contrast, in the fission yeast *Schizosaccharomyces pombe*, the SPB resides outside the nuclear envelope in interphase, embeds into the nuclear envelope in mitosis, and dissociates from the nuclear envelope afterwards; thus, the *S. pombe* nuclear envelope partially breaks down and then re-forms at the site of the SPB during mitosis (Ding et al. 1997).

In *Caenorhabditis elegans*, the nuclear envelope seems to be regulated in a more complex manner. During development the nuclear envelope partially disassembles near the spindle poles during metaphase, but most of the nuclear envelope remains intact until early anaphase, and does not fully disassemble until mid-late anaphase (Lee et al. 2000). The timing of NPC disassembly, however, depends on the embryonic stage. In the early embryonic stages (2- to 24-cell embryos) the NPCs remain intact during metaphase, while in the late embryonic stages (>30-cell embryos) the NPCs disassemble during metaphase.

In *Drosophila* syncytial blastoderm embryos the nuclear envelope persists until metaphase (Paddy et al. 1996). Fixed specimens show that nuclear envelope breakdown and mitotic spindle formation are coordinated: invagination of the nuclear envelope is observed in prometaphase, followed by partial breakdown of the nuclear envelope at the site where the mitotic spindle is forming (Fuchs et al. 1983; Hiraoka et al. 1990; Paddy et al. 1996). Thus, in both *C. elegans* and *Drosophila* spindle formation precedes nuclear envelope dissociation.

In human cells formation of a mitotic spindle in the presence of an intact nuclear envelope has also been observed (Georgatos et al. 1997; Beaudouin et al. 2002). It was originally held that in open mitosis the nuclear envelope breaks down before the mitotic spindle is formed, but these findings indicate that the nuclear envelope persists after the mitotic spindle is formed in higher eukaryotic cells, and thus there is a limited period when the nuclear envelope and the mitotic spindle are both present. Taken together, these observations suggest a mechanism, discussed below, in which microtubules or centrosomes trigger envelope breakdown (see also Fig. 3).

Nuclear Envelope Breakdown in Human Cells

Live-cell imaging technology based on fluorescence microscopy has revealed the process of nuclear envelope breakdown in living human cells. Assembly and disassembly of the nuclear envelope can be followed by monitoring the behavior of inner membrane proteins. For example, lamin B fused with green fluorescence protein (LBR–GFP) can be visualized throughout mitosis in living cells. Coupled with simultaneous observation of chromosomes stained with Hoechst 33342, a DNA-specific fluorescent dye, a comprehensive picture of envelope dynamics can be developed (Haraguchi et al. 2000). Figure 2A shows an example of nuclear envelope breakdown in human HeLa cells. In late prophase, the nuclear envelope becomes deformed and invaginates into the nucleus (4–8 min in Fig. 2A). These invaginations occur at sites that correspond to the positions of the centrosomes. Following invagination, complete breakdown of the nuclear envelope occurs (10 min in Fig. 2A). After nuclear envelope breakdown, chromosomes migrate to form the metaphase plate (12 min in Fig. 2A).

Nuclear invagination can also observed when fixed cells are used. Figure 2B–E shows the results of indirect immunofluorescence staining using an anti-lamin B antibody with fixed HeLa cells. In interphase and early prophase the nuclear envelope was roughly spherical (Fig. 2B). In late prophase, which is characterized by highly condensed chromosomes, the nuclear envelope is invaginated into the nucleus (Fig. 2C); at this stage, centrosomes are located at the both ends of the nuclear invagination (Fig. 2E). From prometaphase to metaphase the nuclear envelope cannot be detected (Fig. 2D).

Observation of fluorescently labeled nuclear proteins shows that the nuclear envelope is still intact when the mitotic spindle first forms. Figure 2F shows simultaneous observation of rhodamine-conjugated tubulins (to visualize the mitotic spindle) and a fluorescently labeled nuclear protein in living HeLa cells. In this example, the mitotic spindle began to form while the nuclear protein remained in the nucleus, showing that the nuclear envelope was still intact. Shortly after spindle formation began, the nuclear envelope was disassembled and the nuclear protein diffused throughout the cell.

From the observations described above a more detailed description of nuclear envelope breakdown is being formed. In human cells, the mitotic spindle forms in the cytoplasm on an intact nuclear envelope in late prophase. As spindle formation progresses, the spindle pushes against the nuclear envelope causing invagination (Fig. 3A,B).

Such an image of invagination seems reminiscent of observations of early *Drosophila* embryos, but it was not clear whether the breakage point of the nuclear envelope was in the invaginated region proximal to the centrosomes, like that observed in *Drosophila* and *C. elagans* embryos, or elsewhere. To clarify the region of breakage of the nuclear envelope in human cells, live and high-resolution three-dimensional observation of the nuclear envelope using lamin B-GFP was performed (Beaudouin et al. 2002). Surprisingly the observations made suggested the unexpected model that the nuclear envelope was torn at the site of most tension in the nuclear lamina, and not in the invaginated region (Beaudouin et al. 2002) (see Fig. 3D). This model is supported by findings that dynein–dynactin complexes on the nuclear envelope facilitates nuclear envelope breakdown by pulling the nuclear envelope (Salina et al. 2002). It is also supported by the finding that in starfish oocytes nuclear envelope breakdown does not take place uniformly but starts at a specific site distant from the centrosomes (Terasaki et al. 2001; Lenart et al. 2003). These findings contrast with what is observed in *Drosophila* and yeasts, in which the nuclear envelope starts to break down at the sites of the centrosomes (see Fig. 3C). Thus, further investigations are needed to elucidate common mechanisms of nuclear envelope breakage among eukaryotes; for example, in addition to mechanical force models, biochemical modification of lamin B1 is also proposed to cause nuclear envelope breakdown (Panorchan et al. 2004).

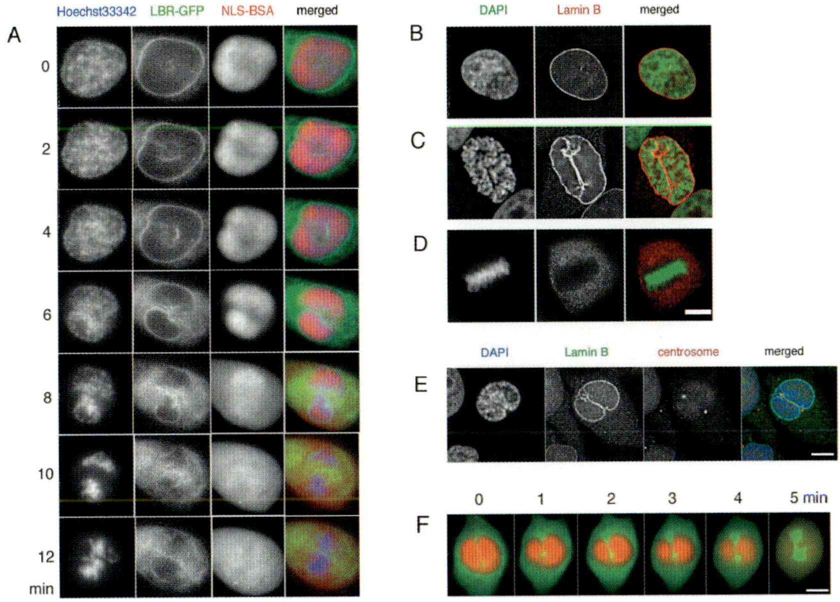

Fig. 2A-F. Invagination and breakdown of the nuclear envelope in prophase. The dynamic behavior of the nuclear envelope was observed by time-lapse observation of fluorescently stained living cells (**A** and **F**) or indirect immunofluorescence staining of fixed cells (**B**–**E**). **A** Fluorescently stained living HeLa cells were prepared as described previously (Haraguchi et al. 1997, 1999, 2000). Briefly, cells plated on a 35-mm glass-bottom culture dish were transfected with a lamin B receptor–green fluorescence protein (*LBR–GFP*) fusion construct to stain the nuclear envelope, treated with Hoechst 33342 to stain the chromosomes, and then microinjected with rhodamine-conjugated nuclear localization sequence–bovine serum albumin (*NLS–BSA*) (used as a nuclear protein) to monitor the timing of nuclear envelope breakdown. Time-lapse images were obtained at 37°C using an Olympus oil immersion objective lens (PlanApo 40×/NA=1.35) on the DeltaVision microscope system (Applied Precision, Seattle, WA, USA; Haraguchi et al. 1999). *Numbers on the left* indicate time in minutes. **B**–**D** Fixed HeLa cells stained with 4′,6-diamidino-2-phenylindole (*DAPI*) (a DNA stain) and anti-lamin B (to stain the nuclear envelope) as described previously (Haraguchi et al. 2000): **B** interphase, **C** prophase, **D** metaphase. **E** A fixed prophase cell stained with DAPI, anti-lamin B, and anti-γ-tubulin (to stain centrosomes). **F** Time-lapse images of a living cell stained with rhodamine-conjugated tubulins (to stain microtubules: *green*) and NLS–GFP (used as a nuclear protein: *red*). Images were taken every 1 min. *Numbers on the top* indicate the time in minutes

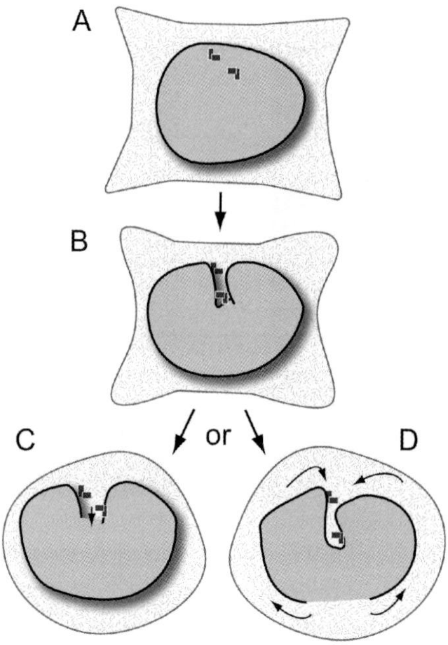

Fig. 3A-D. Diagrams of invagination and breakdown of the nuclear envelope. Centrosomes on the nuclear envelope (**A**) become localized within a nuclear invagination (**B**). The nuclear envelope is broken either in the invaginated region proximal to the centrosomes (**C**) or at other sites where stress is accumulated (**D**). The former case has been observed in *Drosophila* (Fuchs et al. 1983; Hiraoka et al. 1990; Paddy et al. 1996), fission yeast (Ding et al. 1997), and the nematode (Lee et al. 2000); the latter case has been observed in starfish (Terasaki et al. 2001; Lenart et al. 2003) and humans (Georgatos et al. 1997; Beaudouin et al. 2002)

4.4 Reformation of the Nuclear Envelope

Temporal Sequences of Nuclear Envelope Reformation

The nuclear envelope reforms around chromosomes during telophase to re-establish functional nuclear structures for the subsequent interphase. Reformation of the nuclear envelope begins with attachment of vesicular membranes to chromosomes. This is followed by fusion of the membranes and re-composition of integral membrane proteins, the NPCs, and nuclear lamina into the nuclear envelope.

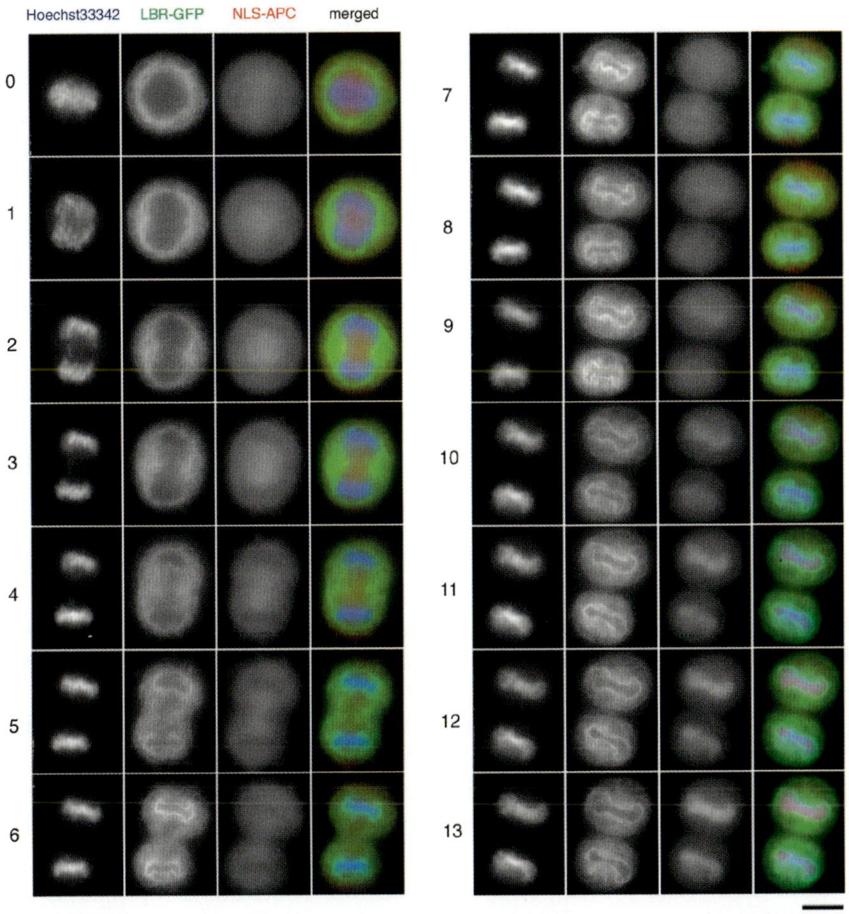

Fig. 4. Reformation of the nuclear envelope in telophase. The dynamic behavior of the reforming nuclear envelope during telophase was observed in fluorescently stained living cells using the DeltaVision microscope system as described in the legend to Fig. 2A (Haraguchi et al. 2000). Cells were transfected with LBR–GFP to stain the nuclear envelope, treated with Hoechst 33342 to stain chromosomes, and then microinjected with NLS–allophycocyanin (*NLS–APC*) to monitor nuclear transport activity. Images were taken every 1 min. *Numbers on the left* indicate time in minutes. *Scale bar* 10 μm

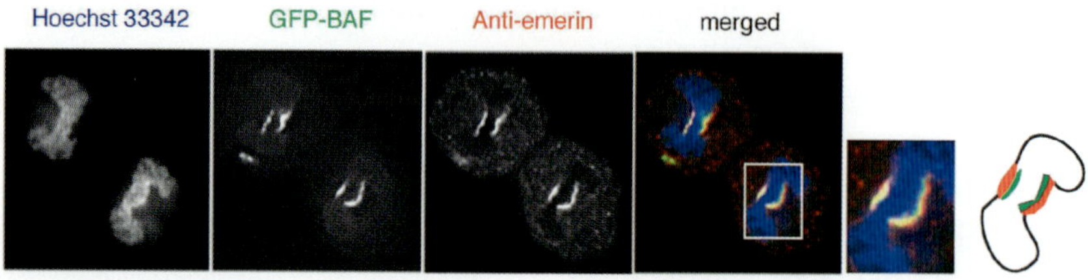

Fig. 5. Intracellular localization of barrier-to-autoregulation factor (*BAF*) and emerin in telophase. HeLa cells expressing GFP–BAF were fixed and stained with antibodies against emerin. Chromosomes were stained with Hoechst 33342. An enlarged image of the indicated area (*white square*), together with a schematic diagram, is shown at the right. BAF is shown in *green* and emerin in *red*. Reproduced from Haraguchi et al. (2001)

To understand the molecular mechanism of reformation of the nuclear envelope, the temporal sequence of the reconstruction of envelope components onto the chromosome surface has been determined using fixed cells (Bodoor et al. 1999) and, more recently, by time-lapse observation of nuclear proteins in living cells (Haraguchi et al. 2000, 2001). Figure 4 shows an example of a series of observations in living cells. Live cells stained with Hoechst 33342, LBR–GFP, and a fluorescently labeled nuclear protein were observed every 1 min from metaphase to telophase to monitor the behavior of the chromosomes, nuclear membrane, and nuclear protein, respectively. These experiments reveal when the nuclear membrane reforms and when nuclear import function recovers. Nuclear membranes started to reassemble 5 min after the metaphase-anaphase transition and nuclear import function recovered at 8 min (Fig. 4). Another nuclear integral membrane protein, emerin (one of the LEM-domain proteins; see Fig. 1), has also been examined, using GFP–emerin, to confirm the behavior of the nuclear membrane (Haraguchi et al. 2001). Interestingly, the results of the observations using emerin-GFP were different from those using LBR–GFP. Both emerin and LBR first attached to the telophase chromosome mass at a site distant from the mitotic spindle at 4–5 min after the onset of anaphase. Emerin then accumulated at the central core region of the telophase chromosome mass close to the mitotic spindle (designated the "core" region; Fig. 5) from 6 to 8 min after the onset of anaphase (Haraguchi et al. 2001). In contrast, LBR accumulated at chromosomal regions flanking the "core" region. This difference suggests that emerin and

LBR reassemble by different mechanisms (see the next section). Two other LEM-domain proteins, LAP2β and MAN1, also accumulate at the "core" region. After 8 min, LBR and emerin become uniformly distributed, suggesting that vesicular membranes that contain different nuclear integral membrane proteins fuse at this time (Fig. 6).

NPC assembly has also been examined by immunofluorescence staining of cells fixed at precise times after the onset of anaphase (Bodoor et al. 1999; Haraguchi et al. 2000). RanBP2, p62, and Nup153 re-assemble in early telophase, whereas Tpr reassembles in late telophase. Nuclear transport function is recovered 2 min later, coinciding with the time when localization of emerin and LBR becomes uniform as observed in living cells. Thus, the uniform distribution of emerin and LBR is an indication of the completion of a nuclear envelope bearing nuclear import activity.

Molecular Mechanisms of Nuclear Envelope Reformation

There are at least three independent pathways, involving Ran, barrier-to-autointegration factor (BAF), and A-kinase anchoring protein AKAP149, involved in envelope reformation.

Involvement of Ran has been demonstrated by in vitro experiments using *Xenopus* oocyte extracts, in which the nuclear envelope re-forms around plastic beads coated with Ran (Zhang and Clarke 2000), and by microscopic observation in living *C. elegans* embryos (Askjaer et al. 2002). Importins α and β modulate Ran-dependent assembly of the nuclear envelope (Harel et al. 2003; Hachet et al. 2004) by mediating the assembly of nuclear pore complexes (Walther et al. 2003).

Involvement of BAF has been demonstrated by live observation of nuclear components in living human cells (Haraguchi et al 2001). BAF was first discovered as a cellular protein that prevented retroviral DNA from suicidal auto-integration and ensured its integration into host DNA (Lee and Craigie 1998). It was later described as a binding partner of the LEM domain nuclear envelope proteins (Furukawa 1999; Lee et al. 2001). BAF binds nonspecifically to double-stranded DNA in vitro (Lee and Craigie 1998) and colocalizes with emerin at the "core" region of chromosomes during telophase (Haraguchi et al. 2001, see Figs. 5 and 6). In HeLa cells, both emerin and BAF accumulate briefly at the "core" region of telophase chromosomes and later distribute over the entire nuclear rim (Fig. 5). Recruitment of emerin is mediated by BAF: in HeLa cells expressing the dominant BAF mutant G25E, which does not show "core" localization, the mutated BAF disrupts emerin localization at the core region during telophase and dissociates emerin from the nuclear envelope in the subsequent interphase (Haraguchi et al. 2001). The mutation also dissociates LAP2β and lamin A from the nuclear envelope, but has no effect on lamin B. Thus, BAF may be required for the assembly of LAP2β and lamin A as well as emerin, when the nuclear envelope re-forms during telophase (Haraguchi et al. 2001).

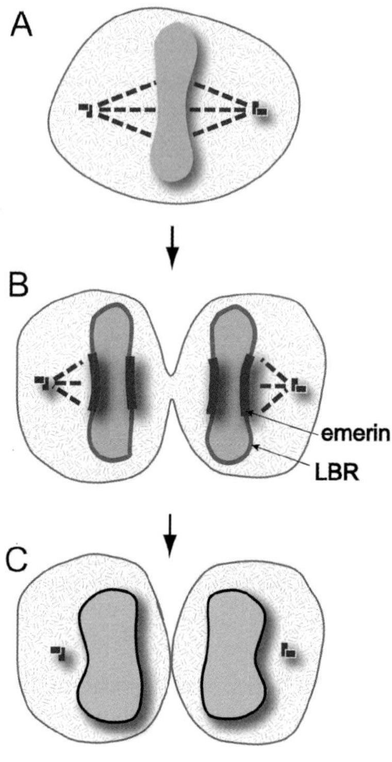

Fig. 6A-C. Diagram of the assembly of the nuclear envelope during telophase. Metaphase chromosomes with no nuclear envelope (A). Emerin and LBR initially accumulate at distinct, separate locations of the telophase chromosome mass (**B**), but later become uniformly distributed (**C**). Emerin accumulates at the "core" region, and LBR accumulates at regions outside the core regions. BAF, LAP2β, and MAN1 also accumulate at the core region

AKAP149, A kinase-anchoring protein with a molecular weight of 149 kDa, is a factor responsible for reassembly of lamin B into the nuclear envelope at the end of mitosis (Steen et al. 2003). The AKAP149 protein targets protein phosphatase 1 (PP1) to the nuclear envelope, and this targeting is necessary for assembly of lamin B. In the presence of an inhibitory peptide that inhibits PP1 binding to the nuclear envelope, lamin B failed to assemble into the re-forming nuclear envelope at the end of mitosis (Steen et al. 2003).

Lamin A and lamin B seem to assemble into the nuclear lamina by different pathways. For example, the timing of assembly is different; lamin B assembles early in telophase, before lamin A. Also, AKAP149 and PP1 are involved in assembly of lamin B, but not of lamin A: while the inhibitory peptide that inhibited PP1 binding to the nuclear envelope caused failure of lamin B to reassemble into the reforming nuclear envelope at the end of mitosis, it had no effect on lamin A assembly (Moir et al. 2000; Steen et al. 2003). Moreover, assembly of lamin B is not affected by the BAF mutant G25E, whereas lamin A assembly is inhibited (Haraguchi et al. 2001). These results indicate that lamin A and lamin B form separate filaments, unlike the current model, in which lamin A and lamin B form mixed filaments.

4.4 Dynamics of Nuclear Envelope Components in Interphase

The interphase nuclear envelope may seem to be static, but its protein components are continuously moving around within it, and there is also some movement of proteins into and out of the inner membrane of the nuclear envelope. Studies by fluorescence recovery after photobleaching (FRAP) have measured the mobility or exchange rates of various integral membrane proteins, such as LBR, emerin, LAP2β, and MAN1. LBR is relatively immobile, similar to nuclear lamina components, suggesting that LBR binds stably to the nuclear lamina (Ellenberg et al. 1997). In contrast, emerin, LAP2β, and MAN1 all exhibit significantly greater movement in the nuclear envelope than does LBR (Ostlund et al. 1999; Shimi et al. 2004). Moreover, while the nuclear lamina and NPCs form immobile networks with a very low turnover (Daigle et al. 2001), FRAP studies have shown that the constituent proteins of the NPCs are actually in flux. NPC components could be classified into three groups according to their mobility: slow (retention time >20 h), intermediate (retention time of several hours), and fast (retention time less than a few minutes) (Rabut et al. 2004).

4.5 Nuclear Envelope in Plants

The nuclear envelope in plants has structures similar to that in animals, but the molecular components are quite different (Rose et al. 2004). The entire genome of *Arabidopsis* has now been sequenced (Arabidopsis Genome Initiative 2000), but it contains no amino acid sequences similar to lamin. On the other hand, plant cells may have lamin-like molecules, as antibodies against metazoan lamin recognize the plant nuclear envelope (Li and Roux 1992; McNulty and Saunders 1992; Minguez and Moreno Diaz de la Espina 1993). So far, three components are known to localize to the plant nuclear envelope: nuclear matrix constituent protein 1 (NMCP1), MFP1 attachment factor 1 (MAF1), and Ran GTPase activating protein (RanGAP). NMCP1 is a coiled-coil protein forming intermediate filaments similar to lamin or myosin (Masuda et al. 1997), and thus is a good candidate for a lamin homologue in plants. RanGAP is conserved from yeasts to humans, but mechanisms for localization to the nuclear envelope in plants seem to be different from those in animals. An amino terminal domain unique to plant RanGAP is necessary for its localization to the nuclear envelope (Rose and Meier 2001). MAF1 also has this domain, suggesting that the domain may be a plant-specific localization signal to the nuclear envelope. However, the animal LBR is localized to the nuclear envelope in plant cells that have no lamin B (Irons et al. 2003). Thus, there seems to be common mechanisms for LBR localization to the nuclear envelope independent of lamin B.

4.6 Concluding Remarks

We have described the cell cycle behavior of the nuclear envelope observed by direct observations in living cells. As the protein components of the nuclear envelope are insoluble, it has been difficult to examine those proteins biochemically. Recent developments in fluorescence imaging of proteins in living cells have made it possible to examine their dynamics, mobility, and interactions without biochemical purification. In particular, by using GFP-fusion constructs, mutated proteins can be fluorescently tagged in living cells and their behavior examined. A combination of molecular genetic approaches and live cell imaging will provide new insights into understanding the fundamental functions of the nuclear envelope.

References

Alberts B, Johnson A, Lewis J, Raff M, Roberts K, Walter P (2002) Molecular biology of the cell, 4th edn. Garland, New York

Arabidopsis Genome Initiative (2000) Analysis of the genome sequence of the flowering plant *Arabidopsis thaliana*. Nature 408:796–815

Askjaer P, Galy V, Hannak E, Mattaj IW (2002) Ran GTPase cycle and importins alpha and beta are essential for spindle formation and nuclear envelope assembly in living *Caenorhabditis elegans* embryos. Mol Biol Cell 13:4355–4370 (Erratum appears in Mol Biol Cell 2004;15(1): preceding Table of Contents)

Beaudouin J, Gerlich D, Daigle N, Eils R, Ellenberg J (2002) Nuclear envelope breakdown proceeds by microtubule-induced tearing of the lamina. Cell 108:83–96

Bione S, Maestrini E, Rivella S, Mancini M, Regis S, Romeo G, Toniolo D (1994). Identification of a novel X-linked gene responsible for Emery–Dreifuss muscular dystrophy. Nat Genet 8:323–327

Bodoor K, Shaikh S, Salina D, Raharjo WH, Bastos R, Lohka M, Burke B (1999) Sequential recruitment of NPC proteins to the nuclear periphery at the end of mitosis. J Cell Sci 112:2253–2264

Bonne G, Di Barletta MR, Varnous S, Becane HM, Hammouda EH, Merlini L, Muntoni F, Greenberg CR, Gary F, Urtizberea JA, Duboc D, Fardeau M, Toniolo D, Schwartz K (1999) Mutations in the gene encoding lamin A/C cause autosomal dominant Emery–Dreifuss muscular dystrophy. Nat Genet 21:285–288

Chen L, Lee L, Kudlow BA, Dos Santos HG, Sletvold O, Shafeghati Y, Botha EG, Garg A, Hanson NB, Martin GM, Mian IS, Kennedy BK, Oshima J (2003) LMNA mutations in atypical Werner's syndrome. Lancet 362:440–445

Clements L, Manilal S, Love DR, Morris GE (2000) Direct interaction between emerin and lamin A. Biochem Biophys Res Commun 267:709–714

Cronshaw JM, Krutchinsky AN, Zhang W, Chait BT, Matunis MJ (2002) Proteomic analysis of the mammalian nuclear pore complex. J Cell Biol 158:915–927

Daigle N, Beaudouin J, Hartnell L, Imreh G, Hallberg E, Lippincott-Schwartz J, Ellenberg J (2001) Nuclear pore complexes form immobile networks and have a very low turnover in live mammalian cells. J Cell Biol 154:71–84

De Sandre-Giovannoli A, Chaouch M, Kozlov S, Vallat JM, Tazir M, Kassouri N, Szepetowski P, Hammadouche T, Vandenberghe A, Stewart CL, Grid D, Levy N (2002) Homozygous defects in LMNA, encoding lamin A/C nuclear-envelope proteins, cause autosomal recessive axonal neuropathy in human

(Charcot-Marie-Tooth disorder type 2) and mouse. Am J Hum Genet 70:726–736 (Erratum: Am J Hum Genet 70:1075)

Ding R, West RR, Morphew DM, Oakley BR, McIntosh JR (1997) The spindle pole body of *Schizosaccharomyces pombe* enters and leaves the nuclear envelope as the cell cycle proceeds. Mol Biol Cell 8:1461–1479

Ellenberg J, Siggia ED, Moreira JE, Smith CL, Presley JF, Worman HJ, Lippincott-Schwartz J (1997) Nuclear membrane dynamics and reassembly in living cells: targeting of an inner nuclear membrane protein in interphase and mitosis. J Cell Biol 138:1193–1206

Eriksson M, Brown WT, Gordon LB, Glynn MW, Singer J, Scott L, Erdos MR, Robbins CM, Moses TY, Berglund P, Dutra A, Pak E, Durkin S, Csoka AB, Boehnke M, Glover TW, Collins FS (2003) Recurrent de novo point mutations in lamin A cause Hutchinson-Gilford progeria syndrome. Nature 423:293–298

Fuchs JP, Giloh H, Kuo CH, Saumweber H, Sedat J (1983) Nuclear structure: determination of the fate of the nuclear envelope in Drosophila during mitosis using monoclonal antibodies. J Cell Sci 64:331–349

Furukawa K (1999) LAP2 binding protein 1 (L2BP1/BAF) is a candidate mediator of LAP2-chromatin interaction. J Cell Sci 112:2485–2492

Georgatos SD, Pyrpasopoulou A, Theodoropoulos PA (1997) Nuclear envelope breakdown in mammalian cells involves stepwise lamina disassembly and microtubule-drive deformation of the nuclear membrane. J Cell Sci 110:2129–2140

Hachet V, Kocher T, Wilm M, Mattaj IW (2004) Importin alpha associates with membranes and participates in nuclear envelope assembly in vitro. EMBO J 23:1526–1535

Haraguchi T, Kaneda T, Hiraoka Y (1997) Dynamics of chromosomes and microtubules visualized by multiple-wavelength fluorescence imaging in living mammalian cells: effects of mitotic inhibitors on cell cycle progression. Genes to Cells 2:369–380

Haraguchi T, Ding DQ, Yamamoto A, Kaneda T, Koujin T, Hiraoka Y (1999) Multiple-color fluorescence imaging of chromosomes and microtubules in living cells. Cell Struct Funct 24:291–298

Haraguchi T, Koujin T, Hayakawa T, Kaneda T, Tsutsumi C, Imamoto N, Akazawa C, Sukegawa J, Yoneda Y, Hiraoka Y (2000) Live fluorescence imaging reveals early recruitment of emerin, LBR, RanBP2, and Nup153 to reforming functional nuclear envelopes. J Cell Sci 113:779–794

Haraguchi T, Koujin T, Segura M, Lee KL, Matsuoka Y, Yoneda Y, Wilson KL, Hiraoka Y (2001) BAF is required for emerin assembly into the reforming nuclear envelope. J Cell Sci 114:4575–4585

Harel A, Chan RC, Lachish-Zalait A, Zimmerman E, Elbaum M, Forbes DJ (2003) Importin beta negatively regulates nuclear membrane fusion and nuclear pore complex assembly. Mol Biol Cell 14:4387–4396

Hiraoka Y, Agard DA, Sedat JW (1990) Temporal and spatial coordination of chromosome movement, spindle formation and nuclear envelope breakdown

during prometaphase in *Drosophila melanogaster* embryos. J Cell Biol 111:2815–2828
Holaska JM, Wilson KL, Mansharamani M (2002) The nuclear envelope, lamins and nuclear assembly. Curr Opin Cell Biol 14:357–364
Irons SL, Evans DE, Brandizzi F (2003) The first 238 amino acids of the human lamin B receptor are targeted to the nuclear envelope in plants. J Exp Bot 54:943–950
Lee MS, Craigie R (1998) A previously unidentified host protein protects retroviral DNA from autointegration. Proc Natl Acad Sci USA 95:1528–1533
Lee KK, Gruenbaum Y, Spann P, Liu J, Wilson KL (2000) *C. elegans* nuclear envelope proteins emerin, MAN1, lamin, and nucleoporins reveal unique timing of nuclear envelope breakdown during mitosis. Mol Biol Cell 11:3089–3099
Lee KK, Haraguchi T, Lee RS, Koujin T, Hiraoka Y, Wilson KL (2001) Distinct functional domains in emerin bind lamin A and DNA-bridging protein BAF. J Cell Sci 114:4567–4573
Lenart P, Rabut G, Daigle N, Hand AR, Terasaki M, Ellenberg J (2003) Nuclear envelope breakdown in starfish oocytes proceeds by partial NPC disassembly followed by a rapidly spreading fenestration of nuclear membranes. J Cell Biol 160:1055–1068
Li H, Roux SJ (1992) Purification and characterization of a casein kinase 2-type protein kinase from pea nuclei. Plant Physiol 99:686–692
Masuda K, Xu ZJ, Takahashi S, Ito A, Ono M, Nomura K, Inoue M (1997) Peripheral framework of carrot cell nucleus contains a novel protein predicted to exhibit a long alpha-helical domain. Exp Cell Res 232:173–181
McNulty AK, Saunders MJ (1992) Purification and immunological detection of pea nuclear intermediate filaments: evidence for plant nuclear lamins. J Cell Sci 103:407–414
Minguez A, Moreno Diaz de la Espina S (1993) Immunological characterization of lamins in the nuclear matrix of onion cells. J Cell Sci 106:431–439
Moir RD, Yoon M, Khuon S, Goldman RD (2000) Nuclear lamins A and B1: different pathways of assembly during nuclear envelope formation in living cells. J Cell Biol 151:1155–1168
Mounkes LC, Kozlov S, Hernandez L, Sullivan T, Stewart CL (2003) A progeroid syndrome in mice is caused by defects in A-type lamins. Nature 423:298–301
Nagano A, Arahata K (2000) Nuclear envelope proteins and associated diseases. Curr Opin Neurol 13:533–539
Novelli G, Muchir A, Sangiuolo F, Helbling-Leclerc A, D'Apice MR, Massart C, Capon F, Sbraccia P, Federici M, Lauro R, Tudisco C, Pallotta R, Scarano G, Dallapiccola B, Merlini L, Bonne G (2002) Mandibuloacral dysplasia is caused by a mutation in LMNA-encoding lamin A/C. Am J Hum Genet 71:426–431
Ostlund C, Ellenberg J, Hallberg E, Lippincott-Schwartz J, Worman HJ (1999) Intracellular trafficking of emerin, the Emery–Dreifuss muscular dystrophy protein. J Cell Sci 112:1709–1719

Paddy MR, Saumweber H, Agard DA, Sedat JW (1996) Time-resolved, in vivo studies of mitotic spindle formation and nuclear lamina breakdown in Drosophila early embryos. J Cell Sci 109:591–607

Panorchan P, Schafer BW, Wirtz D, Tseng Y (2004) Nuclear envelope breakdown requires overcoming the mechanical integrity of the nuclear lamina. J Biol Chem 279:43462–43467

Rabut G, Doye V, Ellenberg J (2004) Mapping the dynamic organization of the nuclear pore complex inside single living cells. Nat Cell Biol 6:1114–1121

Rose A, Meier I (2001) A domain unique to plant RanGAP is responsible for its targeting to the plant nuclear rim. Proc Natl Acad Sci USA 98:15377–15382

Rose A, Patel S, Meier I (2004) The plant nuclear envelope. Planta 218:327–336

Rout MP, Aitchison JD, Suprapto A, Hjertaas K, Zhao Y, Chait BT (2000) The yeast nuclear pore complex: composition, architecture, and transport mechanism. J Cell Biol 148:635–651

Salina D, Bodoor K, Eckley DM, Schroer TA, Rattner JB, Burke B (2002) Cytoplasmic dynein as a facilitator of nuclear envelope breakdown. Cell 108:97–107

Schirmer EC, Florens L, Guan T, Yates JR, Gerace L (2003) Nuclear membrane proteins with potential disease links found by subtractive proteomics. Science 301:1380–1382

Shimi T, Koujin T, Segura-Totten M, Wilson KL, Haraguchi T, Hiraoka Y (2004) Dynamic interaction between BAF and emerin revealed by FRAP, FLIP and FRET analyses in living HeLa cells. J Struct Biol 147:31–41

Steen RL, Beullens M, Landsverk HB, Bollen M, Collas P (2003) AKAP149 is a novel PP1 specifier required to maintain nuclear envelope integrity in G1 phase. J Cell Sci 116:2237–2246

Terasaki M, Campagnola P, Rolls MM, Stein PA, Ellenberg J, Hinkle B, Slepchenko B (2001) A new model for nuclear envelope breakdown. Mol Biol Cell 12:503–510

Walther TC, Askjaer P, Gentzel M, Habermann A, Griffiths G, Wilm M, Mattaj IW, Hetzer M (2003) RanGTP mediates nuclear pore complex assembly. Nature 424:689–694

Wilson KL (2000) The nuclear envelope, muscular dystrophy and gene expression. Trends Cell Biol 10:125–129

Wilson KL, Zastrow MS, Lee KK (2001) Lamins and disease: insights into nuclear infrastructure. Cell 104:647–650

Worman HJ, Courvalin JC (2004) How do mutations in lamins A and C cause disease? J Clin Invest 113:349–351

Zhang C, Clarke PR (2000) Chromatin-independent nuclear envelope assembly induced by Ran GTPase in Xenopus egg extracts. Science 288:1429–1432

5. Functional Organization and Dynamic Aspects of Nucleoli During the Cell Cycle

Takuya Saiwaki[1] and Yoshihiro Yoneda[1,2]

[1]Department of Cell Biology and Neuroscience, Graduate School of Medicine, Osaka University, 2-2 Yamada-oka, Suita, Osaka 565-0871, Japan

[2]Department of Frontier Biosciences, Graduate School of Frontier Biosciences, Osaka University, 1-3 Yamada-oka, Suita, Osaka 565-0871, Japan

5.1 Structure of the Nucleolus

The nucleolus is the most prominent structure in the nucleus, and is clearly distinguished from other nuclear substructures (Scheer and Hock 1999; Olson et al. 2000; Spector 2003). Since the 1960s, it has been generally thought to be the site of ribosome biogenesis. Inner nucleolar DNA sequences encode 18S, 5.8S and 28S ribosomal RNAs and in the nucleolus, many ribosomal proteins and nascent ribosomes are detected (Fatica and Tollervey 2002; Raska 2003). In metabolically active animal and plant somatic cells and in yeast, the nucleolus harbors tens to hundreds of active ribosomal genes, which account for roughly one half of the total cellular RNA production, while no other types of active genes have been identified within it.

Structural Basis of the Nucleolus

In human diploid cells, ribosomal genes are located on several chromosomes and form arrays of tandem repeats. These genomic sites can be detected on metaphase chromosomes and are referred to as nucleolar-organizing regions (NORs). NORs are located on the short arm of acrocentric chromosomes (nos. 13, 14, 15, 21, and 22) and can be visualized by silver staining, due to the abundance of associated argyrophilic proteins.

Each NOR is, on average, 3 Mb in length (80 copies of 43 kb repeat), indicating that a NOR dominates more than half the region on the short arm of an acrocentric chromosome. Each ribosomal gene unit is 43 kb in length, which consists of a transcribed segment (13.7 kb) and an external non-transcribed spacer (29.3 kb). The transcribed segment is transcribed by RNA polymerase I (pol I) which synthesizes the long precursor rRNA (pre-rRNA) which, in mammalian cells, contains the 18S, 5.8S, and 28S rRNA sequence and internal and external transcribed spacers (Olson et al. 2000; Raska 2003). Under normal conditions, only a subset of all rRNA genes are ordinarily active, depending on cell growth activity (Carmo-Fonseca et al. 2000). The remaining rRNA genes are silenced by chromatin modification (Grummt and Pikaard 2003). Although it has frequently been observed that chromosomes are doubled up in malignant and immortalized cells, the number of nucleoli are maintained in such cells as well as in normal cells, suggesting that several chromosomes, especially the short arm of acrocentric chromosomes, are assembled on one nucleolus (Sullivan et al. 2001).

The sites of rDNA transcription, rRNA processing, and the assembly of ribosomes are located in separate, morphologically distinct regions within the nucleolus (Fig. 1) such as the fibrillar center (FC), the dense fibrillar component (DFC), and the granular component (GC), as observed at the structural level by transmission electron microscopy (Olson et al. 2000; Raska 2003). Fully active nucleoli are large, with extensive intermingling of FC, DFC, and GC (Raska 2003). In contrast, quiescent cells or those with nucleolar inactivation following induced transcriptional arrest have small and compact nucleoli in which the distinct morphological components tend to be segregated into adjacent blocks (Shaw and Jordan 1995).

Transcription in the Nucleolus

The fibrillar center is usually seen in the nucleoli of most metazoans but is generally not found in lower eukaryotes. It has been established that the FC harbors several hundred copies of rRNA genes, arranged in one or more tandem arrays at chromosomal loci, NOR, where the pol I and its transcription factors involved in rDNA transcription are enriched (Fig. 2). Of all rRNA genes, only a subset is normally active, and these actively transcribed templates appear to have a more peripheral location, extending into the DFC. Therefore, the precise localization of rDNA transcription with respect to the borders between FC and DFC remains a controversial issue (Shaw and Jordan 1995; Huang 2002; Raska 2003).

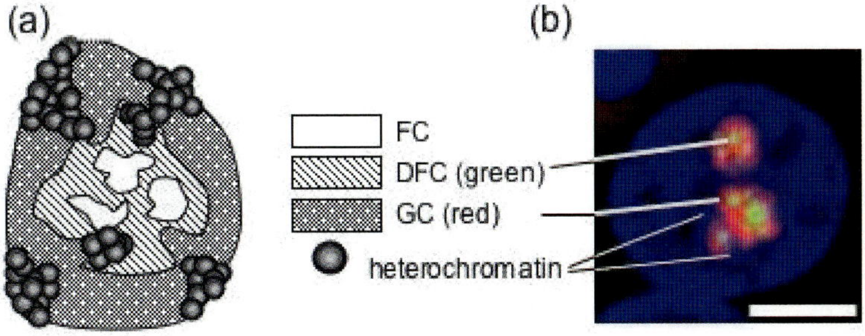

Fig. 1a,b. The structural organization of the nucleolus. **a** The nucleolus is comprised of distinct nucleolar subcompartments, fibrillar center (*FC*; *white area*), dense fibrillar compartment (*DFC*; *lined area*), and granular compartment (*GC*; *dotted area*), and has large heterochromatin domains (*filled circles*) in its periphery, which originate from heterochromatic areas of rDNA genes and the centromere heterochromatin adjacent to rDNA genes. **b** An image of the nucleoli in a HeLa cell stained with antibodies against fibrillarin (*green*) and B23 (*red*). The DNA is stained with Hoechst 33342 (*blue*). Fibrillarin is a member of the pre-rRNA processing complex acting in the DFC, and B23 is involved in the later steps of pre-rRNA processing and the assembly of ribosome subunits in the GC. *Bar* 5 μm

In mammalian cells, the basal pol I transcription machinery is composed of pol I, the upstream binding factor (UBF), and the selectivity factor (SL1), a TBP–TAF_I complex (Comai 2004). It is also known that one of the chromatin remodeling complexes, nucleolar remodeling complex (NoRC), is involved in rDNA transcription and functions as a regulatory target which is related to growth stimuli (Grummt and Pikaard 2003).

RNA Processing in the Nucleolus

The DFC surrounds the FC and functions as a site for pre-rRNA processing (Fig. 2). The DFC contains newly synthesized pre-rRNA, which is a primary large precursor transcript encoded by rDNA, and therefore pre-rRNA requires extensive post-transcriptional processing. The pre-rRNA contains the 18S, 5.8S, and 28S rRNA sequences and internal and external transcribed spacers, is processed by removal of these spacer sequences, and modified by ribose-*O*-2′-methylation and pseudouridylation in the DFC. The processing requires ribonucleoproteins, including a variety of small nucleolar RNAs (snoRNAs) that are responsible not only for the cleavage events that cut the long pre-rRNA, but also for guiding the 2′-*O*-methylation and pseudouridylation of rRNA (Maxwell and Fournier 1995; Smith and Steitz 1997; Tollervey and Kiss 1997). Finally, mature rRNAs (18S, 5.8S, and 28S) are assembled with ribosomal subunits and 5S rRNA, which is synthesized in the nucleoplasm by RNA polymerase III (pol III), to produce 40S and 60S pre-ribosomal particles in the GC (Melese and Xue 1995; Fatica and Tollervey 2002). The GC constitutes the remainder of the nucleolus surrounding the FC and DFC, is enriched in ribosomal proteins and ribosomal subunit assembly factors, and contains nearly completed preribosomal particles destined for the cytoplasm (Tschochner and Hurt 2003). The biogenesis of 40S and 60S ribosomal subunits occurs concurrently with the pre-rRNA maturation process and involves association with 5S rRNA as well as 70–80 different ribosomal proteins.

Nucleoli Have Large Heterochromatin Structures

Highly condensed chromatin structures are observed in the areas surrounding of nucleolus. These structures are formed by several acrocentric chromosomes, which contain heterochromatin blocks derived from centromeres and silent rDNA repeats, since the majority of ribosomal genes are inactivated. It is generally thought that heterochromatin structures accumulate in nucleoli as well as in regions of the nuclear envelope to anchor the chromosomes (Gasser 2002; Spector 2003). Movement of the euchromatin region in the nucleoplasm is constrained, whereas the heterochromatin region is tightly fixed on the nucleoli and nuclear envelope. Consequently, it is believed that the euchromatin structure has an advantage, in that it can associate with transcription factors or chromatin modification enzymes in the nucleoplasm.

The Nucleolus Is a Dynamic Structure Depending on Growth Conditions

In spite of the differences in size and protein composition, nucleoli are constructed in essentially the same structure in all organisms. However, prokaryotes do not have a detectable nucleolus, although ribosome biogenesis occurs actively. This could be because prokaryotic cells have a very low copy number of rRNA genes with a nontandem arrangement. This view was addressed in previous studies, using yeast strains carrying a deletion of chromosomal rDNA repeats. These strains grow depending on the transcription of rRNA driven by RNA polymerase II (pol II), when plasmids with a single rRNA gene are introduced into them. These strains are devoid of typical nucleolar structures; several granules, termed mini-nucleoli, replace the nucleolus (Oakes et al. 1998; Trumtel et al. 2000). Thus it would appear that the construction of a normal nucleolus is basically dependent on the pol I-dependent transcription of rDNA. That is, normal nucleolar formation appears to require a specific rDNA organization, the tandem chromosomal arrangement of rDNA repeats, and transcription activity on the chromosome loci by pol I.

In mammalian cells, the importance of the tandem chromosomal arrangement of rDNA genes has not yet been clarified. However, rRNA transcription plays an important role in the maintenance of nucleolar structures (Scheer and Hock 1999; Olson et al. 2000). In fact, when a pol I transcription inhibitor, actinomycin D, is introduced to mammalian cells, the subnucleolar compartments (FC, DFC, and GC) segregate into distinct regions within the nucleus (Olson and Dundr 2005). In addition, the nucleolar structure is sensitive to the pol II inhibitor, 5,6-dichloro-1β-D-ribofuranosylbenzymidazole (DRB). However, since DRB also functions as a protein kinase inhibitor, the issue of whether the segregation of subnucleolar compartments is due to the suppression of the pol II gene product or nucleolar protein dephosphorylation is currently unclear.

5.2 Dynamic Behavior of Nucleolar Structures During the Cell Cycle

Dynamic Distribution of Nucleolar Subcompartments During Mitosis

Although the nucleolus in yeast remains intact during mitosis, which is separated into daughter cells along the mitotic spindle during the late stages of division, the nucleolus in multicellular organisms has a remarkable characteristic (Fig. 3). That is, in higher eukaryotes, this highly organized structure is disassembled at the beginning of mitosis and various nucleolar

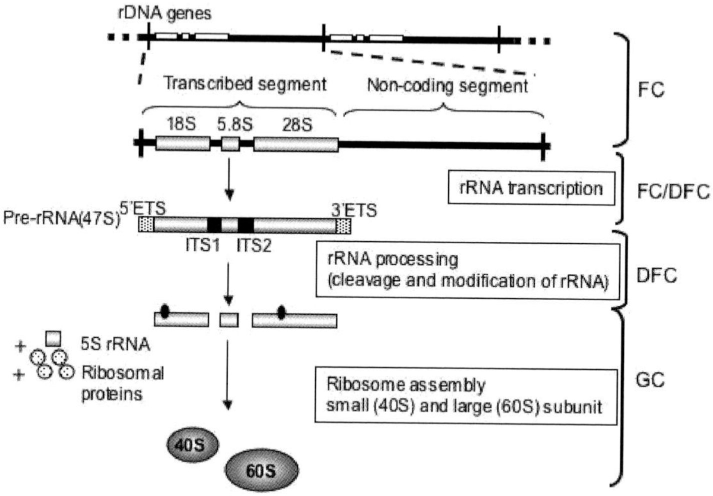

Fig. 2. Schematic representation for the multi-steps in ribosome biogenesis. A single rDNA gene unit is 43 kb in length, and consists of a transcribed segment (13.7 kb) and a noncoding segment (29.3 kb). The rDNA gene is first transcribed as a precursor rRNA (*pre-rRNA*) in the border between the fibrillar center (*FC*) and the dense fibrillar compartment (*DFC*). Pre-rRNAs are then processed in the DFC, where the external transcribed spacer (*ETS*) and internal transcribed spacer (*ITS*) are enzymatically removed. Also, pre-rRNA is methylated and pseudouridylated under the guidance of small nucleolar RNAs, resulting in 18S, 5.8S, and 28S rRNAs. 5S rRNAs are transported out of the nucleolus by RNA polymerase III transcription. These rRNAs are then complexed with ribosomal proteins to form the small (40S) and large (60S) ribosomal subunits in the GC. The nearly mature subunits are exported to the cytoplasm through nuclear pore complexes, and the small and large subunits are eventually incorporated into mature ribosomes in the cytoplasm

proteins leave the subnucleolar structures. The nucleolar proteins are then segregated to daughter cells during mitosis and the nucleolus is reassembled in the resulting daughter cells.

The relative kinetics of the disassembly of nucleolar subcompartments have recently been examined in human cell lines using a quantitative live cell-imaging technique with fluorescent protein-tagged nucleolar proteins (Leung et al. 2004). Nucleolar disassembly starts with the loss of a pol I subunit (RPA39) from the FCs before the nuclear envelope breaks down at the prophase. The dissociation of proteins from outer subcompartments, the GC and the DFC, occurs simultaneously with the breakdown of the nuclear envelope. The NORs, which are chromosome regions containing the repeats of rDNA genes and are morphologically similar to FCs when observed by electron microscopy, then become visible on the mitotic chromosomes, when the interphase chromatin is condensed into mitotic chromosomes. In contrast to RPA39, most of the rDNA transcription factors (the pol I complex, UBF, SL1, and TTF-1) remain bound in the NORs during mitosis. Although this difference in the dissociation kinetics of these proteins remains unclear, rRNA transcription is still detected even after the breakdown of the nuclear envelope, until at least the late prophase (Gebrane-Younes et al. 1997). The disassembly of the nucleolar subcompartments, DFCs and GCs, occurs simultaneously, coinciding with the breakdown of the nuclear envelope and chromosome condensation, followed by the disintegration of the FCs and the formation of NORs (Klein et al. 1998).

During mitosis, various classes of nucleolar components become distributed over several different regions (Hernandez-Verdun et al. 2002). The transcription apparatus, including the pol I complex and its transcription factors, remains located on the NORs of rDNA gene-containing chromosomes (Roussel et al. 1993, 1996; Sirri et al. 1999). In contrast, pre-rRNA processing machinery components become distributed on the surface of all the mitotic chromosomes (Hernandez-Verdun and Gautier 1994; Dundr et al. 1997). The partially processed pre-rRNAs and pre-rRNA processing machinery such as fibrillarin, B23, nucleolin, Nop52, and snoRNAs U3 and U14 also become colocalized mainly on the surface of chromosomes. The colocalization of these different factors that are involved in pre-rRNA processing suggests that the processing complexes remain intact, at least to some extent, during mitosis. Eventually, these pre-rRNA processing factors appear as numerous small dot-like structures, prenucleolar bodies (PNBs) at the telophase and in the early G1 phase, and are involved in the reformation of nucleoli (Jimenez-Garcia et al. 1994; Hernandez-Verdun et al. 2002). This perichromosomal distribution of nucleolar machinery during mitosis may reflect the equal separation of nucleolar proteins to daughter cells as chromosome passenger proteins (Earnshaw and Mackay 1994), although how the chromosome periphery is involved in the separation remains unclear.

In addition, these processing machineries are also distributed to large cytoplasmic spherical particles termed nucleolus derived foci (NDF), which first appear in the prometaphase, diminish in number during the telophase, and eventually disappear in the G1 phase (Dundr and Olson 1998). The NDF do not contain components of the pol I transcription machinery, assuring the spatial separation of rRNA transcription machinery from pre-rRNA processing components during mitosis.

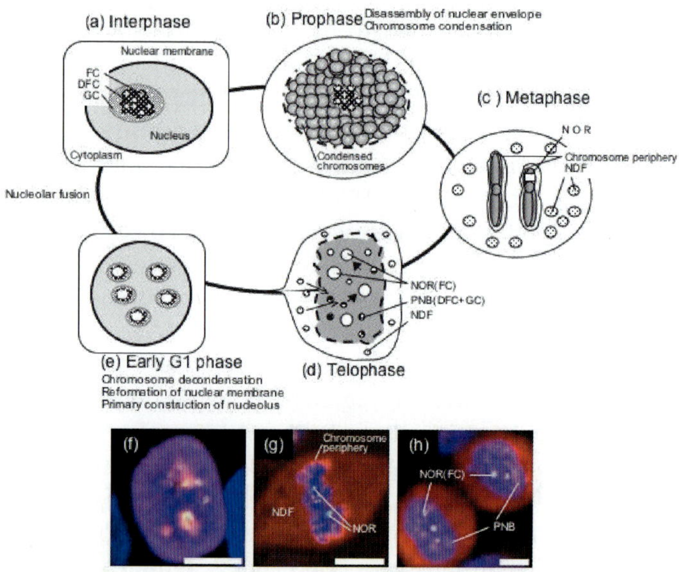

Fig. 3a-h. A model for nucleolar disassembly and reassembly during mitosis. **a** The nucleolus during interphase is organized from distinct subcompartments: the fibrillar center (*FC*) containing the rDNA; the dense fibrillar component (*DFC*) corresponding to the site of pre-rRNA processing; and the granular component (*GC*) corresponding to the assembly of the ribosome subunit. **b** When cells enter the mitotic prophase, the destruction of the nuclear envelope begins through the phosphorylation of proteins (e.g., lamin A) by cyclin B–CDK1, and the chromosomes condense at the same time. A small portion of the protein complexes in DFC and GC are dispersed simultaneously before the destruction of nuclear envelope, and are distributed to the chromosome periphery. After the completion of the nuclear envelope breakdown and chromosome condensation, the FC is replaced by nucleolar-organizing regions on mitotic chromosomes. Although cyclin B–CDK1 also inactivates a part of the rRNA transcription factors, rRNA transcription is still active on some chromosomes during the prometaphase. Nucleolar disassembly appears to take place in the order of DFC/GC and then FC.

Fig. 3. ***(continued)*** **c** At the metaphase, rRNA processing complexes are distributed to the chromosome periphery and cytoplasmic structures, referred to as nucleolar derived foci (*NDF*). In contrast, transcription complexes persist on the nucleolar-organizing region (*NOR*). After the metaphase–anaphase transition, the chromosomes move rapidly to the poles. **d** rRNA transcription is reinitiated during the telophase. Concurrently, the rRNA processing complexes in the chromosome periphery form prenucleolar bodies (*PNB*), and the NDF fuse to PNBs. PNBs are then transported to the NORs, where rRNA transcription is reinitiated. Eventually, nucleolar subcompartments start to reform around the active NORs in the order of FC, DFC, and GC. **e** After the early G1 phase, the nucleoli become mature via association with each other and again become prominent structures in the nucleus. **f–h** The immunofluorescence images show the nucleolar subcompartments at the prophase (**f**), metaphase (**g**), and early G1 phase (**h**), which are stained with antibodies against upstream binding factor (*green*) and B23 (*red*). In **f**, B23 (a GC component) is dissociated from the nucleolus and is distributed to the chromosome periphery. In **g**, NORs are clearly recognized as green punctate signals on mitotic chromosomes, while the processing machinery is distributed to the chromosome periphery and NDF. From the telophase to the early G1 phase (**h**), the rRNA processing machinery forms PNBs in the nucleoplasm (*red*), and PNBs move to active NORs (*green*). *Bars* 5 μm

Molecular Mechanisms of Nucleolar Disassembly

It is well known from previous studies that when cells enter mitosis, RNA transcription is globally repressed in higher eukaryotic cells. Interestingly, the pol I transcriptional machinery (the pol I complex, UBF, SL1, and TTF-1) remains associated with the NORs. However, rDNA transcription is silenced from the prophase to the telophase (Szentirmay and Sawadogo 2000; Spector 2003). Pre-rRNA processing activity is concurrently suppressed at the beginning of mitosis and restored at the exit from mitosis (Dundr et al. 1997).

It is noteworthy that the modification of transcription factors in the pol I complex by phosphorylation appears to play a prominent role in the inhibition of rRNA transcription (Gebrane-Younes et al. 1997; Hernandez-Verdun et al. 2002). Some components of the pol I transcription machinery, such as SL1 (Heix et al. 1998) and TTF-1 (Sirri et al. 1999), are mitotically phosphorylated by cyclin B–CDK1, a mitosis-specific kinase. The phosphorylation of SL1 inhibits its interaction with UBF, which prevents the formation of the transcription initiation complex in vitro (Heix et al. 1998), and cyclin B–CDK1 is necessary for not only the establishment but also the maintenance of the repression from the prophase to the anaphase.

In fact, the in vivo inhibition of cyclin B–CDK1 activity leads to the dephosphorylation of the components of the pol I transcription machinery and restores pol I transcription in mitotic cells (Sirri et al. 2000). Thus, the repression of pol I transcription is regulated by the activity of cyclin B–CDK1.

On the other hand, the in vivo inhibition of cyclin B–CDK1 leads to the accumulation of pre-rRNAs, which are not processed to mature rRNAs, indicating that the repression of rRNA processing activity is not dependent on cyclin B–CDK1. It is reasonable to speculate that pol I transcription and rRNA processing might be regulated by distinct mechanisms.

Nucleolar Reassembly After Mitosis

The process of nucleolar reassembly mainly consists of two steps. In the first step, in the telophase, the pol I transcription machinery is released from mitotic inhibition. Nucleolar reassembly begins in the late anaphase or early telophase when pol I transcription is reinitiated, depending on the decrease in cyclin B–CDK1 activity (Sirri et al. 2000). Cyclin B–CDK1 activity is decreased by the ubiquitin-mediated proteolysis of cyclin B, when chromosomes align on the metaphase plate, just after the spindle-assembly checkpoint is inactivated (Vorlaufer and Peters 1998; Clute and Pines 1999; Kraft et al. 2003). The nucleolar components that are present in the chromosome periphery are released from the decondensing chromosomes and form PNBs together with rRNA processing complexes (Jimenez-Garcia et al. 1994). The population of nucleolar components present in the NDF associates with PNBs in the telophase at around the same time as the formation of the nuclear envelope. Collectively, the inactivation of cyclin B–CDK1, the reactivation of pol I, and the formation of PNBs are critical for the early phase of nucleolar reassembly.

In the second step, the processing machinery is recruited to the newly growing nucleoli. The PNBs appear to be involved in pre-RNA processing, because they contain a number of nucleolar proteins, snoRNAs, and partially processed pre-rRNAs (Azum-Gelade et al. 1994; Jimenez-Garcia et al. 1994). It is important to note that PNBs do not contain any transcriptional components, and that transcription is never initiated within them (Bell et al. 1992; Gebrane-Younes et al. 1997). The PNBs then migrate toward the NORs to deliver the nucleolar components and approach each other to reassemble the mature nucleolus. Investigations of the dynamic behavior of PNBs in living cells, using fluorescent proteins, have demonstrated that the delivery of pre-rRNA processing machinery from the PNBs to the NORs occurs via the spatial exchange of components

(Dundr et al. 2000; Savino et al. 2001), as evidenced by the observation that dense fibrils transiently connect PNBs to NORs. That is, the dynamic behavior of PNBs is required for the transfer of protein from the PNBs to the reforming nucleoli in the late phase of nucleolar reassembly. Previous investigations have shown that a cyclin B–CDK1 inhibitor induces transcription by pol I and the formation of PNBs in mitotic cells, but that the inhibitor does not induce the transport of pre-rRNA processing machinery from the PNBs to the NORs (Sirri et al. 2002). This suggests that an alternate activation mechanism for the components in PNBs or NORs is operative in the late phase of the nucleolar reformation process. Thus, the structure of the complete nucleolus is reformed in the mid G1 phase.

Moreover, a recent study indicated that nucleolar reassembly consists of sequential reconstruction processes of the nucleolar subcompartments in the order FC, DFC, and finally, GC (Leung et al. 2004). The temporal order of nucleolar reassembly reflects the root of ribosome biogenesis; the reactivation of pol I induces the formation of FC, and DFC is then constructed as compartments of rRNA processing. Finally, GC is constructed for ribosome assembly.

The Activity of rRNA Transcription and rRNA Processing: Which Activity Governs the Late Stage of Nucleolar Reassembly?

As described above, the degradation of cyclin B–CDK1 leads to the initiation of pol I transcription in the telophase. This event appears to trigger the early stage of nucleolar reassembly. However, a transcription factor, UBF, is still inactivated during this stage. In addition, it has been shown that phosphorylation by cyclin D1–CDK4 and cyclin E–CDK2 is required for the activation of UBF (Voit et al. 1999). From this evidence, it is conceivable that during the early G1 phase, pol I transcription activity is not so high compared with the following stages of the cell cycle. The late stage of nucleolar reassembly is therefore regulated by activity other than rRNA transcription. In support of this conclusion, the inhibition of pol I transcription during the nucleolar reassembly process does not impair the recruitment of nucleolar proteins such as components of the DFC or GC in the vicinity of the NORs (Dousset et al. 2000). The nucleoli that are formed under these conditions contain partially processed pre-rRNAs synthesized in the early prophase of the last cell cycle, suggesting that pre-rRNAs may play some important role in the late stage of nucleolar reassembly.

Partially processed pre-rRNAs are produced in the prophase when the rRNA processing complex leaves the nucleoli, since rRNA transcription is still active at that time (Gautier et al. 1992). The pre-rRNAs are maintained in a stable state during mitosis in the NDFs or PNBs in which the pre-rRNAs form processing complexes with numerous nucleolar proteins and snoRNAs (Dundr and Olson 1998; Pinol-Roma 1999; Dousset et al. 2000). This suggests that the processing complexes formed in the NDFs and PNBs are nucleated by the pre-rRNAs. Since the late stage of nucleolar reassembly occurs in the absence of pol I transcription, pre-rRNA processing activity has an important role in the late stage of nucleolar reassembly. In fact, during the early stage of *Xenupus lavevis* embryogenesis, rRNA transcription is established after the transition to the mid-blastula stage. Until the mid-blastula transition, nucleoli are repeatedly disassembled and assembled, and are devoid of rRNA transcription during several cell cycles. However, rRNA processing machinery is recruited to the NORs to associate with pre-rRNA of maternal origin (Verheggen et al. 1998, 2000). Thus, the late stage of nucleolar reassembly is accomplished by the formation of DFC and GC, depending on the protein transport from dynamic PNBs to the NORs through rRNA processing activity.

5.3 Conclusions

Based on live cell-imaging studies, a model for cell-cycle-dependent nucleolar dynamics has been established. When cyclin B–CDK1 is transported to the nucleus during the G2/M transition, the pol I transcription complex is inactivated and the kinase triggers the disassembly of the nucleolus as well as the breakdown of the nuclear envelope. The nucleolar subcompartments, DFC and GC, are disassembled more rapidly than FC so that partially processed pre-rRNAs are released from the nucleolus and associate with processing components. This complex is distributed to the periphery of chromosome and cytoplasmic particles (NDF), while most of the components of the pol I complex are located in the region of the rDNA sequences on several chromosomes, referred to as NOR. Such a discriminative distribution is not observed in the pol II or pol III transcription machinery that is dispersed within the cytoplasm, and has a great advantage in terms of the equipartition of nucleolar proteins and the immediate supply of ribosomes during the G1 phase. During the formation of the nuclear envelope in the telophase, the rRNA processing machineries are distributed to the PNB from the chromosome periphery. In contrast, a processing complex in the NDF is dissociated and transported to the nucleus through the newly reassembled nuclear envelope and are eventually incorporated into the PNB. rRNA transcription is reactivated by the inactivation of cyclin B–CDK1. The components of the PNB are then recruited to the NOR to form the mature nucleolus (Leung et al. 2004). The transfer of protein to the NOR is dependent on the rRNA processing rather than rRNA transcription activity (Verheggen et al. 1998, 2000; Dousset et al. 2000). Another interesting discovery is that, even after the reassembly of nucleolus, the nucleolus is under the control of CDK throughout the interphase (Sirri et al. 2002). This will be a clue to understanding how the nucleolus is maintained in a cell cycle-dependent manner (Hernandez-Verdun et al. 2003).

A statistical evaluation shows that the number of FC foci and functional nucleoli after mitosis in daughter cells is mutually similar and nearly equivalent to that of their mother cell, compared with unrelated cells (Leung et al. 2004), raising another interesting unanswered question of how the number and distribution of nucleoli are accurately maintained in the daughter cell nuclei. NORs may participate in the structural and functional preservation of the nucleolar states in daughter cell lineages, since it has been shown that several NORs originating from several chromosomes fuse with each other to participate in the formation of one nucleolus (Hernandez-Verdun et al. 2002). Alternatively, transcriptionally silent heterochromatic regions adjacent to rDNA genes, which are bound with Net1 and Sir2 in yeast, may play crucial roles in the maintenance of nucleolar integrity (Carmo-Fonseca et al. 2000). In addition, the fact that global chromosome positioning is maintained after mitosis (Gerlich et al. 2003; Walter et al. 2003) may lead to an understanding of the spatial positioning and the number of nucleoli in daughter cells.

References

Azum-Gelade MC, Noaillac-Depeyre J, Caizergues-Ferrer M, Gas N (1994) Cell cycle redistribution of U3 snRNA and fibrillarin. Presence in the cytoplasmic nucleolus remnant and in the prenucleolar bodies at telophase. J Cell Sci 107:463–475

Bell P, Dabauvalle MC, Scheer U (1992) In vitro assembly of prenucleolar bodies in Xenopus egg extract. J Cell Biol 118:1297–1304

Carmo-Fonseca M, Mendes-Soares L, Campos I (2000) To be or not to be in the nucleolus. Nat Cell Biol 2:E107–E112

Clute P, Pines J (1999) Temporal and spatial control of cyclin B1 destruction in metaphase. Nat Cell Biol 1:82–87

Comai L (2004) Mechanism of RNA polymerase I transcription. Adv Protein Chem 67:123–155

Dousset T, Wang C, Verheggen C, Chen D, Hernandez-Verdun D, Huang S (2000) Initiation of nucleolar assembly is independent of RNA polymerase I transcription. Mol Biol Cell 11:2705–2717
Dundr M, Olson MO (1998) Partially processed pre-rRNA is preserved in association with processing components in nucleolus-derived foci during mitosis. Mol Biol Cell 9:2407–2422
Dundr M, Meier UT, Lewis N, Rekosh D, Hammarskjold ML, Olson MO (1997) A class of nonribosomal nucleolar components is located in chromosome periphery and in nucleolus-derived foci during anaphase and telophase. Chromosoma 105:407–417
Dundr M, Misteli T, Olson MO (2000) The dynamics of postmitotic reassembly of the nucleolus. J Cell Biol 150:433–446
Earnshaw WC, Mackay AM (1994) Role of nonhistone proteins in the chromosomal events of mitosis. FASEB J 8:947–956
Fatica A, Tollervey D (2002) Making ribosomes. Curr Opin Cell Biol 14:313–318
Gasser SM (2002) Visualizing chromatin dynamics in interphase nuclei. Science 296:1412–1416
Gautier T, Robert-Nicoud M, Guilly MN, Hernandez-Verdun D (1992) Relocation of nucleolar proteins around chromosomes at mitosis. A study by confocal laser scanning microscopy. J Cell Sci 102:729–737
Gebrane-Younes J, Fomproix N, Hernandez-Verdun D (1997) When rDNA transcription is arrested during mitosis, UBF is still associated with non-condensed rDNA. J Cell Sci 110:2429–2440
Gerlich D, Beaudouin J, Kalbfuss B, Daigle N, Eils R, Ellenberg J (2003) Global chromosome positions are transmitted through mitosis in mammalian cells. Cell 112:751–764
Grummt I, Pikaard CS (2003) Epigenetic silencing of RNA polymerase I transcription. Nat Rev Mol Cell Biol 4:641–649
Heix J, Vente A, Voit R, Budde A, Michaelidis TM, Grummt I (1998) Mitotic silencing of human rRNA synthesis: inactivation of the promoter selectivity factor SL1 by cdc2/cyclin B-mediated phosphorylation. EMBO J 17:7373–7381
Hernandez-Verdun D, Gautier T (1994) The chromosome periphery during mitosis. Bioessays 16:179–185
Hernandez-Verdun D, Roussel P (2003) Regulators of nucleolar functions. Prog Cell Cycle Res 5:301–308
Hernandez-Verdun D, Roussel P, Gebrane-Younes J (2002) Emerging concepts of nucleolar assembly. J Cell Sci 115:2265–2270
Huang S (2002) Building an efficient factory: where is pre-rRNA synthesized in the nucleolus? J Cell Biol 157:739–741
Jimenez-Garcia LF, Segura-Valdez ML, Ochs RL, Rothblum LI, Hannan R, Spector DL (1994) Nucleologenesis: U3 snRNA-containing prenucleolar bodies move to sites of active pre-rRNA transcription after mitosis. Mol Biol Cell 5:955–966
Klein C, Cheutin T, O'Donohue MF, Rothblum L, Kaplan H, Beorchia A, Lucas L, Heliot L, Ploton D (1998) The three-dimensional study of chromosomes

and upstream binding factor-immunolabeled nucleolar organizer regions demonstrates their nonrandom spatial arrangement during mitosis. Mol Biol Cell 9:3147–3159

Kraft C, Herzog F, Gieffers C, Mechtler K, Hagting A, Pines J, Peters JM (2003) Mitotic regulation of the human anaphase-promoting complex by phosphorylation. EMBO J 22:6598–6609

Leung AK, Gerlich D, Miller G, Lyon C, Lam YW, Lleres D, Daigle N, Zomerdijk J, Ellenberg J, Lamond AI (2004) Quantitative kinetic analysis of nucleolar breakdown and reassembly during mitosis in live human cells. J Cell Biol 166:787–800

Maxwell ES, Fournier MJ (1995) The small nucleolar RNAs. Annu Rev Biochem 64:897–934

Melese T, Xue Z (1995) The nucleolus: an organelle formed by the act of building a ribosome. Curr Opin Cell Biol 7:319–324

Oakes M, Aris JP, Brockenbrough JS, Wai H, Vu L, Nomura M (1998) Mutational analysis of the structure and localization of the nucleolus in the yeast *Saccharomyces cerevisiae*. J Cell Biol 143:23–34

Olson MO, Dundr M (2005) The moving parts of the nucleolus. Histochem Cell Biol 123:203–216

Olson MO, Dundr M, Szebeni A (2000) The nucleolus: an old factory with unexpected capabilities. Trends Cell Biol 10:189–196

Pinol-Roma S (1999) Association of nonribosomal nucleolar proteins in ribonucleoprotein complexes during interphase and mitosis. Mol Biol Cell 10:77–90

Raska I (2003) Oldies but goldies: searching for Christmas trees within the nucleolar architecture. Trends Cell Biol 13:517–525

Roussel P, Andre C, Masson C, Geraud G, Hernandez-Verdun D (1993) Localization of the RNA polymerase I transcription factor hUBF during the cell cycle. J Cell Sci 104:327–337

Roussel P, Andre C, Comai L, Hernandez-Verdun D (1996) The rDNA transcription machinery is assembled during mitosis in active NORs and absent in inactive NORs. J Cell Biol 133:235–246

Savino TM, Gebrane-Younes J, De Mey J, Sibarita JB, Hernandez-Verdun D (2001) Nucleolar assembly of the rRNA processing machinery in living cells. J Cell Biol 153:1097–1110

Scheer U, Hock R (1999) Structure and function of the nucleolus. Curr Opin Cell Biol 11:385–390

Shaw PJ, Jordan EG (1995) The nucleolus. Annu Rev Cell Dev Biol 11:93–121

Sirri V, Roussel P, Hernandez-Verdun D (1999) The mitotically phosphorylated form of the transcription termination factor TTF-1 is associated with the repressed rDNA transcription machinery. J Cell Sci 112:3259–3268

Sirri V, Roussel P, Hernandez-Verdun D (2000) In vivo release of mitotic silencing of ribosomal gene transcription does not give rise to precursor ribosomal RNA processing. J Cell Biol 148:259–270

Sirri V, Hernandez-Verdun D, Roussel P (2002) Cyclin-dependent kinases govern formation and maintenance of the nucleolus. J Cell Biol 156:969–981

Smith CM, Steitz JA (1997) Sno storm in the nucleolus: new roles for myriad small RNPs. Cell 89:669–672

Spector DL (2003) The dynamics of chromosome organization and gene regulation. Annu Rev Biochem 72:573–608

Sullivan GJ, Bridger JM, Cuthbert AP, Newbold RF, Bickmore WA, McStay B (2001) Human acrocentric chromosomes with transcriptionally silent nucleolar organizer regions associate with nucleoli. EMBO J 20:2867–2874

Szentirmay MN, Sawadogo M (2000) Spatial organization of RNA polymerase II transcription in the nucleus. Nucleic Acids Res 28:2019–2025

Tollervey D, Kiss T (1997) Function and synthesis of small nucleolar RNAs. Curr Opin Cell Biol 9:337–342

Trumtel S, Leger-Silvestre I, Gleizes PE, Teulieres F, Gas N (2000) Assembly and functional organization of the nucleolus: ultrastructural analysis of *Saccharomyces cerevisiae* mutants. Mol Biol Cell 11:2175–2189

Tschochner H, Hurt E (2003) Pre-ribosomes on the road from the nucleolus to the cytoplasm. Trends Cell Biol 13:255–263

Verheggen C, Le Panse S, Almouzni G, Hernandez-Verdun D (1998) Presence of pre-rRNAs before activation of polymerase I transcription in the building process of nucleoli during early development of *Xenopus laevis*. J Cell Biol 142:1167–1180

Verheggen C, Almouzni G, Hernandez-Verdun D (2000) The ribosomal RNA processing machinery is recruited to the nucleolar domain before RNA polymerase I during *Xenopus laevis* development. J Cell Biol 149:293–306

Voit R, Hoffmann M, Grummt I (1999) Phosphorylation by G1-specific cdk-cyclin complexes activates the nucleolar transcription factor UBF. EMBO J 18:1891–1899

Vorlaufer E, Peters JM (1998) Regulation of the cyclin B degradation system by an inhibitor of mitotic proteolysis. Mol Biol Cell 9:1817–1831

Walter J, Schermelleh L, Cremer M, Tashiro S, Cremer T (2003) Chromosome order in HeLa cells changes during mitosis and early G1, but is stably maintained during subsequent interphase stages. J Cell Biol 160:685–697

6. Dynamics, Roles, and Diseases of the Nuclear Membrane, Lamins, and Lamin-binding Proteins

Tsuneyoshi Horigome[1,2], Yasuhiro Hirano[3], and Kazuhiro Furukawa[1]

[1]Department of Chemistry, Faculty of Science, [2]Center for Transdisciplinary Research, and [3]Course of Biosphere Science, Graduate School of Science and Technology, Niigata University, Igarashi-2, Niigata 950-2181, Japan

6.1 Introduction

The nuclear envelope is the boundary between the nucleus and cytoplasm. The nuclear envelope consists of two lipid bilayers, the nuclear lamina and nuclear pore complexes (NPCs) (Fig. 1). The transfer of materials between the nucleoplasm and cytoplasm is regulated by NPCs. The nuclear envelope is also the basis of the nuclear architecture and functions. Inner nuclear membrane proteins connect the nuclear lamina and the nuclear membrane (Fig. 1). The nuclear envelope provides a platform for chromatin. The participation of inner nuclear membrane proteins in gene replication and expression has been demonstrated. The nuclear envelope also dynamically changes in structure during the cell cycle. In vertebrates, it is disassembled in the prometaphase, and reassembled at the transition from the anaphase to the telophase. Some control systems for these dynamic changes, i.e., microtubule-dynein, Ran-importinβ and phosphorylation–dephosphorylation systems, were partially revealed recently. On the other hand, it has become clear that when some proteins supporting such nuclear envelope functions are mutated, unexpected diseases, i.e., muscular dystrophies, familial partial lipodystrophy, cardiomyopathy, progeria, and others, are caused.

6.2 Nuclear Membrane and Lamina Proteins

Intrinsic membrane proteins specific to the inner nuclear membrane are summarized in Table 1. Many novel proteins were recently found. The

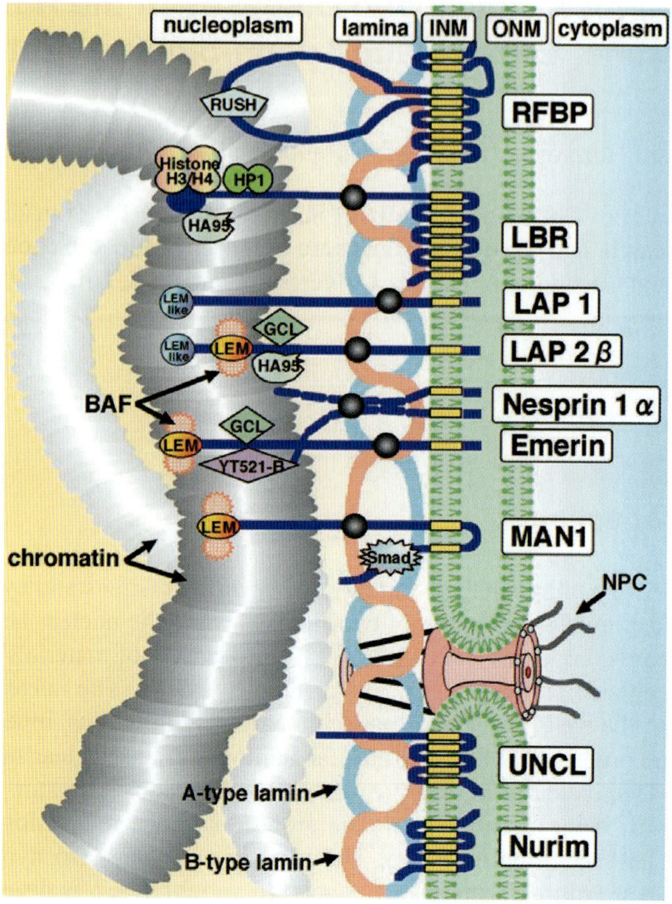

Fig. 1. Schematic view of the nuclear envelope and chromatin. The inner and outer nuclear membranes (*INM* and *ONM*) are joined by the NPC. A- and B-type lamins are shown as *blue* and *red thick-curved lines*, respectively. Selected proteins of the inner nuclear membrane and proteins that interact with INM proteins in the nuclear interior are shown. The LEM domain and LEM-like domain are depicted. The *gray circles* represent interactions with lamins. Chromatin at the nuclear periphery is structurally condensed, since it is mostly transcriptionally silent

development of new visual assay systems contributed to their finding (Rolls et al. 1999; Ding et al. 2000). Proteome analysis of the nuclear envelope is also a powerful method for finding novel proteins. Nineteen unknown or uncharacterized proteins were identified in the nuclear envelope fraction on proteome analysis (Dreger et al. 2001). Cross-contamination by other organelles causes many problems in proteomics. Schirmer improved this point, and found 67 uncharacterized open reading frames in a liver nuclear membrane fraction (Schirmer et al. 2003). Interestingly, 23 of them were mapped within chromosome regions linked to a variety of dystrophies (Schirmer et al. 2003).

The major proteins comprising the nuclear lamina are lamins. Vertebrates have three major A-type lamin proteins (lamins A, ΔA10, and C) derived from the LMNA gene, and two major B-type lamins (lamins B1 and B2) derived from the LMNB1 and LMNB2 genes. All vertebrate cells constitutively express at least one B-type lamin, whereas A-types are developmentally regulated and expressed in differentiated cells.

Table 1. Properties of intrinsic membrane proteins specific to the inner nuclear membrane

Protein	Poly-pepti-de mass (kDa)	Lamin binding	Chro-matin binding	Other binding proteins	Com-ments	References[a]
LAP1A	75	A/B	No	–	–	Senior A and Gerace L 1988
LAP1B	68	A/B	No	–	Splice variant of LAP1A	Senior A and Gerace L 1988
LAP1C	57	A/B?	No	–	Splice variant of LAP1A	Martin L et al. 1995
LAP2β	50	B	Yes	BAF, GCL, HA95	LEM protein	Foisner R and Gerace L 1993; Furukawa K et al. 1995
LAP2ε,δ,γ	38–46	Most likely B	Pro-bable	–	Splice variants of LAP2β	Berger R et al. 1996; Harris CA et al. 1995

Table 1 (*continued*).

Protein	Polypeptide mass (kDa)	Lamin binding	Chromatin binding	Other binding proteins	Comments	References[a]
Emerin	29	A/B	Probable	BAF, GCL, YT521-B, Actin, Nesprin 1α	LEM protein	Bione S et al. 1994
MAN1	82	B	Probable	BAF, Smads 1, 5 and 8	LEM protein	Lin F et al. 2000
Lamin B receptor	70	B	Yes	Histone H3/H4, HP1, p18, HA95, LBR-kinase	Sterol C-14 reductase activity	Woman HJ et al. 1990; Silve S et al. 1998
Nurim	29	–	–	–	–	Rolls MM et al. 1999
Unc-84	125	B	–	Unc-83	–	Malone CJ et al. 1999
Unc-83	118	–	–	Unc-84	–	Starr DA et al. 2001
RFBP	126	–	–	RUSH	ATPase	Mansharamani M et al. 2001
LUMA	45	–	–	–	–	Dreger M et al. 2001
UNCL	27	–	–	–	RNA binding	Fitzgerald J et al. 2000
Nesprin1 α (Syne1B, Myne1)	112	A	–	Emerin, Self-association	Spectrin repeats, Interrupted LEM domain	Apel ED et al. 2000; Mislow JMK et al. 2002; Zhang Q et al. 2001

Table 1 (*continued*).

Protein	Polypeptide mass (kDa)	Lamin binding	Chromatin binding	Other binding proteins	Comments	References[a]
Nesprin2 α,β,γ (Syne2)	377 87 61	–	–	–	Spectrin repeats	Apel ED et al. 2000; Zhang Q et al. 2001

[a]Most of these references are not included in the references at the end of this chapter

6.3 Dynamics of the Nuclear Envelope

The major known regulation systems for assembly and disassembly of the nuclear envelope are the microtubule–dynein system, Ran–importinβ system, p97 system, and phosphorylation–dephosphorylation system. The microtubule–dynein system functions in the breakdown of the nuclear envelope on entering mitosis (Beaudouin et al. 2002; Salina et al. 2002). The Ran–importinβ system functions in the formation of the nuclear envelope on exiting mitosis. The Ran–importinβ system indicates the site where nuclear membrane fusion (Harel et al. 2003) and NPC formation (Walther et al. 2003) should occur. Ran-GDP is converted into Ran-GTP by RCC1 on chromatin, and consequently Ran-GTP exists on or near the surface of chromatin at a high concentration. The suppression of nuclear envelope formation by importinβ is abolished by Ran-GTP, and thereby nuclear membrane fusion (Harel et al. 2003) and NPC formation are restricted to the surface of chromatin. Nuclear membrane fusion is caused by machineries involving the p97–Ufd1–Npl4 and p97–p47 complexes (Hetzer et al. 2001). The participation of a multifunctional enzyme, glyceraldehyde-3-phosphate dehydrogenase, in the nuclear membrane fusion step has also been demonstrated with a *Xenopus* egg extract system (Nakagawa et al. 2003).

The dynamics of the assembly and disassembly of the nuclear envelope are precisely regulated by the phosphorylation–dephosphorylation system. The phosphorylation and dephosphorylation of nuclear membrane and nuclear lamina proteins are summarized in Table 2. Most nuclear membrane proteins lose their binding ability as to chromatin on specific phosphorylation at the beginning of mitosis. A- and B-type lamins are also depolymerized and solubilized through phosphorylation (Table 2). Lamin B and lamin B receptor (LBR) recover the ability to assemble the nuclear lamina and to bind to chromatin through dephosphorylation (Steen et al. 2000; Table 2). Recruitment of protein phosphatase 1 (PP1) to the inner nuclear membrane by AKAP149 and the resulting dephosphorylation of lamin B are prerequisites for the lamina formation at the end of the metaphase (Steen et al. 2000). PP1 is also required to maintain the cell nucleus integrity during the interphase (Steen et al. 2003).

Table 2. Phosphorylation and dephosphorylation of nuclear membrane and lamina proteins

Protein	R	Sites	Reaction system	Enzymes	Cell cycle	Comments	References[a]
Nuclear membrane proteins							
LAP1A, 1B, 1C	P	–	Cell-free	–	M	–	Foisner R and Gerace L 1993
LAP2β	P	–	In vivo, In vitro	–	M, I	Inhibition of binding to lamin B1 and chromosomes	Foisner R and Gerace L 1993
	P	T74, T159, S176, S179	In vivo	–	I	–	Dreger M et al. 1999

R, reaction type; P, phosphorylation; D, dephosphorylation. Sites: S, serine; T, threonine. Enzymes: SRPK, SR protein-specific kinase; cdc2, cdc2 kinase; PKA, protein kinase A; CaMKII, calmodulin dependent protein kinase II; PKC, protein kinase C; PP1, protein phosphatase 1. Cell cycle: I, interphase; S, synthetic phase; M, mitotic phase

Table 2 (continued).

Protein	R	Sites	Reaction system	Enzymes	Cell cycle	Comments	References[a]
Nuclear membrane proteins (continued)							
LBR	P	S76, S78, S80, S82, S84	In vivo	SRPK	I	–	Nikolakaki E et al. 1997
	P	S71, S76, S78, S80, S82, S84	In vivo	SRPK, cdc2	M	–	Nikolakaki E et al. 1997
	P	S71	Cell-free	cdc2	M	Suppression of binding to chromatin	Takano M et al. 2004
	D	S71	Cell-free	PP1	I	Recovery of binding to chromatin	Ito H et al. in preparation
Emerin	P	–	In vivo	–	I/S/M		Ellis JA et al. 1998
AKAP 149	P	S	In vivo	–	S	Dissociation of PP1	Steen RL et al. 2003

R, reaction type; P, phosphorylation; D, dephosphorylation. Sites: S, serine; T, threonine. Enzymes: SRPK, SR protein-specific kinase; cdc2, cdc2 kinase; PKA, protein kinase A; CaMKII, calmodulin dependent protein kinase II; PKC, protein kinase C; PP1, protein phosphatase 1. Cell cycle: I, interphase; S, synthetic phase; M, mitotic phase

Table 2 (continued).

Protein	R	Sites	Reaction system	En-zymes	Cell cycle	Comments	References[a]
Lamins							
Lamin B	D	–	Cell-free	PP1	I	–	Thompson LJ et al. 1997
	P	–	Cell-free	PKC	I	–	Collas P et al. 1997
	p	S395, S405	In vitro	PKCβII	–	–	Goss VL et al. 1994
	p	S23	In vitro	cdc2	–	–	Goss VL et al. 1994
	P	S405	In vivo	PKCβII	M	–	Goss VL et al. 1994
	P	–	In vivo		–	–	Hocevar BA et al. 1993
	P	S395, S405	In vitro	PKC	–	Solubilization of lamin B	Hocevar BA et al. 1993
Lamin A/C	P	S22, S392	In vitro	–	–	Responsible for depolymerization	Heald R and Mckeon F 1990
Lamin C	P	S22, S404, S406(?)	Cell-free	–	I	–	Ward GE and Kirschner MW 1990
		S22, S392, S404, S406(?)	Cell-free	–	M	–	Ward GE and Kirschner MW 1990

R, reaction type; P, phosphorylation; D, dephosphorylation. Sites: S, serine; T, threonine. Enzymes: SRPK, SR protein-specific kinase; cdc2, cdc2 kinase; PKA, protein kinase A; CaMKII, calmodulin dependent protein kinase II; PKC, protein kinase C; PP1, protein phosphatase 1. Cell cycle: I, interphase; S, synthetic phase; M, mitotic phase

[a]Most of these references are not included in the references at the end of this chapter

6.4 Roles of Intrinsic Nuclear Membrane Proteins and Lamins

The major roles of intrinsic inner nuclear membrane proteins and lamins are classified into three groups, i.e., roles regarding the nuclear architecture, roles as systems for the assembly and disassembly of nuclear envelope during the cell cycle, and roles as regulation systems for chromatin function. In this section, the roles of selected nuclear envelope proteins are summarized.

Lamina-Associated Polypeptide 2β

Lamina-associated polypeptide 2β (LAP2β) binds to lamin B, barrier-to-autointegration factor (BAF) (Furukawa 1999), and HA95 (Martins et al. 2003b). LAP2β belongs to the "LEM domain" family, which mediates direct binding to BAF. BAF is known to play crucial roles in cell cycle progression and nuclear organization in *Drosophila* (Furukawa et al. 2003). The association of LAP2β with HA95 is regulated by PKA (Martins et al. 2003a), and disruption of this association prevents the initiation step, but not the elongation step, of DNA replication (Martins et al. 2003b). These observations suggested that the binding of LAP2β to chromatin through HA95 is important for initiation of DNA replication. In addition, it has also been shown that LAP2β interacts with a transcriptional repressor, germ cell-less (GCL), in the interphase nucleus and thereby mediates transcriptional repression (Nili et al. 2001). The roles of LAP2β in membrane–chromatin attachment, lamina assembly, and nuclear envelope formation have also been demonstrated using a *Xenopus* egg extract system with truncated LAP2β proteins (Gant et al. 1999).

Emerin

Emerin binds to lamins A and B, BAF, and a transcriptional repressor, GCL (Holaska et al. 2003). Emerin belongs to the "LEM domain" family. BAF competes with GCL as to binding to emerin in vitro (Holaska et al. 2003). Further emerin-binding proteins were searched for recently by means of a yeast two-hybrid method in human heart and HeLa cell cDNA libraries, a splicing factor, YT521-B, a transcription factor, vav-1, and a death-promoting transcriptional repressor, Btf, being found (Wilkinson et al. 2003; T. Haraguchi, personal communication). Emerin bound YT521-B at the same site as that for GCL in vitro and the binding inhibited YT521-B-dependent splice site selection in vivo. These results suggested that emerin participates in the regulation of transcription and splicing.

MAN1

MAN1 of *Caenorhabditis elegans* is known to interact directly with B-type lamin and BAF in vitro like emerin in the N-terminal region (Liu et al. 2003). MAN1 is a LEM domain protein, and at least one of its functions overlap those of emerin in *C. elegans* (Liu et al. 2003). On the other hand, MAN1 was recently identified as a neutralizing factor in *Xenopus* embryos (Osada et al. 2003). The neutralizing and bone morphogenetic protein (BMP)-antagonizing activities of MAN1 reside in the C-terminal region, and the C-terminal region binds to Smad1, Smad5, and Smad8, which are intracellular mediators of the BMP pathway (Osada et al. 2003). Thus, MAN1 acts as a Smad-interacting protein that antagonizes BMP signaling during *Xenopus* embryogenesis.

LBR

LBR provides an anchorage site in the nuclear envelope for chromatin in interphase cells (Pyrpasopoulou et al. 1996). LBR is also suggested to function in the membrane targeting to chromatin on exiting mitosis (Collas et al. 1996). LBR has been shown to bind directly to the DNA part of chromatin (Duband-Gouleta and Courvalin 2000, Takano et al. 2002, Takano et al. 2004), HP1 (Ye et al. 1997), and histone H3/H4 complexes including HP1 (Polioudaki et al. 2001) by in vitro experiments. The binding to HP1 was also confirmed in vivo by means of a yeast two-hybrid method (Ye et al. 1997). LBR seems to be important for maintenance of the nuclear architecture because nuclear abnormalities in the karyotype of neutrophils are caused by LBR deficiency (see below; Hoffmann et al. 2002). LBR exhibits 3β-hydroxysterol Δ^{14}-reductase activity and thereby contributes to sterol metabolism (Silve et al. 1998; Waterham et al. 2003).

Lamins

The lamins are the major structural components of the nuclear lamina (Gruenbaum et al. 2003; Hutchison 2002). In cells that do not exhibit A-type lamin expression, highly elongated or irregularly shaped nuclei with herniation of the nuclear membrane are often observed (Sullivan et al.

1999). These observations are consistent with the idea that the lamina provides strength and support for the nuclear membrane. The lamina plays an important role in anchoring the elements of the nuclear membrane in their correct positions. Other than these structural functions of lamins, it is widely accepted that lamins play roles in the regulation of gene expression and replication. When lamin B3 was depleted from *Xenopus* egg extracts with specific antibodies, small nuclei that failed to show initiation of DNA replication were assembled, and readdition of purified lamin B3 restored the DNA replication activity (Goldberg et al. 1995). On the other hand, when dominant-negative lamin mutants were added to assembled nuclei in a *Xenopus* egg extract to disrupt the lamin organization, the elongation step but not the initiation step of DNA replication was prevented (Moir et al. 2000). Both studies suggest the direct participation of lamins in DNA replication, although their precise roles are not yet clear. In both somatic cells and transcriptionally active *Xenopus* embryonic nuclei, disruption of the lamina by dominant negative mutants inhibits RNA polymerase II activity, which provides direct evidence of the involvement of lamins in transcription (Spann et al. 2002). A-type lamins have also been reported to be colocalized with RNA splicing-factor speckles in interphase cells (Jagatheesan et al. 1999). A-type lamins might also support nuclear compartments containing proteins involved in RNA splicing.

6.5 Nuclear Envelope and Genetic Diseases

Considerable interest has been focused on the nuclear envelope in recent years following the realization that several human diseases are linked to defects in genes encoding nuclear envelope-specific proteins, i.e., A-type lamins, emerin, LBR, and ALADIN. These diseases, described collectively as the "nuclear envelopathies," are summarized in Table 3. Studies on the molecular and cellular mechanisms underlying these diseases have only just started.

Table 3. The nuclear envelopathies

Disease	Type of mutation	Principal phenotypes	References[a]
A-type lamins			
Autosomal-dominant Emery–Dreifuss muscular dystrophy (EDMD2)	Throughout the coding sequence. Includes amino-acid substitutions, codon deletions, frameshifts, and premature stop codons	Early contractures of tendons, slowly progressive muscle weakness, and cardiac conduction defects	Bonne G et al. 2000; Genschl J and Schmidt HH 2000: Brown CA et al. 2001
Autosomal-recessive Emery–Dreifuss muscular dystrophy (EDMD3)	A single homozygous mutation at C664T, causing amino-acid substitution H222Y	Early contractures of tendons, slowly progressive muscle weakness, and cardiac conduction defects	Raffaele di Barletta M et al. 2000
Limb-girdle muscular dystrophy type 1 (LGMD1B)	Amino-acid substitution, codon deletion and splice donor mutations	Progressive pelvic girdle weakness but sparing of the lower leg muscles	Murchir A et al. 2000; Kitaguchi T et al. 2001
Dilated cardiomyopathy with conduction system defect (CMD1A)	Throughout the coding sequence. Includes amino-acid substitutions, frameshifts, and premature stop codons	Impaired systolic function and ventricular dilation. Variable skeletal muscle involvement	Brodsky GL et al. 2000
Dilated cardiomyopathy with conduction system defect (CMD1A)	Throughout the coding sequence. Includes amino-acid substitutions, frameshifts, and premature stop codons	Impaired systolic function and ventricular dilation. Variable skeletal muscle involvement	Brodsky GL et al. 2000
Dunnigan-type familial partial lipodystrophy (FPLD)	Majority of missense mutations clustered in exon 8 with a few mutations in exon 11	Loss of subcutaneous fat in the limb and trunk with excessive fat accumulation around the neck	Cao H et al. 2000; Shackleton S et al. 2000

Table 3 (continued).

Disease	Type of mutation	Principal phenotypes	References[a]
A-type lamins			
Charcot–Marie–Tooth neuropathy type 2B1 (CMT2B1)	A homozygous recessive mutation at R298C in the rod domain	Reduced axon density, demyelinated axons, and wasting of the peripheral muscles	De Sandre-Giovannoli A et al. 2002
Hutchison–Gilford progeria syndrome (HGP)	Point mutation in exons 2 and 11. Incorrect splicing generates a shorter product	Accelerated aging, including shortened stature, craniofacial disproportion, alopecia, and osteoporosis	Eriksson M et al. 2003; De Sandre-Giovannoli A et al. 2003
Emerin			
X-linked Emery–Dreifuss muscular dystrophy (X-EDMD)	Throughout the coding sequence. Includes frameshift and missense mutations, and premature stop codons	Early contractures of tendons, slowly progressive muscle weakness, and cardiac conduction defects	Bione S et al. 1994
LBR			
Autosomal dominant Pelger–Huet anomaly (PHA)	Splice donor and acceptor mutations, frameshifts, and nonsense mutations	Abnormal nuclear shape and chromatin organization in blood granulocytes	Hoffmann K et al. 2002; Best S et al. 2003
Autosomal recessive HEM/Greenberg skeletal dysplasia	Substitution in exon 13 results in a truncated LBR protein	Chondrodystrophy with a lethal course	Waterham HR et al. 2003
ALADIN			
Autosomal recessive Triple A syndrome	Throughout the gene. Nonsense, frameshifts, and splice-site mutations	Adrenal insufficiency, achalasia, and alacrima	Tullio-Pelet A et al. 2000

[a]Most of these references are not included in the references at the end of this chapter

A-Type Lamins

Mutations of the A-type lamin gene, LMNA, cause as many as seven diseases (Table 3). How do mutations in the lamin gene promote these particular disease phenotypes, and why do certain mutations can give rise to tissue-specific effects? To explain these phenotypes and tissue specificities, several hypotheses have been put forward. The "structural hypothesis" proposes that mutations weaken the lamina structure, and thereby lead to fragility of the nuclear envelope and its breakage in patients. The "gene-expression hypothesis" proposes that the tissue-specific changes in gene expression that are associated with some mutations promote disease.

More than 30 mutations in the LMNA gene have been identified in patients with autosomal dominant Emery–Dreifuss muscular dystrophy (EDMD2) (Krimm et al. 2002). EDMD2 mutations are distributed throughout the gene, including both the amino- and carboxy-terminal globular domains, and the central rod domain, with no clear correlation between the site of the mutation and the severity of the disease. EDMD3 is called autosomal-recessive Emery–Dreifuss muscular dystrophy and the only mutation identified is H222Y. These and other observations on EDMD seem consistent with the structural hypothesis. There has also been an interesting report on the phenotypical variability and tissue specificity among cardiac and muscular disorders. In one family, one member was classified as having dilated cardiomyopathy (CMD1A) and one as having EDMD2, and two displayed symptoms of limb-girdle muscular dystrophy (LGMD1B) (Brodsky et al. 2000). This strongly supports that the phenotypical variability among EDMD2, LGMD1B and CMD1A is due to the influence of environmental factors rather than modified genes.

In contrast to those in individuals with cardiac and skeletal muscular diseases, mutations in familial partial lipodystrophy (FPLD) are localized mostly in exon 8, particularly at codons 482 and 486, with a few mutations in exons 1 and 11 (Ostlund and Worman 2003). Recently, the three-dimensional (3D) structure of the C-terminal domain of lamin A/C, which is mutated in muscular dystrophies, cardiomyopathy and partial lipodystrophy, was determined by nuclear magnetic resonance (Krimm et al. 2002). In the 3D structure, most mutations associated with muscular dystrophies involve buried residues in the core. These mutations may destabilize the structure of the entire carboxyl-terminal tails of lamins A and C. Actually, studies on one lamin mutation in EDMD2, R453W, showed that its higher-order structure is less stable than that of the wild-type or mutants R482Q and R482W, which cause FPLD (Krimm et al. 2002). In contrast, FPLD-associated mutations are clustered within a small surface region and lead to a decrease in positive surface charge. These findings suggest that these mutations may disturb the interaction between A-type lamins and a specific binding partner. Recently this partner was searched for by means of yeast two-hybrid screening, an adipocyte differentiation factor, i.e., sterol response element binding protein 1 (SREBP1), being found (Lloyd et al. 2002). The binding of SREBP1 to lamin A was noticeably reduced by FPLD mutations (Lloyd et al. 2002). Whilst the physiological relevance of this interaction has yet to be elucidated, these results are consistent with the gene-expression hypothesis. Muscular dystrophy and FPLD, which are caused by mutations of the A-type lamin gene, may be partially explained by the "structural hypothesis" and "gene-expression hypothesis," respectively.

The mutations that cause Hutchison–Gilford progeria syndrome (HGP) are localized mostly in exon 11, with a few in exon 2, of LMNA, particularly at codon 608 (Eriksson et al. 2003; De Sandre-Giovannoli et al. 2003). These mutations in HGP cause incorrect splicing, which generates shorter products.

Emerin

The emerin gene, STA, was found to be the gene responsible for X-linked Emery–Dreifuss muscular dystrophy (X-EDMD) (Bione et al. 1994). More than 90 mutations in the STA gene have been identified in subjects with X-EDMD, and it was found that the mutations are distributed throughout the gene. To determine whether or not the mutations cause instability of the nuclear envelope, lamin solubility and distribution were examined. It was found that all lamin subtypes exhibited increased solubility in fibroblasts from X-EDMD patients compared to in normal individuals, and that lamin C became redistributed from the nuclear lamina to the nucleoplasm in X-EDMD fibroblasts (Markiewicz et al. 2002). Another study involving cells expressing a X-EDMD emerin mutant, Del236–241 (a deletion in the transmembrane domain and the protein mainly localized to the cytoplasm), showed that the mutant emerin affects the endogenous lamin A/C distribution and the cell cycle length (Fairley et al. 2002). These results suggested that the mutations in X-EDMD cause instability of the nuclear envelope. However, as mentioned above, the binding of GCL, Btf, YT521-B, and other proteins to emerin is known. To clarify the molecular and cellular mechanisms in X-EDMD, further studies are necessary.

LBR

Mutations in the LBR gene cause the Pelger–Huet anomaly (Hoffmann et al. 2002). The Pelger–Huet anomaly (PHA) is an autosomal dominant disorder characterized by an abnormal nuclear shape and chromatin organization in blood granulocytes. Affected individuals show hypolobulated neutrophil nuclei with coarse chromatin. These phenotypes suggest that LBR is responsible for the nuclear shape and chromatin organization. Eleven different LBR mutations have been identified, and it has been shown that these mutations occur throughout the gene, and involve both the transmembrane and nucleoplasmic domains (Hoffmann et al. 2002: Best et al. 2003). The wide range of LBR mutations underlying the remarkable constant phenotype suggest that the Pelger–Huet anomaly is essentially a quantitative defect, and that the nuclear hypolobulation occurs because of the reduced level of LBR protein in the inner nuclear membrane (Best et al. 2003).

A recent study has shown that a homozygous mutation in the LBR gene causes hydrops-ectopic calcification-"moth-eaten" (HEM) or Greenberg skeletal dysplasia (Waterham et al. 2003). HEM is an autosomal recessive chondrodystrophy with a lethal course, characterized by fetal hydrops, short limbs, and abnormal chondro-osseous calcification. The concentration of cholesta-8,14-dien-3β-ol, which is a substrate of LBR, in cultured skin fibroblasts of an 18-week-old fetus with HEM was determined, an elevated level of the sterol compatible with a deficiency of 3β-hydroxysterol Δ^{14}-reductase being observed (Waterham et al. 2003). Most disorders in HEM could be attributed to the enzymatic defect of LBR in the cholesterol biosynthesis pathway (Waterham et al. 2003).

Presumed homozygous individuals with the Pelger–Huet anomaly have ovoid neutrophil nuclei, as well as varying degrees of developmental delay, epilepsy, and skeletal abnormalities (Hoffmann et al. 2002). HEM and homozygous PHA can be considered as allelic disorders that exhibit a wide clinical spectrum, with nonviable fetuses with HEM representing the severe end and minor limb defects representing the mild end of the spectrum.

ALADIN

ALADIN is a component of the NPC and is localized to the cytoplasmic surface of the complex (Cronshaw and Matunis 2003). The protein belongs to the WD-repeat family of regulatory proteins. Mutations of the protein cause the Triple A syndrome. The triple A syndrome is a human autosomal recessive disorder characterized by adrenal insufficiency, achalasia and alacrima. Mutations in Triple A syndrome individuals are spread throughout the ALADIN gene. With a variety of disease-associated mutations in the ALADIN gene, the mutant proteins fail to be localized to NPC. Microscopic analysis of cells from a patient suggested that defects in NPC function, rather than structure, give rise to the Triple A syndrome (Cronshaw and Matunis 2003). ALADIN may play a cell type-specific role in the regulation of nucleocytoplasmic transport, and this function is essential for proper maintenance and/or development of certain tissues (Cronshaw and Matunis 2003).

6.6 Perspectives

During the last few years, many novel nuclear envelope proteins have been reported and many inherited diseases have been found to be related to specific nuclear envelope proteins. However, the identification and characterization of further novel nuclear membrane proteins by means of proteomic analysis (Schirmer et al. 2003) and other methods are indispensable for elucidation of the function of the nuclear envelope and the pathogenic mechanisms of nuclear envelopathies. To understand the dynamics of nuclear envelope assembly/disassembly, further studies on kinase/phosphatase systems, especially phosphatase systems, regulating the dynamics are important. Inherited diseases, i.e., "nuclear envelopathies," offer unique opportunities to define the functions of nuclear envelope proteins in normal cells, and an opportunity to understand how defects in nuclear envelope function contribute to disease pathogenesis. cDNA microarray analysis (Tsukahara et al. 2002), and proteome analysis of nuclear envelope proteins of normal and disease individuals could provide new information for understanding the diseases. In the future, studies involving knockout or mutated animals (De Sandre-Giovannoli et al. 2002; Mounkes et al. 2003) could provide further valuable information for elucidation of the mechanisms underlying these diseases, and pave the way for the development of novel treatment strategies.

References

Beaudouin J, Gerlich D, Daigle N, Eils R, Ellenberg J (2002) Nuclear envelope breakdown proceeds by microtubule-induced tearing of the lamina. Cell 108:83–96

Best S, Salvati F, Kallo J, Garner C, Height S, Thein SL, Rees DC (2003) Lamin B-receptor mutations in Pelger-Huet anomaly. Br J Haematol 123:542–544

Bione S, Maestrini E, Rivella S, Mancini M, Regis S, Romeo G, Toniolo D (1994) Identification of a novel X-linked gene responsible for Emery-Dreifuss muscular dystrophy. Nat Genet 8:323–327

Brodsky GL, Muntoni F, Miocic S, Sinagra G, Sewry C, Mestroni L (2000) Lamin A/C gene mutation associated with dilated cardiomyopathy with variable skeletal muscle involvement. Circulation 101:473–476

Collas P, Courvalin J-C, Poccia D (1996) Targeting of membranes to sea urchin sperm chromatin is mediated by a lamin B receptor-like integral membrane protein. J Cell Biol 135:1715–1725

Cronshaw JM, Matunis MJ (2003) The nuclear pore complex protein ALADIN is mislocalized in triple A syndrome. Proc Natl Acad Sci USA 100:5823–5827

De Sandre-Giovannoli A, Chaouch M, Kozlov S, Vallat JM, Tazir M, Kassouri N, Szepetowski P, Hammadouche T, Vandenberghe A, Stewart CL, Grid D, Levy N (2002) Homozygous defects in LMNA, encoding lamin A/C nuclear-envelope proteins, cause autosomal recessive axonal neuropathy in human (Charcot-Marie-Tooth disorder type 2) and mouse. Am J Hum Genet 70:726–736

De Sandre-Giovannoli A, Bernard R, Cau P, Navarro C, Amiel J, Boccaccio I, Lyonnet S, Stewart CL, Munnich A, Le Merrer M, Levy N (2003) Lamin A truncation in Hutchinson-Gilfold progeria. Science 300:2055

Ding D-Q, Tomita Y, Yamamoto A, Chikashige Y, Haraguchi T, Hiraoka Y (2000) Large-scale screening of intracellular protein localization in living fission yeast cells by the use of a GFP-fusion genomic DNA library. Gene Cell 5:169–190

Dreger M, Bengtsson L, Schoneberg T, Otto H, Hucho F (2001) Nuclear envelope proteomics: Novel integral membrane proteins of the inner nuclear membrane. Proc Natl Acad Sci USA 98:11943–11948

Duband-Goulet I, Courvalin J-C (2000) Inner nuclear membrane protein LBR preferentially interacts with DNA secondary structures and nucleosomal linker. Biochemistry 39:6483–6488

Eriksson M, Brown WT, Gordon LB, Glynn MW, Singer J, Scott L, Erdos MR, Robbins CM, Moses TY, Berglund P, Dutra A, Pak E, Durkin S, Csoka AB, Boehnke M, Glover TW, Collins FS (2003) Recurrent de novo point mutations in lamin A cause Hutchinson-Gilford progeria syndrome. Nature 423:293–298

Fairley EA, Riddell A, Ellis JA, Kendrick-Jones J (2002) The cell cycle dependent mislocalisation of emerin may contribute to the Emery-Dreifuss muscular dystrophy phenotype. J Cell Sci 115:341–354

Furukawa K (1999) LAP2 binding protein 1 (L2BP1/BAF) is a candidate mediator of LAP2-chromatin interaction. J Cell Sci 112:2485–2492

Furukawa K, Sugiyama S, Osouda S, Goto H, Inagaki M, Horigome T, Omata S, McConnell M, Fisher PA, Nishida Y (2003) Barrier-to-autointegration factor plays crucial roles in cell cycle progression and nuclear organization in *Drosophila*. J Cell Sci 116, 3811–3823

Gant TM, Harris CA, Wilson KL (1999) Roles of LAP2 proteins in nuclear assembly and DNA replication: truncated LAP2β proteins alter lamina assembly, envelope formation, nuclear size, and DNA replication efficiency in *Xenopus laevis* extracts. J Cell Biol 144:1083–1096

Goldberg M, Jenkins H, Allen T, Whitfield WGF, Hutchison CJ (1995) *Xenopus* lamin B3 has a direct role in the assembly of a replication competent nucleus: evidence from cell-free egg extracts. J Cell Sci 108:3451–3461

Gruenbaum Y, Goldman RD, Meyuhas R, Mills E, Margalit A, Fridkin A, Dayani Y, Prokocimer M, Enosh A (2003) The nuclear lamina and its functions in the nucleus. International Rev Cytol 226:1–62

Harel A, Chan RC, Lachish-Zalait A, Zimmenman E, Elbaum M, Forbes DJ (2003) Importin β negatively regulates nuclear membrane fusion and nuclear pore complex assembly. Mol Biol Cell 14:4387–4396

Hetzer M, Meyer HH, Walther TC, Bilbao-Cortes D, Warren G, Mattaj IW (2001) Distinct AAA-ATPase p97 complexes function in discrete steps of nuclear assembly. Nat Cell Biol 3:1086–1091

Hoffmann K, et al (2002) Mutations in the gene encoding the lamin B receptor produce an altered nuclear morphology in granulocytes (Pelger-Huet anomaly). Nat Genet 31:410–414

Holaska JM, Lee KK, Kowalski AK, Wilson KL (2003) Transcriptional repressor germ cell-less (GCL) and barrier to autointegration factor (BAF) compete for binding to emerin in vitro. J Biol Chem 278:6969–6975

Hutchison CJ (2002) Lamins: building blocks or regulators of gene expression? Nat Rev Mol Cell Biol 3:848–858

Jagatheesan G, Thanumalayan S, Muralikrishna B, Rangaraj N, Karande AA, Parnaik VK (1999) Colocalization of intranuclear lamin foci with RNA splicing factors. J Cell Sci 112:4651–4661

Krimm I, Ostlund C, Gilquin B, Couprie J, Hossenlopp P, Mornon J-P, Bonne G, Courvalin J-C, Worman HJ, Zinn-Justin S (2002) The Ig-like structure of the C-terminal domain of lamin A/C, mutated in muscular dystrophies, cardiomyopathy, and partial lipodystrophy. Structure 10:811–823

Liu J, Lee KK, Segura-Totten M, Neufeld E, Wilson KL, Gruenbaum Y (2003) MAN1 and emerin have overlapping function(s) essential for chromosome segregation and cell division in Caenorhabditis elegans. Proc Natl Acad Sci USA 100:4598–4603

Lloyd DJ, Trembath RC, Shackleton S (2002) A novel interaction between lamin A and SREBP1: implications for partial lipodystrophy and other laminopathies. Human Mol Genet 11:769–777

Markiewicz E, Venables R, Mauricio-Alvarez-Reyes, Quinlan R, Dorobek M, Hausmanowa-Petrucewicz I, Hutchison C (2002) Increased solubility of lamins and redistribution of lamin C in X-linked Emery-Dreifuss muscular dystrophy fibroblasts. J Struct Biol 140:241–253

Martins SB, Marstad A, Collas P (2003a) In vitro modulation of the interaction between HA95 and LAP2β by cAMP signaling. Biochemistry 42:10456–10461

Martins S, Eikvar S, Furukawa K, Collas P (2003b) HA95 and LAP2β mediate a novel chromatin-nuclear envelope interaction implicated in initiation of DNA replication. J Cell Biol 160:177–188

Moir RD, Spann TP, Herrmann H, Goldman RD (2000) Disruption of nuclear lamin organization blocks the elongation phase of DNA replication. J Cell Biol 149:1179–1191

Mounkes LC, Kozlov S, Hernandez L, Sullivan T, Stewart CL (2003) A progeroid syndrome in mice is caused by defects in A-type lamins. Nature 423:298–301

Nakagawa T, Hirano Y, Inomata A, Yokota S, Miyachi K, Kaneda M, Umeda M, Furukawa K, Omata S, Horigome T (2003) Participation of a fusogenic protein, glyceraldehyde-3-phosphate dehydrogenase, in nuclear membrane assembly. J Biol Chem 278:20395–20404

Nili E, Cojocaru GS, Kalma Y, Ginsberg D, Copeland NG, Gilbert DJ, Jenkins NA, Berger R, Shaklai S, Amariglio N, Brok-Simoni F, Simon AJ, Rechavi G (2001) Nuclear membrane protein LAP2β mediates transcriptional repression alone and together with its binding partner GCL (germ-cell-less). J Cell Sci 114:3297–3307

Osada S, Ohmori SY, Taira M (2003) XMAN1, an inner nuclear membrane protein, antagonizes BMP signaling by interacting with Smad1 in *Xenopus* embryos. Development 130:1783–1794

Ostlund C, Worman HJ (2003) Nuclear envelope proteins and neuromuscular diseases. Muscle Nerve 27:393–406

Polioudaki H, Kourmouli N, Drosou V, Bakou A, Theodoropoulos PA, Singh PB, Giannakouros T, Georgatos SD (2001) Histones H3/H4 form a tight complex with the inner nuclear membrane protein LBR and heterochromatin protein 1. EMBO Rep 2:920–925

Pyrpasopoulou A, Meier J, Maison C, Simos G, Georgatos SD (1996) The lamin B receptor (LBR) provides essential chromatin docking sites at the nuclear envelope. EMBO J 15:7108–7119

Rolls MM, Stein PA, Taylor SS, Ha E, McKeon F, Rapoport TA (1999) A visual screen of a GFP-fusion library identifies a new type of nuclear envelope membrane protein. J Cell Biol 146:29–43

Salina D, Bodoor K, Eckley DM, Schroer TA, Rattner JB, Burke B (2002) Cytoplasmic dynein as a facilitator of nuclear envelope breakdown. Cell 108:97–107

Schirmer EC, Florens L, Guan T, Yates JR III, Gerace L (2003) Nuclear membrane proteins with potential disease links found by subtractive proteomics. Science 301:1380–1382

Silve S, Dupuy P-H, Ferrara P, Loison G (1998) Human lamin B receptor exhibits sterol C14-reductase activity in *Saccharomyces cerevisiae.* Biochim Biophys Acta 1392:233–244

Spann TP, Goldman AE, Wang C, Huang S, Goldman RD (2002) Alteration of nuclear lamin organization inhibits RNA polymerase II-dependent transcription. J Cell Biol 156:603–608

Steen RL, Martins SB, Tasken K, Collas P (2000) Recruitment of protein phosphatase 1 to the nuclear envelope by A-kinase anchoring protein AKAP149 is a prerequisite for nuclear lamina assembly. J Cell Biol 150:1251–1261
Steen RL, Beullens M, Landsverk HB, Bollen M, Collas P (2003) AKAP149 is a novel PP1 specifier required to maintain nuclear envelope integrity in G1 phase. J Cell Sci 116:2237–2246
Sullivan T, Escalante-Alcalde D, Bhatt H, Anver M, Bhat N, Nagashima K, Stewart CL, Burke B (1999) Loss of A-type lamin expression compromises nuclear envelope integrity leading to muscular dystrophy. J Cell Biol 147:913–919
Takano M, Takeuchi M, Ito H, Furukawa K, Sugimoto K, Omata S, Horigome T (2002) The binding of lamin B receptor to chromatin is regulated by phosphorylation in the RS region. Eur J Biochem 269:943–953
Takano M, Koyama Y, Ito H, Hoshino S, Onogi H, Hagiwara M, Furukawa K, Horigome T (2004) Regulation of binding of lamin B receptor to chromatin by SR protein kinase and cdc2 kinase in *Xenopus* egg extracts. J Biol Chem 279:13265-13271
Tsukahara T, Tsujino S, Arahata K (2002) cDNA microarray analysis of gene expression in fibroblasts of patients with X-linked Emery-Dreifuss muscular dystrophy. Muscle Nerve 25:898–901
Walther TC, Askjaer P, Gentzel M, Habermann A, Griffiths G, Wilm M, Mattaj IW, Hetzer M (2003) RanGTP mediates nuclear pore complex assembly. Nature 424:689–694
Waterham HR, Koster J, Mooyer P, van Noort G, Kelley RI, Wilcox WR, Wanders RJA, Hennekam RCM, Oosterwijk JC (2003) Autosomal recessive HEM/Greenberg skeletal dysplasia is caused by 3β-hydroxysterol Δ^{14}-reductase deficiency due to mutations in the lamin B receptor gene. Am J Hum Genet 72:1013–1017
Wilkinson FL, Holaska JM, Zhang Z, Sharma A, Manilal S, Holt I, Stamm S, Wilson KL, Morris GE (2003) Emerin interacts in vitro with the splicing-associated factor, YT521-B. Eur J Biochem 270:2459–2466
Ye Q, Callebaut I, Pezhman A, Courvalin J-C, Worman HJ (1997) Domain-specific interactions of human HP1-type chromodomain proteins and inner nuclear membrane protein LBR. J Biol Chem 272:14983–14989

7. Gene Selectors Consisting of DNA-Binding Proteins, Histones, and Histone-Binding Proteins Regulate the Three Major Stages of Gene Expression

Shinsuke Muto[1,2] and Masami Horikoshi[1,2]*

[1]Laboratory of Developmental Biology, Institute of Molecular and Cellular Biosciences, University of Tokyo, 1-1-1 Yayoi, Bunkyo-ku, Tokyo 113-0032, Japan

[2]Horikoshi Gene Selector Project, Exploratory Research for Advanced Technology (ERATO), Japan Science and Technology Corporation (JST), 5-9-6 Tokodai, Tsukuba, Ibaraki 300-2635, Japan

7.1 Eukaryotic Gene Regulation: Three Major Stages and Three Main Types of Gene Selectors

Gene expression is the process whereby DNA sequence information is converted into a functional transmitter or player, namely, mRNA, and then a major functional player, namely, protein. Transcription is the first step in gene expression. Since the temporal and spatial regulation of gene expression define cellular identity, transcription is the most critical and fundamental step in the cellular functions of a gene. We have classified transcriptional regulation into three functional stages on the basis of the complexity of the DNA structures involved. The first level concerns the activation/inactivation of promoters on naked DNA, the second level entails activation/inactivation of nucleosomes, while the third level involves the activation/inactivation of chromosomal regions (Fig. 1). We denote the components that determine which genes are activated or repressed at each of these levels as "gene selectors." We have categorized the gene selectors into three main groups, namely, DNA-binding proteins, histones (non-specific DNA-binding proteins), and histone-binding proteins. These three types of gene selectors work in cooperation to select the genes that are to be expressed (Fig. 2).

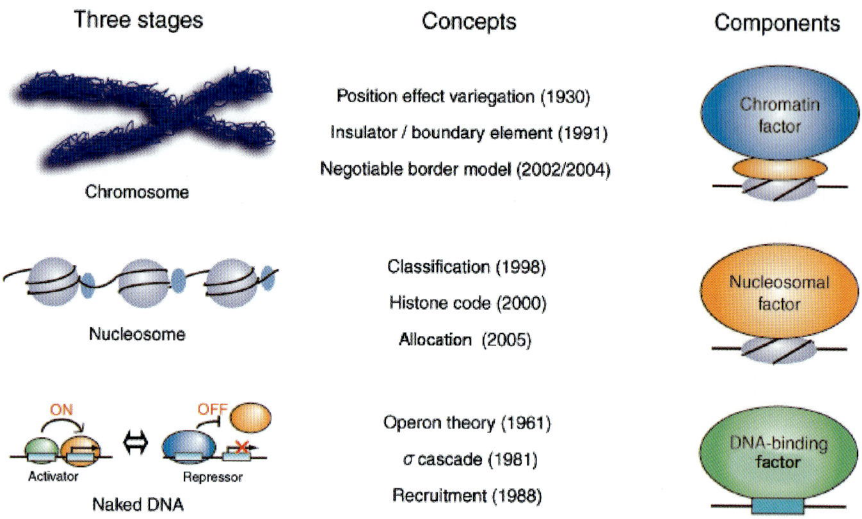

Fig. 1. Three stages of eukaryotic gene expression. Eukaryotic gene regulation can be classified into three functional stages. The first level is the activation/inactivation of promoters on naked DNA. The main factors involved at this stage are DNA and DNA-binding proteins. The second level concerns activation/inactivation of nucleosomes. Nucleosomal factors regulate the activities of DNA-binding proteins. The third level entails the activation/inactivation of chromosomal regions. At this level, chromatin factors are further involved in gene regulation. Three representative concepts are shown according to each stage.

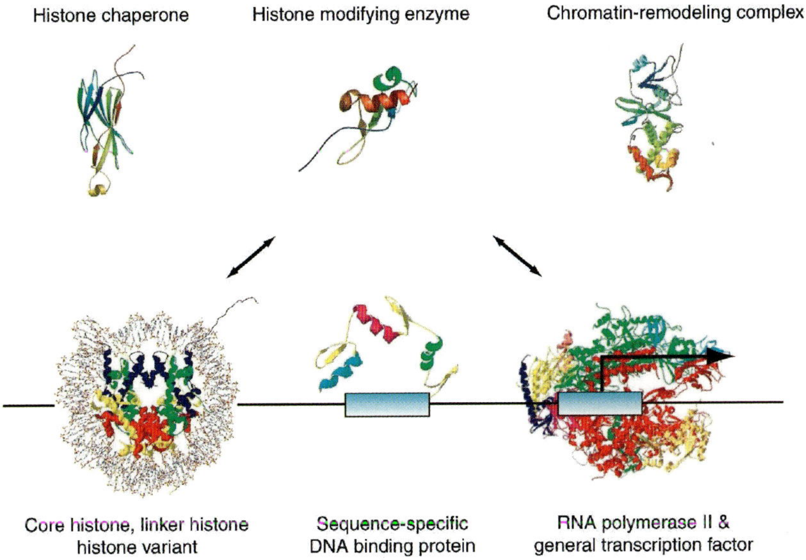

Fig. 2. Three main types of gene selectors. We denote the factors that determine which genes are activated or repressed as gene selectors. Gene selectors can be categorized into three main groups: DNA-binding proteins, histones, and histone-binding proteins.

Since the structure of DNA in eukaryotes is complex, the regulation of eukaryotic transcription initiation is more complicated and sophisticated than that of bacterial transcription (Cold Spring Harbor Laboratory 1999). However, the process of transcription itself on the naked DNA is basically the same in both prokaryotes and eukaryotes, as both require RNA polymerase and additional transcription initiation factors (Cold Spring Harbor Laboratory 1999). Nevertheless, there are differences in the molecules or machinery architecture employed by each case: eukaryotes have three RNA polymerases (pol I, II, and III) and several initiation factors called general transcription factors (GTFs) whereas bacteria have only one RNA polymerase and one initiation factor (Roeder and Rutter 1969; Matsui et al. 1980; Roeder 2003). Pol II and the GTFs (TFIIA, TFIIB, TFIID, TFIIE, TFIIF, and TFIIH) are believed to regulate the basal level of mRNA expression (Roeder 2003). To regulate the level of transcription initiation in response to various biological phenomena, additional regulatory transcriptional regulatory proteins that act on DNA directly and indirectly are required (Chambon 2004). We denote this stage as "the activation/inactivation of promoter stage." This stage, which determines the frequency of transcription initiation, involves the interaction of DNA-binding proteins with specific DNA elements (promoters) followed by the assembly of transcription machineries onto the DNA near the transcription initiation site through the interaction with these DNA-binding proteins (Horikoshi et al. 1988a,b; Hai et al. 1988). Thus, gene selectors like the DNA-binding proteins modulate transcription initiation at this stage.

The majority of eukaryotic DNA is packaged into nucleosomes, which are composed of four different basic proteins called core histones and the DNA that wraps around them (Kornberg 1974; Luger et al. 1997). This nucleosome structure is generally thought to inhibit DNA-binding proteins (activators/repressors) and transcription machineries from accessing the DNA (Workman and Roeder 1987). Consequently, modulation of nucleosome structure that allows these factors to associate with particular nucleosomal DNA regions is critical for the activation/inactivation of specific genes (Workman and Kingston 1998). The modulation of nucleosome structure involves three distinct processes: (i) assembly/disassembly of the nucleosome by histone chaperones (Laskey et al. 1978), (ii) sliding or transfer of the nucleosome by nucleosome remodeling complexes (Tsukiyama et al. 1994; Cote et al. 1998; Peterson and Tamkun 1995), and (iii) alteration of nucleosome accessibility to chromatin-related factors due to the modification of histones by histone-modifying enzymes (Brownell et al. 1996; Taunton et al. 1996). The three different chromatin-related factors involved in modulating nucleosome structure (i.e. histone chaperones, nucleosome remodeling complexes, and histone-modifying enzymes) are now known to act in collaboration with each other (Wolffe 1998; Tyler 2002). We denote this stage as "the activation/inactivation of nucleosomes stage." Thus, gene selectors like the histones and various chromatin-related factors control transcription initiation through the remodeling of the nucleosome structure at this stage.

Finally, the third stage is "the activation/inactivation of chromosomal regions stage." Cytologically, the eukaryotic genomes can be roughly divided into euchromatin and heterochromatin (Turner 2001; Grewal and Elgin 2002). Heterochromatin was originally defined as being the portion of the genome that remains condensed when the cell transits from metaphase to interphase. Generally, telomeres, centromeres, HM loci, and rDNA loci are considered to be the heterochromatin regions in yeast (Grunstein 1998; Kimura and Horikoshi 2004). Higher eukaryotes also bear many additional heterochromatin regions. There have been two important findings indicating that there is a link between the formation of heterochromatin structure and the inactivation of genes. First, in *Drosophila*, it was observed that clonally inherited silencing followed chromosome rearrangements with one breakpoint within the heterochromatin (this is known as position effect variegation [PEV]) (Schotta et al. 2003). Second, X-chromosome inactivation results in one X chromosome becoming the Barr body, which is essentially a heterochromatin aggregate (Chadwick and Willard 2003). While the choice of which X chromosome to inactivate appears to be random in most mammalian species, the decision is clonally inherited (Latham 2005). How are the boundaries between heterochromatin domains and euchromatin domains established? Two major models have been proposed. One of the prevailing models involves insulator activity (West et al. 2002). In this model, specific DNA elements called insulators control activators by inhibiting the spread of chromatin modifications or by acting as DNA markers that bind to perinuclear substrates by forming loops. In other words, an insulator acts as a wall and blocks the effects of activators or repressors, thereby defining the boundaries between heterochromatin and euchromatin. Recently, however, we advocated a rather different model, namely, the "negotiable border model" (Kimura et al. 2002; Kimura and Horikoshi 2004). In this model, the boundaries between inactive chromosomal regions (silenced regions) and active chromosomal regions (nonsilenced regions) are decided by the balance between two opposing histone modifications. This model therefore postulates that the boundary is flexible and negotiable rather than fixed.

In this chapter, we will focus only on the mRNA-synthesizing initiation reaction in eukaryotes. Thus, bacterial or archaeal transcription, the pol I or pol III systems, and other steps than initiation will not be covered here. In the next section, we summarize the mechanism of eukaryotic transcription initiation on the naked DNA and how it is regulated. We then describe the nucleosome structure and review the three important factors/enzymes that alter nucleosome structure and function. The fourth section details the mechanism by which chromosomal domains regulate activation and repression of transcription. In the final section, we briefly summarize our observations and suggest prospective research avenues regarding eukaryotic gene regulation.

7.2 Promoter Activation/Repression from Naked DNA

Transcription Initiation

The basal apparatus used to initiate mRNA transcription in eukaryotes always consists of pol II (RNA polymerase II) and various GTFs (Matsui et al. 1980; Roeder 2003). To accurately start to transcribe from a specific site, core promoters are also necessary. The eukaryotic core promoter consists of a minimal set of DNA sequences that extend either upstream or downstream of the transcription initiation site (Buratowski 2000). Pol II core promoters are known to involve four distinct consensus sequences, namely, the TATA box, the TFIIB recognition element (BRE), the initiator (Inr), and the downstream promoter element (DPE) (Kadonaga 2002). The TATA box is recognized by the general transcription initiation factor called TBP (TATA box-binding protein) (Horikoshi et al. 1989a,b). TBP is a part of the multi-subunit complex TFIID. The other subunits of TFIID are denoted as TBP-associated factors (TAFs) (Green 2000).

Upon the start of transcription, TBP binds to the TATA box and extensively distorts its sequence (Horikoshi et al. 1992; Kim et al. 1993). The resulting TBP–DNA complex forms a platform that recruits pol II and other GTFs (Horikoshi et al. 1988b; Buratowski 2000). These proteins are thought to assemble at the promoter in a sequential manner, namely, TFIIA, TFIIB, TFIIF and pol II first followed by TFIIE and TFIIH (Buratowski et al. 1989). After the formation of this so-called preinitiation closed complex, the promoter melts and a distinct complex called the open complex is formed (Buratowski 2000). The promoter melting requires ATP hydrolysis mediated by TFIIH (Holstege et al. 1996). In eukaryotes, before pol II leaves the promoter, the carboxyl terminal domain (CTD) of the largest subunit of pol II is phosphorylated by TFIIH (Feaver et al. 1991; Lu et al. 1992; Serizawa et al. 1992). This event enables pol II to leave most of the GTFs behind and to proceed with transcription. This is the process by which pol II initiates transcription from a naked DNA template in vitro. The efficiency of this basal transcription process is controlled by various regulatory proteins (Horikoshi et al. 1988b; Roeder 1996).

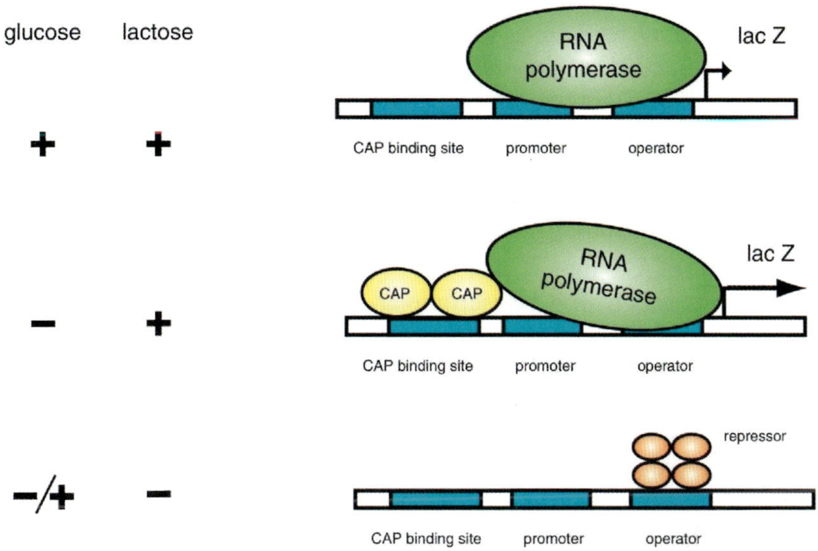

Fig. 3. Operon theory. CAP mediates the effect of glucose, whereas Lac repressor mediates the lactose signal. Lac repressor can bind to operator and repress transcription only in the absence of lactose. In the presence of glucose, the repressor is inactive and the transcription is active. CAP can bind to DNA and activate the genes only in the absence of glucose. Thus, only when lactose is present and glucose is absent, the gene is expressed at high levels.

Transcriptional Regulation: Activators and Repressors

Some regulators control gene expression in response to extracellular signals. Regulatory proteins are called activators when they increase the basal transcription level and repressors when they decrease it (Ptashne and Gann 2001). These regulatory proteins typically show sequence-specific DNA-binding activity, and can be both activators and repressors depending on the physiological context (Ma 2005). The basic ON/OFF idea of how these regulatory proteins control gene expression was first elegantly proposed by Jacob and Monod in 1961 on the basis of their genetic work with the simple organism *Escherichia coli* and additional biochemical analyses (Fig. 3) (Jacob and Monod 1961). Their so-called operon theory states that a repressor inhibits transcription by binding to an operator adjoining the promoter; this then blocks the binding of polymerase to the promoter. After this great work, regulators were found to act by a variety of distinct mechanisms. For example, activators often simultaneously interact with a site on the DNA near the promoter and act as a stimulator that connects the enzyme and the promoter (Horikoshi et al. 1988b). Some activators stimulate the transition from the closed to the open complex by allostery, namely, by triggering a conformational change in either polymerase or DNA. Other activators can act from a distance by forming a loop and/or bending the DNA (Rippe et al. 1995).

As research into transcriptional regulation progressed, two other important ideas were proposed. The first is that most regulators have distinct activating and DNA-binding surfaces. The activating surface interacts with core polymerase while the DNA-binding surface recognizes specific DNA elements (Ptashne and Gann 2001; Ptashne 2004). In a typical case, the DNA-binding domain is a helix-turn-helix that consists of two α-helices separated by a short turn, one of which fits into the major groove of the DNA and recognizes the specific sequence (Ptashne 2004). Other DNA-binding motifs have also been discovered in eukaryotes, including the zinc finger motif, the leucine zipper motif, and the helix-loop-helix motif. The second important idea is that many regulators interact to form dimers that bind in a cooperative fashion to adjacent sites on the DNA (Murre et al. 1989; Blackwood and Eisenman 1991). This enhances the specificity of DNA-binding regulators, as it increases the affinity of the regulators for the correct sites without facilitating their interactions with incorrect sites (Ptashne and Gann 2001).

In eukaryotes, as in bacteria, the most stringently regulated stage in transcription is the initiation step (Fig. 4) (Roeder 2003). However, there are two striking differences between eukaryotes and bacteria in terms of their promoters and regulators. The first is that the eukaryotic genome is wrapped around histones to form nucleosomes (Luger et al. 1997). This will be discussed further in the next section. The other striking difference is that many eukaryotic genes have more regulatory binding sites and are controlled by more regulatory proteins (McKnight and Tjian 1986). The greater number of regulatory binding sites reflects extensive signal integration and more complicated temporal- and spatial-specific regulation (Evans et al. 1990; Weintraub et al. 1991; Gehring 1992). Moreover, the eukaryotic regulatory factors, unlike their bacterial counterparts, bear structurally and functionally distinct domains that possess on the one hand DNA-binding activity and regulatory activity (including activating activity) on the other (Ptashne 1988). Interestingly, analysis of these domains as they appear throughout evolution from yeast to mammals suggests that they are separable and interchangeable (Ptashne 1988). The DNA-binding

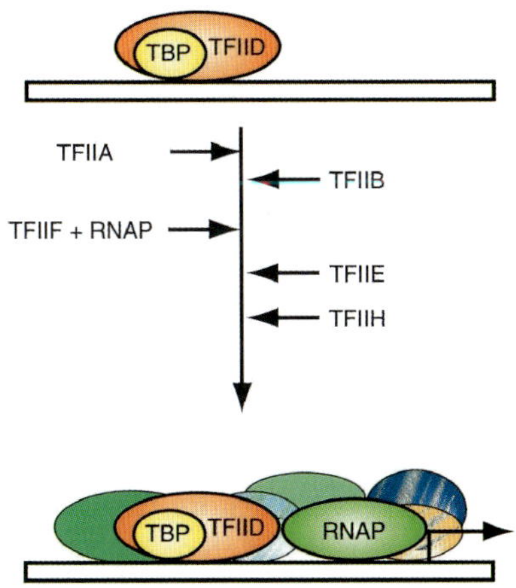

Fig. 4. Transcription initiation by RNA polymerase II. The general transcription factors help polymerase II bind to the promoter and melt the DNA. First, the TATA DNA sequence is recognized by TBP (TATA-box binding protein), one of the subunits of the general transcription initiation factor called TFIID. TBP extensively distorts the TATA sequence. The resulting TBP–DNA complex provides a platform to recruit other general transcription initiation factors and RNA polymerase II to the promoter. In vitro, these factors assemble at the promoter in order: TFIIA, TFIIB, TFIIF together with RNA polymerase, and then TFIIE and TFIIH. TFIIE and TFIIH bind upstream of RNA polymerase II. Promoter melting comes after the formation of the preinitiation complex.

domain has also been shown to bear a binding surface for other proteins and to play another regulatory role (Wagner and Green 1994).

Unlike the DNA-binding domains, the regulatory domains of transcriptional regulators do not always have stable structures and it is believed that their stable (or fixed) structure is induced by binding with other factors, including the transcriptional machinery (Ptashne 2004). In eukaryotes, the regulatory domains may mediate the direct interaction between activators and Pol II, as is the case with bacteria (Horikoshi et al. 1988a). In addition, the activators indirectly recruit Pol II by inducing the association between the transcription initiation machinery (general transcription factors), cofactors, and chromatin-related factors (Fig. 5) (Roeder 2003). In eukaryotes, the transcriptional machinery contains many factors that form complexes, such as mediator or TFIID complexes (Kornberg and Lorch 1999). Activators interact with one or more components of these complexes and thereby recruit them to the gene promoter (Horikoshi et al. 1988b). Activators also associate with chromatin-related factors that help the transcriptional machinery to bind to the targeted promoter (Kadam and Emerson 2002). Three types of chromatin-related factors are known: histone chaperones, chromatin-remodeling complexes, and histone-modifying enzymes (Tyler 2002). There are at least two ways by which these factors regulate transcription initiation. First, they can uncover the promoter region, which would otherwise remain inaccessible within the nucleosome (Wolffe 1998). Second, they can create specific binding sites on the nucleosome for other regulators or the TFIID complex (Workman and Kingston 1998). When a gene is controlled by many regulatory proteins, it acts synergistically. For example, two activators can recruit different components of a single complex, or recruit different complexes. For gene expression to be the center of signal integration, such synergy is essential (Ptashne and Gann 2001).

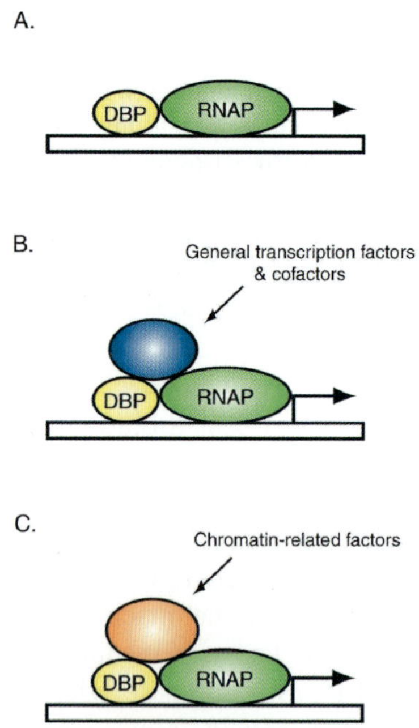

Fig. 5A–C. Activation by DNA-binding proteins. In eukaryotes, DNA-binding proteins recruit RNA polymerase II to the promoter (**A**), as is the case with bacteria. In addition, the DNA-binding protein indirectly recruits polymerase II by inducing the association between the general transcription factors, cofactors (**B**), and chromatin-related factors (**C**).

7.3 Alteration of Chromatin Structure

Nucleosome Structure

The DNA in eukaryotic cells is packaged into nucleosomes, which are composed of eight core histone proteins and the DNA that wraps around them (Luger et al. 1997). The core histone proteins are H2A, H2B, H3, and H4. The DNA between each nucleosome is called linker DNA. The histone protein H1 is referred to as a linker histone because it binds to the linker DNA (Wolffe 1997). In the presence of DNA, the core histones form a tuna can-shaped structure. H3 and H4 first form heterodimers that then come together to form a tetramer. H2A and H2B form heterodimers but not tetramers. The assembly of the nucleosome involves the ordered association of these complexes with DNA, namely, the binding of the H3–H4 tetramer to the DNA followed by binding of the two H2A–H2B dimers to the H3–H4–DNA complex (Wolffe 1998; Khorasanizadeh 2004). This process needs the help of other nuclear proteins called histone chaperones, which will be discussed later in this section.

Each core histone is made up of a conserved region, called the histone-fold domain, and an N-terminal tail (Luger et al. 1997). The histone-fold domain mediates nucleosome assembly while the tail is subjected to extensive chemical modifications that regulate nuclear events (Khorasanizadeh 2004).

The determination of a series of nucleosome structures has markedly improved our understanding of nucleosome and chromatin structure (Luger 2003). Given the high degree of evolutionary conservation of the histone proteins, it is not surprising that the structure of *Xenopus* nucleosome is highly similar to that of chicken (Luger et al. 1997; Harp et al. 2000). In contrast, the yeast histones are little different from those of higher eukaryotes (White et al. 2001). Minor protein sequence differences between yeast and higher eukaryotes alter the internucleosomal contacts in crystal packing (White et al. 2001).

Three types of chromatin-related factors discussed below are known to be involved in the alteration of chromatin structure (Fig. 6). The localized introduction of histone variants into chromatin may also change the equilibrium between different types of nucleosome-nucleosome interactions and nucleosome-chromatin factor interactions, which affect various nuclear events.

Histone Chaperones

Histones and DNA form aggregates when mixed directly in physiological conditions. To prevent improper and premature interactions between DNA and histones, acidic proteins (histone chaperones) bind to free histones (Mello and Almounzi 2001). These histone chaperones act in concert with each other or other chromatin factors to enable the correct and rapid assembly of the nucleosomes in synchronization with DNA replication, DNA repair, and transcription (Tyler et al. 2001; Mello et al. 2002; Akey and Luger 2003).

The first identified histone chaperone is nucleoplasmin, which is also the first characterized chromatin-related factor (Laskey et al. 1978). Since this discovery, many additional histone-binding proteins have been found to be histone chaperones. Examples are NAP1, CAF1, TAF1, and CIA (see below). The histone chaperones can be classified into two groups on the basis of the histones they target, namely, the H2A–H2B dimer-binding group and the $(H3–H4)_2$ tetramer-binding group. For example, nucleoplasmin and N1, which are the most abundant proteins in the nuclei of *Xenopus* oocytes, bind to H2A–H2B dimers and $(H3–H4)_2$ tetramers, respectively (Kleinschmidt et al. 1985; Dilworth et al. 1987). Although both nucleoplasmin and N1 can mediate nucleosome assembly in the absence of the other, it has been suggested that they may sequentially deposit their histones onto the DNA (Dilworth et al. 1987). Nucleoplasmin and N1 homologues have not been found in yeast, *Caenorhabditis elegans*, or plants and appear to exist only in higher organisms, often in germinal cells. This indicates histone chaperones play specialized roles.

The structure of the nucleoplasmin decamer indicated that this and other histone chaperones might participate in nucleosome assembly. The nucleoplasmin monomer is composed of a conserved N-terminal domain and a flexible acidic C-terminal domain. The crystal structure of the N-terminal domain has been resolved (Dutta et al. 2001). Two pentamers of the nucleoplasmin monomer are packed face-to-face to form a decamer. These observations led to a model of the nucleoplasmin-histone complex wherein five histone octamers dock around the periphery of a nucleoplasmin decamer. This model is consistent with previous experiments that suggested (i) the H2A–H2B dimer end of the histone octamer docks into the concave lateral surface of nucleoplasmin and (ii) nucleosome assembly at DNA replication forks requires a nucleoplasmin decamer complex (Dutta et al. 2001).

Nucleosome assembly protein 1 (NAP1) was originally identified as an activity that facilitates the in vivo reconstitution of nucleosomes in HeLa cell extracts (Ishimi et al. 1984). In *Saccharomyces cerevisiae*, NAP1 participates in the control of mitotic events through an interaction with Clb2, one of the cyclin B family members (Kellogg and Murray 1995). NAP1 is located in the cytoplasm during the G2 phase of the cell cycle and in the nucleus in S phase, suggesting that NAP1 may shuttle histones between the cytoplasm and the nucleus (Ito et al. 1996; Miyaji-Yamaguchi et al. 2003). Therefore, NAP1 may be involved in the control of the cell cycle and the transfer of histones.

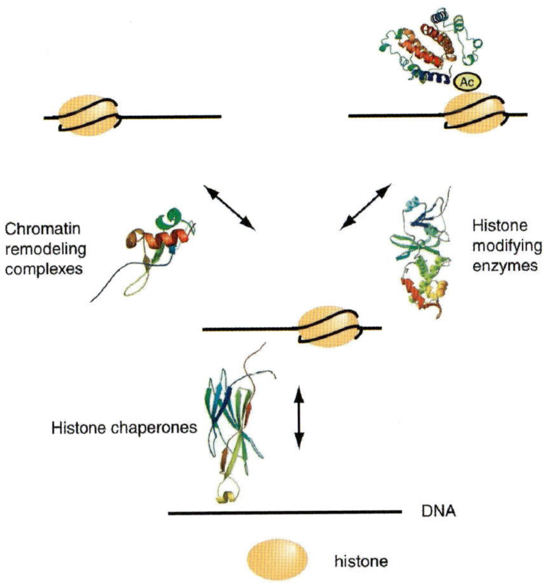

Fig. 6. Alteration of nucleosome structure. Three types of chromatin-related factors are known to be involved in the alteration of chromatin structure. Histone chaperones dynamically assemble and disassemble nucleosomes. Chromatin remodeling complexes may facilitate sliding of a histone octamer and transferring it from one DNA to another or may remodel locally a nucleosome structure. Histone-modifying enzymes create new binding surfaces for protein recognition. The bromodomain and chromodomain recognize acetylated and methylated lysines, respectively. Distinct modifications may take place sequentially or concurrently to specify downstream events.

The deposition of histones onto newly replicated DNA is controlled by the chromatin assembly factor 1 (CAF-1) complex, which is recruited to the DNA by interacting with proliferating cell nuclear antigen (PCNA) (Smith and Stillman 1989; Shibahra and Stillman 1999). CAF-1 is also tightly associated with DNA repair (Gaillard et al. 1996). In cells, CAF-1 interacts with those H4 histones whose lysine 5 and 12 residues have been acetylated, which is the acetylation pattern of newly synthesized histones (Verreault et al. 1996). Therefore, CAF-1 seems regulate histone metabolism by depositing histones onto newly synthesized DNA during DNA replication and DNA repair (Smith and Stillman 1989).

Template activating factor 1 (TAF1) was first identified as a positive regulator of replication and transcription by in vitro chromatin studies (Matsumoto et al. 1993, 1995). Later, TAF1 was shown to bear histone chaperone activity (Okuwaki and Nagata 1998) and to inhibit histone acetylation caused by p300 (Seo et al. 2001). TAF1 also negatively regulates transcription by inhibiting the DNA-binding activity of zinc finger type DNA-binding proteins and their p300-mediated acetylation (Suzuki et al. 2003; Miyamoto et al. 2003). However, the detailed mechanisms by which TAF1 acts in replication and transcription remain to be determined.

CCG1 interacting factor A (CIA) was isolated as a binding factor of CCG1, the largest subunit of TFIID (Munakata et al. 2000; Chimura et al. 2002). CIA was found to have histone chaperone activity; moreover, it is the most conserved histone chaperone (Munakata et al. 2000). There are two highly identical CIA family members in higher eukaryotes (Umehara and Horikoshi 2003). CIA-I is ubiquitously expressed while CIA-II is specifically expressed in tissues such as the testis (Umehara and Horikoshi 2003). CIA binds to the entry/exit site on the nucleosome, where DNA begins to wind around the core histones (Munakata et al. 2000; Mousson et al. 2005). The yeast homologue of CIA, Asf1, was initially identified as a gene that de-repressed the silent mating type loci upon overexpression (Le et al. 1997).

Biochemical studies in *Drosophila* identified a complex that facilitated efficient nucleosome assembly on replicated DNA when CAF-1 activity was limited (Tyler et al. 1999). This complex, replication-coupling assembly factor (RCAF), contains CIA/Asf1, H3, and H4. The *Drosophila* CIA/Asf1 functions synergistically with the CAF-1 complex to assemble chromatin during DNA replication and DNA repair (Tyler et al. 2001). Thus, histone chaperones may act in part as heterochaperones, including as heterodimers of two family forms such as CIA-I and CIA-II. Many studies have revealed CIA to be a highly versatile player in transcription (Chimura et al. 2002; Adkins et al. 2004). CIA may be involved in various cell activities by transferring histones to specialized histone deposition factors and monitoring the fluctuation in histone pools coupled with DNA replication, DNA repair, and transcription.

Recently, a novel type of histone chaperone was identified. FKBP, a member of the peptidyl prolyl *cis-trans* isomerase (PPIase) family, was shown to be required for the silencing of gene expression at the rDNA locus (Kuzuhara and Horikoshi 2004). It was also shown to possess histone chaperone activity. Moreover, it was suggested on the basis of research on FKBP that histone chaperones might be involved in aging via chromatin remodeling and/or alteration of protein folding.

Chromatin Remodeling Complexes

Chromatin remodeling complexes change the chromatin structure by altering the contacts between the DNA and histones in an ATP-dependent manner (Flaus and Owen-Hughes 2004). Chromatin remodeling complexes can be grouped on the basis of their ATPase subunits into three classes, namely, Swi2/Snf2, Isw1, and Mi-2/NuRD (Becker and Horz 2002; Narlikar et al. 2002; Martens and Winston 2003). Of these, the Swi2/Snf2-related groups have been particularly well studied. Swi2/Snf2 complexes have been detected in *S. cerevisiae*, *Drosophila*, and humans, and the Swi2/Snf2 subunit possesses both ATPase and remodeling activities in vitro (Martens and Winston 2002). The other subunits have been relatively poorly characterized but genetic analysis in *S. cerevisiae* has shown that most are required for the remodeling activity of the complex in vivo (Geng et al. 2001). The other subunit(s) also appear to have another role, namely, the interaction with DNA-binding proteins (Kal et al. 2000). Previous studies have shown that the Swi2/Snf2 complex is recruited to the promoter by an activator or a repressor, and that interactions between specific Swi2/Snf2 component(s) and DNA-binding proteins may determine when, where and how transcription is regulated by the Swi2/Snf2 complex (Martens and Winston 2002). Many reports indicate that the Swi2/Snf2 complex plays a role in transcriptional activation and repression (Belandia et al. 2002; Zhang et al. 2002). With regard to its repressive role, Swi2/Snf2 complex was shown to be brought to the promoter by the CoREST corepressor, which is itself also recruited to the promoter by the DNA-binding protein (Battaglioli et al. 2002). A recent study has demonstrated the recruitment of Swi2/Snf2 complex by the heterochromatin protein HP1α suggesting Swi2/Snf2 participates in heterochromatin formation (Nielsen et al. 2002).

Swi2/Snf2 complexes have another striking characteristic. All Swi2/Snf2-type remodeling complexes contain a motif called bromodomain that recognizes acetylated lysine residues in the histone N-terminus in vitro (Hassan et al. 2002). It is still unclear whether this domain is required for the stable interaction of Swi2/Snf2 with the nucleosome, but recognition of histone epitopes by Swi2/Snf2-type complexes may partly determine the specificity of promoters that they work.

Alteration of the nucleosome structure involve (1) sliding of a histone octamer along the DNA, (2) transfer of a histone octamer from one DNA to another, and (3) local remodeling of a nucleosome structure (Tsukiyama and Wu 1997; Becker and Horz 2002; Peterson 2002; Flaus and Owen-Hughes 2004). All chromatin remodeling complexes can facilitate sliding but only the Swi2/Snf2 complex can transfer or remodel nucleosomes (Martens and Winston 2003). However, these different means by which the chromatin remodeling complex alters the nucleosome may reflect, at least in part, the experimental substrate and conditions used. Therefore, it is not clear whether remodeling events occur by a single mechanism or by combined mechanisms. Several recent studies have focused on another possible mechanism. Sth1, a subunit of RSC complex, a Swi2/Snf2 complex, was shown to have ATP-dependent DNA translocation activity (Laurent et al. 1993). Sth1 is a length-dependent ATPase whose activity is greater on circular DNA than on linear DNA. Sth1 was also found to displace the third strand in a triple helix, which suggests that it can track along DNA (Saha et al. 2002). These results together suggest Sth1 can twist and translocate DNA. In a novel model that has been proposed, Swi2/Snf2 first binds at the entry/exit sites on a nucleosome and its translocation activity breaks DNA-histone interactions (Saha et al. 2002).

Histone-Modifying Enzymes

The four core histones are subject to a wide range of covalent modifications, including lysine acetylation, lysine and arginine methylation, serine phosphorylation, lysine ubiquitination, and lysine sumoylation (Cheung et al. 2000a,b; Roth et al. 2001). These modifications may affect chromosome functions through two distinct yet related mechanisms. First, since all modifications may change the electrostatic charge of the histones, they may affect the structure of the nucleosome and its interaction with DNA- and nucleosome-associated factors. Second, since the modifications could create new binding surfaces for protein recognition, they may be responsible for recruiting specific functional proteins such as DNA-binding proteins or mediators to their proper sites of action. This notion is exemplified by the case of the bromodomain that recognizes acetylated lysine, and by the chromodomain that can recognize methylated lysine (Yang 2004). One prevailing idea is that a "histone code" exists wherein distinct modifications take place sequentially or in combination to specify downstream events (Strahl and Allis 2000). Some examples of such combinations of histone chemical modifications and the regulatory roles they play are described below.

The acetylation of lysines in histone tails plays diverse roles in regulating nucleosome structure and function (Jenuwein and Allis 2001). Acetylated lysines may decrease histone–DNA interactions and promote the accessibility of the DNA to chromatin-related factors. Therefore, specific acetylation may regulate DNA replication and DNA repair as well as transcription by recruiting proteins that have an acetylated lysine-recognizing module, namely, the bromodomain (Peterson and Laniel 2004). There is evidence that the lysines in histones H3 and H4 are sequentially acetylated by HAT and that this leads to the ordered recruitment of bromodomain-containing complexes (Jenuwein and Allis 2001).

Lysine methylation is believed to play a role in epigenetic inheritance (Lachner and Jenuwein 2002; van Leeuwen and Gottschling 2002). The establishment and maintenance of mitotic and meiotic transcription patterns are crucial for cell fate determination, and the methylation of H3–K9 and H3–K27 are linked to epigenetic silencing mechanisms (Grewal and Elgin 2002). By utilizing their chromodomains, the HP1 protein binds to methylated H3–K9 while the polycomb protein recognizes methylated H3–K27; this mediates gene silencing (Nakayama et al. 2001; Cao et al. 2002). Chromodomains are also found in several chromatin factors that also bear HAT domains, ATPase domains, and SET domains (Brehm et al. 2004). SET domains are responsible for lysine methylation (Marmorstein 2003). The chromodomain binds to methylated histone tails in a site-specific manner with an affinity that is the strongest for trimethylated lysine and the weakest for monomethylated lysine (Rice and Allis 2001).

The methylation of arginines is believed to be involved in the active state of transcription since methylation of H4–R3 facilitates H4 acetylation and enhances transcription (Wang et al. 2001). Methylated H4–R3 may well become a recognition site for other proteins such as HAT; however, the details remain to be elucidated.

The phosphorylation of H3–S10 or H3–S28 correlates with mitosis and chromosome condensation, as well as with transcription activation (Wei et al. 1998; Goto et al. 1999). The acetyltransferase GCN5/PCAF, CBP/p300, and SAGA prefer to acetylate lysine by interacting with the adjacent phosphorylated serine (Cheung et al. 2000a,b; Lo et al. 2000).

The binding of ubiquitin or the small ubiquitin-related modifier SUMO to a specific lysine plays an important role in regulating transcription (Zhang 2003). The ubiquitin is attached by three steps, namely, E1 activation, E2 conjugation, and E3 ligation (Finley and Chau 1991). The ubiquitination and sumoylation of the lysines may induce proteosome-dependent degradation of transcription factors or modification-dependent recruitment of chromatin factors (Gill 2004). Several studies suggest that Rad6-mediated H2B ubiquitination is linked to gene silencing. Deletion of RAD6 in budding yeast resulted in defects of telomeric and HML silencing (Huang et al. 1997). The silencing function of Rad6 seems to be conserved as its fission yeast homologue is also required for mating-type silencing (Naresh et al. 2003). However, some studies link Rad6-mediated H2B ubiquitination to activation of specific genes. H2B ubiquitination is important for the methylation of H3–K4 and H3–K79 (Sun and Allis 2002; Briggs et al. 2002), while a subunit of the SAGA complex can remove the ubiquitin from the H2B lysine (Henry et al. 2003). These results suggest sequential ubiquitination and deubiquitination may occur in concert with lysine methylation, thereby regulating transcription (Zhang 2003). Further, the binding of ubiquitin to H2A occurs near the entry/exit sites of DNA and the binding site of H1 (Bonner and Stedman 1979). Thus, this post-translational modification is likely to affect the stability of the higher order structure of chromatin.

7.4 Silencing/Antisilencing

Heterochromatin Structure

Gene silencing is dependent on where a gene is located. Silencing is usually associated with a dense form of chromatin called heterochromatin. Heterochromatin was named for its dense appearance as compared to the other type of chromatin (euchromatin) (Fig. 7) (Grewal and Elgin 2002). Heterochromatic regions are gene-poor and mainly consist of repetitive DNA, including satellite sequences and middle repetitive sequences related to transposable elements (Craig 2005; Dimitri et al. 2005). If a gene is experimentally moved to this region, it typically enters a transcriptionally silent state. The well-known phenomenon related to such silencing in heterochromatic regions is the PEV that is observed after rearrangements of the *Drosophila* chromosome that bears a breakpoint within heterochromatin (Henikoff 1990). Silencing and its release, antisilencing, are critical for the maintenance and the inheritance of life because any disorder in this system induces anarchic gene expression (Delaval and Feil 2004).

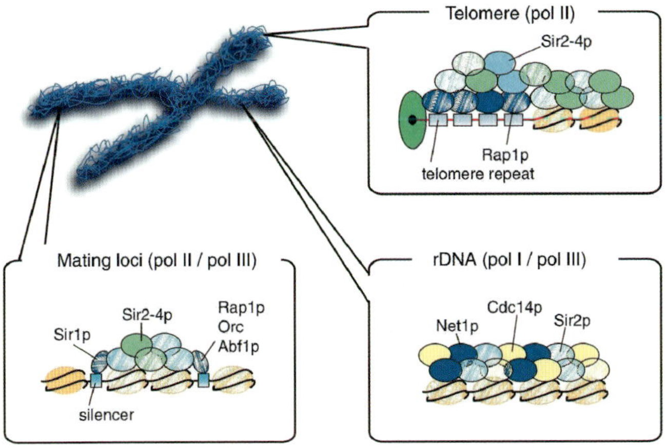

Fig. 7. Heterochromatic regions in *Saccharomyces cerevisiae*. There are three typical heterochromatic regions in *S. cerevisiae*: telomere, rDNA, and mating loci. Many DNA-binding proteins and chromatin-related factors are known to function to maintain heterochromatic conditions.

Establishment and Maintenance of Heterochromatin Domains

Two mechanisms are utilized to establish and maintain heterochromatin, namely, the chemical modification of histones and the methylation of DNA (Craig 2005). Gene silencing by DNA methylation has not been found in yeast but is common in plants and mammals (Rangwala and Richards 2004). Silencing in yeast is mediated by the deacetylation and methylation of histones (Dhillon and Kamakaka 2002). In *S. cerevisiae*, the telomeres, the silent mating-type locus, the centromeres, and the rDNA genes are silent regions (Kimura and Horikoshi 2004). Of these heterochromatin regions, the telomere is the best studied. Telomeres comprise the last 1–5 kb of each chromosome, which is folded and has a dense structure, and many repeated sequences are found in these sequences. A silencing complex that consists of Sir2, Sir3, and Sir4 is recruited to the telomere by a DNA-binding protein Rap1p that recognizes these repeated sequences (Cockell and Gasser 1999). Sir stands for silent information regulators while Sir2 is a histone deacetylase (Imai et al. 2000). The recruitment of the silencing complex initiates the local deacetylation of histone tails. The deacetylated histones are in turn directly recognized by the silencing complex (Hecht et al. 1995; Grunstein 1998). In this way, the local deacetylation spreads along the chromosome in a self-perpetuating manner (Guarente 1999). The extent to which chromatin deacetylation spreads is thought to be limited by two mechanisms. One is histone reverse acetylation and the other is another chemical modification, namely, methylation (Craig 2005). We will see how the balance of histone acetylation forms a boundary between heterochromatin and euchromatin in the next section. Histone methylases have been better characterized in higher eukaryotes and in *Schizosaccharomyces pombe* than in *S. cerevisiae*. From various studies using these organisms, it appears that modifications of histone tails by deacetylase and methylase act in concert to establish and maintain the heterochromatin domains (Craig 2005).

Boundaries of Heterochromatin Domains

That heterochromatin domains have distinct boundaries has long been inferred from the consequences of removing these boundaries, as seen by PEV in *Drosophila* (Schotta et al. 2003). The previously prevailing theory of how these boundaries are formed is based on the insulator model. In this model, insulators or boundary elements act as walls that prevent gene activation or repression mechanisms from spreading to a particular side of the wall (Labrador and Corces 2002; West et al. 2002). The model proposes that the components of these walls are recruited by the insulators, and it was initially suggested that the wall components may be sequence-specific DNA-binding proteins (Bi and Broach 1999; Donze and Kamakaka 2001; Fourel et al. 2004). However, the latter notion has been challenged by experimental observations showing that insulators decide the position of a marked transition in histone chemical modifications (Litt et al. 2001; Mutskov et al. 2002). For example, deletion of the insulators induced spreading of methylated H3–K9 and Swi6 into adjacent euchromatic regions, concomitant with a decrease in methylated H3–K4 (Nakayama et al. 2001). This indicates differential methylation of histone H3 and may serve as a marker for specific euchromatin and heterochromatin domains that are separated by insulators.

Recently, we proposed a novel boundary-formation model that stated the boundaries between euchromatic and heterochromatic regions are not strictly fixed; rather, they are defined in a passive manner by the balance between two opposing activities (Fig. 8) (Kimura et al. 2002). We have denoted this model as the “negotiable border model” and the previous model as the "fixed border model" (Kimura and Horikoshi 2004). There are two important differences between the two models. In the negotiable border model, the transition of the chemical modification status is gradual rather than definite, and a specific DNA element is not required at the border; rather, two DNA elements that recruit histone-modifying enzymes on both sides of the border are essential (Kimura and Horikoshi 2004). It should be noted that the negotiable border and fixed border models are not mutually exclusive. Some of the insulators, however, may be involved in establishing a negotiable border. The introduction of the negotiable border model may help us to better understand complex transcriptional mechanisms in higher eukaryotes, such as the transcription at homeotic or globin gene clusters.

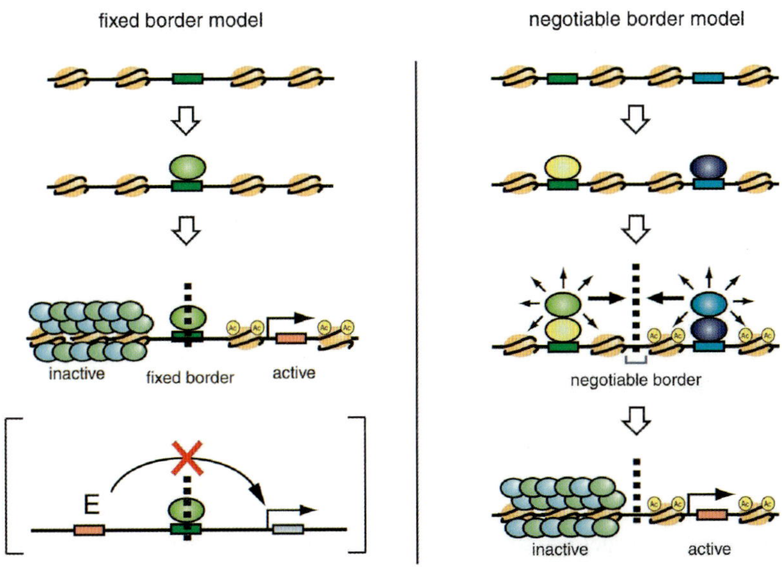

Fig. 8. Establishment of heterochromatic domains. The previously prevailing theory of how boundaries between euchromatic and heterochromatic domains are established is based on the insulator model, or fixed-border model. In this model, insulators or boundary elements act as solid walls that prevent gene activation or repression mechanisms from spreading over them. Enhancer cannot effect gene regulation beyond this fixed wall. In a novel model, the negotiable-border model, the boundaries are not strictly fixed but are defined by the balance between two opposing enzymatic activities. The position of the border is unstable, because transition of the chemical modification status is gradual rather than definite.

7.5 Summary of Eukaryotic Gene Regulation

In this chapter, we have presented a brief overview of the three major stages and three main types of gene selectors that regulate transcription. The last decade has seen significant progress in our knowledge in the field of chromatin transcription. As a result, we have realized that not only acetylation but methylation are linked to the control of transcriptional activation and repression, and how a border between heterochromatin and euchromatin domains may be established. However, the more we learn about the regulation of transcription, the further the point at which we will comprehensively understand it seems to get. This is because transcription seems to involve vast, perhaps infinite, numbers of combinations of chemical modifications and interactions of a variety of components. To simplify our goal, we should continue to search for the basic rules that operate at each step of transcription. While the study of the regulatory mechanisms that govern temporal and spatial gene expression is important, the extraction of such basic principles is vital to molecular and cellular biology.

References

Adkins MW, Howar SR, Tyler JK (2004) Chromatin disassembly mediated by the histone chaperone Asf1 is essential for transcriptional activation of the yeast PHO5 and PHO8 genes. Mol Cell 14:657–666

Akey CW, Luger K (2003) Histone chaperones and nucleosome assembly. Curr Opin Struct Biol 13:6–14

Battaglioli E, Andres ME, Rose DW, Chenoweth JG, Rosenfeld MG, Anderson ME, Mandel G (2002) REST repression of neuronal genes requires components of the hSWI.SNF complex. J Biol Chem 277:41038–41045

Becker PB, Horz W (2002) ATP-dependent nucleosome remodeling. Annu Rev Biochem 71:247–273

Belandia B, Orford RL, Hurst HC, Parker MG (2002) Targeting of SWI/SNF chromatin remodeling complexes to estrogen-responsive genes. EMBO J 21:4094–4103

Bi X, Broach JR (1999) UASrpg can function as a heterochromatin boundary element in yeast. Genes Dev 13: 1089–1101

Blackwood EM, Eisenman RN (1991) Max: a helix-loop-helix zipper protein that forms a sequence-specific DNA-binding complex with Myc. Science 251:1211–1217

Bonner WM, Stedman JD (1979) Histone 1 is proximal to histone 2A and to A24. Proc Natl Acad Sci 76:2190–2194

Brehm A, Tufteland KR, Aasland R, Becker PB (2004) The many colours of chromodomains. Bioessays 26:133–40

Briggs SD, Xiao T, Sun ZW, Caldwell JA, Shabanowitz J, Hunt DF, Allis CD, Strahl BD (2002) Gene silencing: trans-histone regulatory pathway in chromatin. Nature 418:498

Brownell JE, Zhou J, Ranalli T, Kobayashi R, Edmondson DG, Roth SY, Allis CD (1996) Tetrahymena histone acetyltransferase A: a homolog to yeaset Gcn5p linking histone acetylation to gene activation. Cell 84:843–851

Buratowski S (2000) Snapshots of RNA polymerase II transcription initiation. Curr Opin Cell Biol 12:320–325

Buratowski S, Hahn S, Guarente L, Sharp PA (1989) Five intermediate complexes in transcription initiation by RNA polymerase II. Cell 56:549–561

Cao R, Wang L, Wang H, Xia L, Erdjument-Bromage H, Tempst P, Jones RS, Zhang Y (2002) Role of histone H3 lysine 27 methylation in Polycomb-group silencing. Science 298:1039–1043

Chadwick BP, Willard HF (2003) Barring gene expression after XIST: maintaining facultative heterochromatin on the inactive X. Semin Cell Dev Biol 14:359–367

Chambon P (2004) How I became one of the fathers of a superfamily. Nat Med 10:1027–1031

Cheung WL, Briggs SD, Allis CD (2000a) Acetylation and chromosomal functions. Curr Opin Cell Biol 12:326–333

Cheung P, Tanner KG, Cheung WL, Sassone-Corsi P, Denu JM, Allis CD (2000b) Synergistic coupling of histone H3 phosphorylation and acetylation in response to epidermal growth factor stimulation. Mol Cell 5:905–915

Chimura T, Kuzuhara T, Horikoshi M (2002) Identification and characterization of CIA/ASF1 as an interactor of bromodomains associated with TFIID. Proc Natl Acad Sci USA 99:9334–9339

Cockell M, Gasser SM (1999) Nuclear compartments and gene regulation. Curr Opin Genet Dev 9:199–205

Cote J, Peterson CL, Workman JL (1998) Perturbation of nucleosome core structure by the SWI/SNF complex persists after its detachment, enhancing subsequent transcription factor binding. Proc Natl Acad Sci USA 95:4947–4952

Craig JM (2005) Heterochromatin—many flavours, common themes. Bioessays 27:17–28

Delaval K, Feil R (2004) Epigenetic regulation of mammalian genomic imprinting. Curr Opin Genet Dev 14:188–195

Dhillon N, Kamakaka RT (2002) Breaking through to the other side: silencers and barriers. Curr Opin Genet Dev 12:188–192

Dilworth SM, Black SJ, Laskey RA (1987) Two complexes that contain histones are required for nucleosome assembly in vitro: role of nucleoplasmin and N1 in Xenopus egg extracts. Cell 51:1009–1018

Dimitri P, Corradini N, Rossi F, Verni F (2005) The paradox of functional heterochromatin. Bioessays 27:29–41

Donze D, Kamakaka RT (2001) RNA polymerase III and RNA polymerase II promoter complexes are heterochromatin barriers in *Saccharomyces cerevisiae*. EMBO J 20:520–531

Dutta S, Akey IV, Dingwall C, Hartman KL, Laue T, Nolte RT, Head JF, Akey CW (2001) The crystal structure of nucleoplasmin-core: implications for histone binding and nucleosome assembly. Mol Cell 8:841–853
Evans T, Felsenfeld G, Reitman M (1990) Control of globin gene transcription. Annu Rev Cell Biol 6:95–124
Feaver WJ, Gileadi O, Li Y, Kornberg RD (1991) CTD kinase associated with yeast RNA polymerase II initiation factor b. Cell 67:1223–1230
Finley D, Chau V (1991) Ubiquitination. Annu Rev Cell Biol 7:25–69
Fourel G, Magdinier F, Gilson E (2004) Insulator dynamics and the setting of chromatin domains. Bioessays 26:523–532
Flaus A, Owen-Hughes T (2004) Mechanisms for ATP-dependent chromatin remodelling: farewell to the tuna-can octamer? Curr Opin Genet Dev 14:165–173
Gaillard PH, Martini EM, Kaufman PD, Stillman B, Moustacchi E, Almouzni G (1996) Chromatin assembly coupled to DNA repair: a new role for chromatin assembly factor I. Cell 86:887–896
Gehring WJ (1992) The homeobox in perspective. Trends Biochem Sci 17:277–280
Geng F, Cao Y, Laurent BC (2001) Essential roles of Snf5p in Snf-Swi chromatin remodeling in vivo. Mol Cell Biol 21:4311–4320
Gill G (2004) SUMO and ubiquitin in the nucleus: different functions, similar mechanisms? Genes Dev 18:2046–2059
Goto H, Tomono Y, Ajiro K, Kosako H, Fujita M, Sakurai M, Okawa K, Iwamatsu A, Okigaki T, Takahashi T, Inagaki M (1999) Identification of a novel phosphorylation site on histone H3 coupled with mitotic chromosome condensation. J Biol Chem 274:25543–25549
Green MR (2000) TBP-associated factors (TAFIIs): multiple, selective transcriptional mediators in common complexes. Trends Biochem Sci 25:59–63
Grewal SI, Elgin SC (2002) Heterochromatin: new possibilities for the inheritance of structure. Curr Opin Genet Dev 12:178–187
Grunstein M (1998) Yeast heterochromatin: regulation of its assembly and inheritance by histones. Cell 93:325–328
Guarente L (1999) Diverse and dynamic functions of the Sir silencing complex. Nat Genet 23:281–285
Hai TW, Horikoshi M, Roeder RG, Green MR (1988) Analysis of the role of the transcription factor ATF in the assembly of a functional preinitiation complex. Cell 54:1043–1051
Harp JM, Hanson BL, Timm DE, Bunick GJ (2000) Asymmetries in the nucleosome core particle at 2.5A resolution. Acta Crystallogr D Biol Crystallogr. 56:1513–1534
Hassan AH, Prochasson P, Neely KE, Galasinski SC, Chandy M, Carrozza MJ, Workman JL (2002) Function and selectivity of bromodomains in anchoring chromatin-modifying complexes to promoter nucleosomes. Cell 111:369–379
Henikoff S (1990) Position-effect variegation after 60 years. Trends Genet 6:422–426

Henry KW, Wyce A, Lo WS, Duggan LJ, Emre NC, Kao CF, Pillus L, Shilatifard A, Osley MA, Berger SL (2003) Transcriptional activation via sequential histone H2B ubiquitylation and deubiquitylation, mediated by SAGA-associated Ubp8. Genes Dev 17:2648–2663

Hecht A, Laroche T, Strahl-Bolsinger S, Gasser SM, Grunstein M (1995) Histone H3 and H4 N-terminal interact with SIR3 and SIR4 protein: a molecular model for the formation of heterochromatin in yeast. Cell 80:583–592

Holstege FC, van der Vliet PC, Timmers HT (1996) Opening of an RNA polymerase II promoter occurs in two distinct steps and requires the basal transcription factors IIE and IIH. EMBO J 15:1666–1677

Horikoshi M, Carey MF, Kakidani H, Roeder RG (1988a) Mechanism of action of a yeast activator: direct effect of GAL4 derivatives on mammalian TFIID-promoter interactions. Cell 54:665–669

Horikoshi M, Hai T, Lin YS, Green MR, Roeder RG (1988b) Transcription factor ATF interacts with the TATA factor to facilitate establishment of a preinitiation complex. Cell 54:1033–1042

Horikoshi M, Wang CK, Fujii H, Cromlish JA, Weil PA, Roeder RG (1989a) Purification of a yeast TATA box-binding protein that exhibits human transcription factor IID activity. Proc Natl Acad Sci USA 86:4843–4847

Horikoshi M, Wang CK, Fujii H, Cromlish JA, Weil PA, Roeder RG (1989b) Cloning and structure of a yeast gene encoding a general transcription initiation factor TFIID that binds to the TATA box. Nature 341:299–303

Horikoshi M, Bertuccioli C, Takada R, Wang J, Yamamoto T, Roeder RG (1992) Transcription factor TFIID induces DNA bending upon binding to the TATA element. Proc Natl Acad Sci USA 89:1060–1064.

Huang H, Kahana A, Gottschling DE, Prakash L, Liebman SW (1997) The ubiquitin-conjugating enzyme Rad6 (Ubc2) is required for silencing in *Saccharomyces cerevisiae*. Mol Cell Biol 17:6693–6699

Imai S, Armstrong CM, Kaeberlein M, Guarente L (2000) Transcripational silencing and longevity protein Sir2 is an NAD-dependent histone deacetylase. Nature 403:795–800

Ishimi Y, Hirosumi J, Sato W, Sugasawa K, Yokota S, Hanaoka F, Yamada M (1984) Purification and initial characterization of a protein which facilitates assembly of nucleosome-like structure from mammalian cells. Eur J Biochem 142:431–439

Ito T, Bulger M, Kobayashi R, Kadonaga JT (1996) Drosophila NAP-1 is a core histone chaperone that functions in ATP-facilitated assembly of regularly spaced nucleosomal arrays. Mol Cell Biol 16:3112–3124

Jacob F, Monod J (1961) Genetic regulatory mechanisms in the synthesis of proteins. J Mol Biol 3:318–356

Jenuwein T, Allis CD (2001) Translating the histone code. Science 293:1074–1080

Kadam S, Emerson BM (2002) Mechanisms of chromatin assembly and transcription. Curr Opin Cell Biol 14:262–268

Kadonaga JT (2002) The DPE, a core promoter element for transcription by RNA polymerase II. Exp Mol Med 34:259–264

Kal AJ, Mahmoudi T, Zak NB, Verrijzer CP (2000) The Drosophila brahma complex is an essential coactivator for the trithorax group protein zesta. Genes Dev 14:1058–1071

Kellogg DR, Murray AW (1995) NAP1 acts with Clb1 to perform mitotic functions and suppress polar bud growth in budding yeast. J Cell Biol 130:675–685

Khorasanizadeh S (2004) The nucleosome: from genomic organization to genomic regulation. Cell 116:259–272

Kim JL, Nikolov DB, Burley SK (1993) Co-crystal structure of TBP recognizing the minor groove of a TATA element. Nature 365:520–527

Kimura A, Horikoshi M (2004) Partition of distinct chromosomal regions: negotiable border and fixed border. Genes Cell 9:499–508

Kimura A, Umehara T, Horikoshi M (2002) Chromosomal gradient of histone acetylation established by Sas2p and Sir2p functions as a shield against gene silencing. Nat Genet 32:370–377

Kleinschmidt JA, Fortkamp E, Krohne G, Zentgraf H, Franke WW (1985) Coexistence of two different types of soluble histone complexes in nuclei of Xenopus laevis oocytes. J Biol Chem 260:1166–1176

Kornberg RD (1974) Chromatin structure: a repeating unit of histones and DNA. Science 184:868–871

Kornberg RD, Lorch Y (1999) Twenty-five years of the nucleosome, fundamental particle of the eukaryote chromosome. Cell 98:285–294

Kuzuhara T, Horikoshi M (2004) A nuclear FK506-binding protein is a histone chaperone regulating rDNA silencing. Nat Struct Mol Biol 11:275–283

Labrador M, Corces VG (2002) Setting the boundaries of chromatin domains and nuclear organization. Cell 111:151–154

Lachner M, Jenuwein T (2002) The many faces of histone lysine methylation. Curr Opin Cell Biol 14:286–298

Laskey RA, Honda BM, Mills AD, Finch JT (1978) Nucleosomes are assembled by an acidic protein which binds histones and transfers them to DNA. Nature 275:416–420

Latham KE (2005) X chromosome imprinting and inactivation in preimplantation mammalian embryos. Trends Genet 21:120–127

Laurent BC, Treich I, Carlson M (1993) The yeast SNF2/SWI2 protein has DNA-stimulated ATPase activity required for transcriptional activation. Genes Dev 7:583–591

Le S, Davis C, Konopka JB, Sternglanz R (1997) Two new S-phase-specific genes from *Saccharomyces cerevisiae*. Yeast 13:1029–1042

Litt MD, Simpson M, Recillas-Targa F, Prioleau MN, Felsenfeld G (2001) Transitions in histone acetylation reveal boundaries of three separately regulated neighboring loci. EMBO J 20:2224–2235

Lo WS, Trievel RC, Rojas JR, Duggan L, Hsu JY, Allis CD, Marmorstein R, Berger SL (2000) Phosphorylation of serine 10 in histone H3 is functionally linked in vitro and in vivo to Gcn5-mediated acetylation at lysine 14. Mol Cell 5:917–926

Lu H, Zawel L, Fisher L, Egly JM, Reinberg D (1992) Human general transcription factor IIH phosphorylates the C-terminal domain of RNA polymerase II. Nature 358:641–645

Luger K (2003) Structure and dynamic behavior of nucleosomes. Curr Opin Genet Dev 13:127–135

Luger K, Mader AW, Richmond RK, Sargent DF, Richmond TJ (1997) Crystal structure of the nucleosome core particle at 2.8A resolution. Nature 389:251–260

Ma J (2005) Crossing the line between activation and repression. Trends Genet 21:54–59

Marmorstein R (2003) Structure of SET domain proteins: a new twist on histone methylation. Trends Biochem Sci. 28:59–62

Martens JA, Winston F (2002) Evidence that Swi/Snf directly represses transcription in *S. cerevisiae.* Genes Dev 16:2231–2236

Martens JA, Winston F (2003) Recent advances in understanding chromatin remodeling by Swi/Snf complexes. Curr Opin Genet Dev 13:136–142

Matsui T, Segall J, Weil PA, Roeder RG (1980) Multiple factors required for accurate initiation of transcription by RNA polymerase II. J Biol Chem 255:11992–11996

Matsumoto K, Nagata K, Ui M, Hanaoka F (1993) Template activating factor I, a novel host factor required to stimulate the adenovirus core DNA replication. J Biol Chem 268:10582–10587

Matsumoto K, Okuwaki M, Kawase H, Handa H, Hanaoka F, Nagata K (1995) Stimulation of DNA transcription by the replication factor from the adenovirus genome in a chromatin-like structure. J Biol Chem 270:9645–9650

McKnight S, Tjian R (1986) Transcriptional selectivity of viral genes in mammalian cells. Cell 46:795–805

Mello JA, Almouzni G (2001) The ins and outs of nucleosome assembly. Curr Opin Genet Dev 11:136–141

Mello, JA, Sillje HH, Roche DM, Kirschner DB, Nigg EA, Almouzni G (2002) Human Asf1 and CAF-1 interact and synergize in a repair-coupled nucleosome assembly pathway. EMBO Rep 3:329–334

Miyaji-Yamaguchi M, Kato K, Nakano R, Akashi T, Kikuchi A, Nagata K (2003) Involvement of nucleocytoplasmic shuttling of yeast Nap1 in mitotic progression. Mol Cell Biol 23:6672–6684

Miyamoto S, Suzuki T, Muto S, Aizawa K, Kimura A, Mizuno Y, Nagino T, Imai Y, Adachi N, Horikoshi M, Nagai R (2003) Positive and negative regulation of the cardiovascular transcription factor KLF5 by p300 and the oncogenic regulator SET through interaction and acetylation on the DNA-binding domain. Mol Cell Biol 23:8528–8541

Mousson F, Lautrette A, Thuret JY, Agez M, Courbeyrette R, Amigues B, Becker E, Neumann JM, Guerois R, Mann C, Ochsenbein F (2005) Structural basis

for the interaction of Asf1 with histone H3 and its functional implications. Proc Natl Acad Sci USA 102:5975–5980
Munakata T, Adachi N, Yokoyama N, Kuzuhara T, Horikoshi M (2000) A human homologue of yeast anti-silencing factor has histone chaperone activity. Genes Cells 5:221–233
Murre C, McCaw PS, Vaessin H, Caudy M, Jan LY, Jan YN, Cabrera CV, Buskin JN, Hauschka SD, Lassar AB Weintraub H. Baltimore D (1989) Interactions between heterologous helix-loop-helix proteins generate complexes that bind specifically to a common DNA sequence. Cell 58:537–544
Mutskov VJ, Farrell CM, Wade PA, Wolffe AP, Felsenfeld G (2002) The barrier function of an insulator couples high histone acetylation levels with specific protection of promoter DNA from methylation. Genes Dev 16:1540–1554
Nakayama J, Rice JC, Strahl BD, Allis CD, Grewal SI (2001) Role of histone H3 lysine 9 methylation in epigenetic control of heterochromatin assembly. Science 292:110–113
Naresh A, Saini S, Singh J (2003) Identification of Uhp1, a ubiquitinated histone-like protein, as a target/mediator of Rhp6 in mating-type silencing in fission yeast. J Biol Chem 278:9185–9194
Narlikar GJ, Fan HY, Kingston RE (2002) Cooperation between complexes that regulate chromatin structure and transcription. Cell 108:475–487
Nielsen AL, Sanchez C, Ichinose H, Cervino M, Lerouge T, Chambon P, Losson R (2002) Selective interaction between the chromatin-remodeling factor BRG1 and the heterochromatin-associated protein HP1 alpha. EMBO J 21:5797–5806
Okuwaki M, Nagata K (1998) Template activating factor-I remodels the chromatin structure and stimulates transcription from the chromatin template. J Biol Chem 273:34511–34518
Peterson CL, Tamkun JW (1995) The SWI-SNF complex: a chromatin remodeling machine?. Trends Biochem Sci. 22:143-146
Peterson CL (2002) Chromatin remodeling: nucleosomes bulging at the seams. Curr Biol 12:R245–247
Peterson CL, Laniel MA (2004) Histones and histone modifications. Curr Biol 14:R546–551
Ptashne M (1988) How eukaryotic transcriptional activators work. Nature 335:683–689
Ptashne M (2004) A genetic switch: phage lambda revisited, 3rd edn. Blackwell Science, Malden, MA and Cell Press, Cambridge, MA
Ptashne M, Gann A (2001) Genes and signals. Cold Spring Harbor Laboratory Press, Cold Spring Harbor, NY
Rangwala SH, Richards EJ (2004) The value-added genome: building and maintaining genomic cytosine methylation landscapes. Curr Opin Genet Dev 14:686–691
Rice JC, Allis CD (2001) Histone methylation versus histone acetylation: new insights into epigenetic regulation. Curr Opin Cell Biol 13:263–273
Rippe K, von Hippel PH, Langowski J (1995) Action at a distance: DNA-looping and initiation of transcription. Trends Biochem Sci 20:500–506

Roeder RG (1996) The role of general initiation factors in transcription by RNA polymerase II. Trends Biochem Sci. 21:327–335

Roeder RG (2003) Lasker Basic Medical Research Award. The eukaryotic transcriptional machinery: complexities and mechanisms unforeseen. Nat Med 9:1239–1244

Roeder RG, Rutter WJ (1969) Multiple forms of DNA-dependent RNA polymerase in eukaryotic organisms. Nature 224:234–237

Roth SY, Denu JM, Allis CD (2001) Histone acetyltransferases. Annu Rev Biochem 70:81–120

Saha A, Wittmeyer J, Cairns BR (2002) Chromatin remodeling by RSC involves ATP-dependent DNA translocation. Genes Dev 16:2120–2134

Schotta G, Ebert A, Dorn R, Reuter G (2003) Position-effect variegation and the genetic dissection of chromatin regulation in Drosophila. Semin Cell Dev Biol 14:67–75

Seo SB, McNamara P, Heo S, Turner A, Lane WS, Chakravarti D (2001) Regulation of histone acetylation and transcription by INHAT, a human cellular complex containing the set oncoprotein. Cell 104:119–130

Serizawa H, Conaway RC, Conaway JW (1992) A carboxyl-terminal-domain kinase associated with RNA polymerase II transcription factor delta from rat liver. Proc Natl Acad Sci USA 89:7476–7480

Shibahara K, Stillman B (1999) Replication-dependent marking of DNA by PCNA facilitates CAF-I coupled inheritance of chromatin. Cell 96:575–585

Smith S, Stillman B (1989) Purification and characterization of CAF-I, a human cell factor required for chromatin assembly during DNA replication in vitro. Cell 58:15–25

Strahl BD, Allis CD (2000) The language of covalent histone modifications. Nature 403:41–45

Sun ZW, Allis CD (2002) Ubiquitination of histone H2B regulates H3 methylation and gene silencing in yeast. Nature 418:104–108

Suzuki T, Muto S, Miyamoto S, Aizawa K, Horikoshi M, Nagai (2003) Functional interaction of the DNA-binding transcription factor Sp1 through its DNA-binding domain with the histone chaperone TAF-I. J Biol Chem 278:28758–28764

Taunton J, Hassig CA, Schreiber SL (1996) A mammalian histone deacetylase related to the yeast transcriptional regulator Rpd3p. Science 272:408–411

Tsukiyama T, Wu C (1997) Chromatin remodeling and transcription. Curr Opin Genet Dev 7:182–191

Tsukiyama T, Becker PB, Wu C (1994) ATP-dependent nucleosome disruption a heat-shock promoter mediated by binding of GAGA transcription factor. Nature 367:525–532

Turner BM (2001) Chromatin and gene regulation: mechanism in epigenetics. Blackwell Science

Tyler JK (2002) Chromatin assembly. Cooperation between histone chaperones and ATP-dependent nucleosome remodeling machines. Eur J Biochem 269:2268–2274

Tyler JK, Adams CR, Chen SR, Kobayashi R, Kamakaka RT, Kadonaga JT (1999) The RCAF complex mediates chromatin assembly during DNA replication and repair. Nature 402:555–560

Tyler JK, Collins KA, Prasad-Sinha J, Amiott E, Bulger M, Harte PJ, Kobayashi R, Kadonaga JT (2001) Interaction between the Drosophila CAF-1 and ASF1 chromatin assembly factors. Mol Cell Biol 21:6574–6584

Umehara T, Horikoshi M (2003) Transcription initiation factor IID-interactive histone chaperone CIA-II implicated in mammalian spermatogenesis. J Biol Chem 278:35660–35667

van Leeuwen F, Gottschling DE (2002) Genome-wide histone modifications: gaining specificity by preventing promiscuity. Curr Opin Cell Biol 14:756–762

Verreault A, Kaufman PD, Kobayashi R, Stillman B (1996) Nucleosome assembly by a complex of CAF-1 and acetylated histones H3/H4. Cell 87:95–104

Wagner S, Green MR (1994) DNA-binding domains: targets for viral and cellular regulators. Curr Opin Cell Biol 6:410–414

Wang H, Huang ZQ, Xia L, Feng Q, Erdjumnet-Bromage H, Strahl BD, Briggs SD, Allis CD, Wong J, Tempst P, Zhang Y (2001) Methylation of histone H4 at arginine 3 facilitating transcriptional activation by nuclear hormone receptor. Science 293:853–857

Wei Y, Mizzen CA, Cook RG, Gorovsky MA, Allis CD (1998) Phosphorylation of histone H3 at serine 10 is correlated with chromosome condensation during mitosis and meiosis in Tetrahymena. Proc Natl Acad Sci USA 95:7480–7484

Weintraub H, Davis R, Tapscott S, Thayer M, Krause M, Benezra R, Blackwell TK, Turner D, Rupp R, Hollenberg S, Zhuang Y, Lassar A (1991) The MyoD gene family: nodal point during specification of the muscle cell lineage. Science 251:761–766

West AG, Gaszner M, Felsenfeld G (2002) Insulators: many functions, many mechanisms. Genes Dev 16:271–288

White CL, Suto RK, Luger K (2001) Structure of the yeast nucleosome core particle reveals fundamental changes in internucleosome interactions. EMBO J 20:5207–5218

Wolffe AP (1997) Histone H1. Int J Biochem Cell Biol 29:1463–1466

Wolffe AP (1998) Chromatin: structure and function, 3rd edn. Academic Press, San Diego

Workman JL, Kingston RE (1998) Alteration of nucleosome structure as a mechanism of transcriptional regulation. Annu Rev Biochem 67:545–579

Workman JL, Roeder RG (1987) Binding of transcription factor TFIID to the major late promoter during in vitro nucleosome assembly potentiates subsequent initiation by RNA polymerase II. Cell 51:613–622

Yang XJ (2004) Lysine acetylation and the bromodomain: a new partnership for signaling. Bioessays 26:1076–1087

Zhang Y (2003) Transcriptional regulation by histone ubiquitination and deubiquitination. Genes Dev 17:2733–2740

Zhang ZK, Davies KP, Allen J, Zhu L, Pestell RG, Zagzag D, Kalpana GV (2002) Cell cycle arrest and repression of cyclin D1 transcription by INI1/hSNF5. Mol Cell Biol 22:5975–5988

8. Dynamic Chromatin Loops and the Regulation of Gene Expression

Hiroshi Kimura[1] and Peter R. Cook[2]

[1]Nuclear Function and Dynamics Unit, HMRO, Graduate School of Medicine, Kyoto University, Yoshidakonoe-cho, Sakyo-ku, Kyoto 606-8501, Japan

[2]Sir William Dunn School of Pathology, University of Oxford, South Parks Road, Oxford, OX1 3RE, UK

8.1 Introduction

Although we have a draft sequence of the human genome, little is known about how the chromatin fiber is packed in three-dimensional (3D) space, or how packing affects function (Jackson 2003). We know packing plays a major role; the rate of transcription of a typical gene can vary over eight orders of magnitude (Ivarie et al. 1983), but deleting local elements like promoters and enhancers reduces expression by less than 5000-fold in transient transfection assays where the 3D "context" is missing. Common sense suggests the fiber cannot be packed randomly, but elucidating what any underlying order might be has proved difficult. First, the foldings of the chromatin fiber have dimensions below the resolution (~200 nm) of the light microscope (LM) and so can only be seen by electron microscopy (EM), but then the fixation required can distort structure. Second, DNA is so long and packed so tightly it breaks and/or aggregates easily on isolation. Third, chromatin is poised in a metastable state so small charge alterations trigger changes in structure and function, and replacing the natural environment with our buffers often promotes aggregation.

Not surprisingly, biochemists minimize aggregation through the use of hypo- and hypertonic buffers, but then different isolates made in different buffers have quite different structures; for example, "matrices" and "scaffolds" contain different sets of DNA sequences associated with different proteins. Some buffers lead to a stable reduction in contour length of chromatin loops (so new attachments of the chromatin fiber to the underlying structure are generated), others increase it (so some attachments are destroyed; Jackson et al. 1990). Again not surprisingly, observations on the different structures have led to different models including: (i) random packing of the nucleosomal string (Sachs et al. 1995), (ii) helical hierarchies—strings are coiled into solenoids (~30 nm diameter), solenoids into higher-order structures, and so on (Sedat and Manuelidis 1978), (iii) loops (50–150 kbp) attached to the peripheral lamina or internal structures like (iv) skeletons, matrices, scaffolds (e.g., Saitoh and Laemmli 1978), or factories (Cook 1995), and (v) combinations of the above—for example, of helical coils and radial loops (Manuelidis 1990), or helical coils and random folding (Li et al. 1998; Strukov et al. 2003). Since there is so little agreement, outsiders often dismiss all the results seen (Cook 1988; Pederson 2000; Belmont 2002).

Here, we concentrate on results obtained using isotonic buffers. We discuss the idea that most large biological structures are intrinsically unstable, persisting only by exchanging subunits with others in their surroundings. We then go on to review evidence for the existence of transient chromatin loops, and propose a general model for the organization of all genomes that involves such loops. Finally, we suggest how ever-changing attachments of the loops to the underlying structure can explain how genes are regulated.

8.2 Self-Assembly and Self-Organization of Macromolecular Structures

Macromolecular structures are generated in two fundamentally different ways (Misteli 2001a). Many virus particles "self-assemble" to a fixed plan to attain a true thermodynamic equilibrium; the particles are stable, and survive in the absence of a pool of unincorporated subunits when released from the host. Our houses are similar structures, although we direct their construction; when the builder has finished he removes any unused bricks but the house usually remains standing. But most cellular structures are built differently. Thus, the cytoskeleton lacks a rigid architecture. It is "self-organizing," intrinsically unstable, and persists only by exchanging subunits with others in its surroundings; if those subunits are removed, it collapses and disappears. The structures discussed here fall into the latter category. They are ever-changing, with their shape at any particular moment depending on past and present environments. We cannot make precise predictions about the position of any particular molecule or gene within the structure; however, we should eventually be able to predict the probabilities that they will be found in one particular place rather than another. Evidence for the dynamic nature of nuclear structures is reviewed in a number of chapters in this book, and photobleaching studies of the critical components that concern us here—transcription factors and polymerases tagged with green fluorescent protein (GFP)—reveal that they exchange continually with the soluble pool (e.g., Stenoien et al. 2001; Becker et al. 2002; Chen et al. 2002; Dundr et al. 2002; Kimura et al. 2002).

8.3 Chromatin Loops

Evidence

The idea that the chromatin fiber is looped is one of the oldest in cell biology. Images of the lampbrush chromosomes that can be isolated from oocytes are often cited as providing the best evidence for looping. During the first meiotic division, duplicated homologues pair, and loops can be seen extending microns away from axial chromomeres. Unusually, these chromosomes are transcribed, and nascent RNA is attached to both loops and chromomeres (Snow and Callan 1969). Note, however, that these loops only become visible on dispersing chromatin in hypotonic buffers, and none are seen in sections of whole oocytes where chromatin appears as a granular aggregate. Therefore, some transcription units may be stripped off the granules during dispersal, and possible intermediates in such a process—small granules—can be seen scattered around loops (Mott and Callan 1975; Cook 2001).

Supercoiling provides additional evidence. Supercoils cannot be maintained in linear eukaryotic DNA without looping. However, lysing cells in >1 M NaCl releases "nucleoids" containing superhelical DNA (Cook and Brazell 1975; Benyajati and Worcel 1976), and nascent transcripts are associated with attachment points, but not loops (Jackson et al. 1984a). Nucleoids made from all active cells examined (e.g., fibroblasts, erythroblasts, epithelial cells of men, chickens, frogs, insects) contain such supercoils. In contrast, inactive chicken erythrocytes (and human sperm) yield relaxed DNA, with the supercoiling (and so looping) being lost progressively as they develop (Jackson et al. 1984a). But this evidence is also compromised by the unphysiological conditions used.

Experiments involving nuclease digestion in isotonic buffers are also consistent with looping. Cutting an unlooped fiber should release long fragments that are then shortened, but the expected long fragments are not seen; rather, kinetics fit the release of short fragments from loops (Jackson et al. 1990, 1996; Jackson and Cook 1993). Other in vitro evidence is also supportive; EM reveals that pure repressors like Gal, AraC, and λ bind to distant sites on one molecule of DNA to loop it (reviewed by Ptashne 1986), and an enhancer can only influence a promoter on another plasmid if the two make molecular contact, which implies that when they are on the same chromosome they must also do so (Mueller-Storm et al. 1989). Contact between an enhancer on one chromosome and a target promoter on another also underlies the phenomenon of transvection seen in *Drosophila* larvae (Wu 1993; Cook 1997).

The application of two new methods—"chromosome conformation capture" (3C; Dekker et al. 2002) and "RNA tagging and recovery of associated proteins" (RNA TRAP; Carter et al. 2002)—provide excellent support for loops tied through two (or more) active transcription units (Cook 2003). Both methods involve careful fixation, before analysis by polymerase chain reaction of which DNA sequences lie next to each other in 3D space (i.e., after ligation in 3C, or purifying complexes in RNA TRAP). The mouse β-globin locus control region (LCR) lies tens of kilobases away from the β-globin genes that it regulates. This LCR is transcribed, and both methods show it contacts the β-globin gene in erythroid nuclei (where the gene is also transcribed). No contacts are found in brain nuclei, where the gene is inactive (Carter et al. 2002; Tolhuis et al. 2002; Palstra et al. 2003). (Competition between two different genes within the locus for the LCR leads to their alternate transcription, with the nearest one initially being transcribed the most; this is consistent with the transcription frequency being determined by LCR:promoter distance [Hanscombe et al. 1991; Dillon et al. 1997].) Similarly, 3C reveals that two distant "barrier" elements (i.e., *scs* and *scs'*) flanking the *Drosophila* 87A7 heat-shock locus—which are both transcribed—lie together (Blanton et al. 2003).

Although none of this evidence is derived from the analysis of living cells, taken together it provides good evidence for looping.

Ever-Changing Attachments

Many models for looping involve stable interactions between a motif like a MAR with some abundant protein bound to the substructure; this molecular tie would persist from one interphase to the next, be highly conserved, and found in all cells in the population. However, genomic sequencing has failed to uncover any such motifs, and a simple experiment demonstrates that different ties are found in different cells in the population. Cells are permeabilized in an isotonic buffer, treated with a nuclease like *Eco*RI, detached fragments removed, and remaining fragments analyzed. If the same DNA ties were found in all cells (Fig. 1a), removing all but 10% cellular DNA should leave the ties enriched tenfold. However, enrichments are never this high, implying that the same ties are not found in all cells and that they change continually (Fig. 1b; Dickinson et al. 1990; Jackson et al. 1990, 1996; Jackson and Cook 1993).

Constrained Diffusion of DNA Within Loops

Chromatin dynamics can be monitored in living cells containing *lac* operator arrays integrated into a chromosome if they also express the repressor tagged with GFP; the repressors bound to the array appear as a moving spot (Robinett et al. 1996; Gasser 2002; Spector 2003). Analyses of the kinetics are consistent with the array diffusing randomly over short distances. In yeast, the array can sample a considerable fraction of nuclear volume in minutes (Heun et al. 2001). In larger fly and human nuclei, the local neighborhood (diameter ~500 nm) is sampled roughly as rapidly, but diffusion further a field is constrained—presumably by neighboring chromatin (Marshall et al. 1997; Vazquez et al. 2001; Chubb et al. 2002; Chubb and Bickmore 2003).

Active Transcription Units Are Attached Through Engaged Polymerases

Which proteins and DNA motifs constitute the molecular ties? In experiments like that illustrated in Fig. 1, the residual fragments prove to be parts of transcription units associated with engaged polymerases. For example, they hybridize with poly(A)$^+$ RNA (Jackson and Cook 1985) and contain active rDNA cistrons (Dickinson et al. 1990), while cloning and sequencing over 100 randomly selected ones shows they are nearly all parts of transcription units (Jackson et al. 1996). Moreover, they remain associated with engaged polymerases; removing most chromatin in an isotonic buffer does not reduce polymerizing activity, whether it be the total activity or due to polymerases I or II (Jackson and Cook 1985; Dickinson et al. 1990). These polymerases cannot be tethered to the substructure through nascent transcripts as they remain after RNase treatment. (Note that when the experiment illustrated in Fig. 1 is conducted in nonisotonic buffers, different fragments [e.g., MARs and SARs] are found to be attached depending on the buffer used [above].)

A detailed study of a model loop confirms that engaged polymerases mediate attachments (Jackson and Cook 1993). When a few plasmids carrying the SV40 origin of replication are transfected into monkey cells (i.e., cos 7), they replicate over 2–3 days to give hundreds of minichromosomes. After permeabilization in an isotonic buffer, nearly all these model loops (~5 kb) resist electroelution and so must be attached to the substructure. Transfecting in progressively more DNA increases the number of attached plasmids up to a maximum of ~1200, but then higher inputs generate additional unattached ones. This suggests there are a saturable number of attachment sites. Only the attached population is active, as eluting the unattached fraction does not reduce plasmid-specific transcription. Cutting with *Hae*III and removing most of the resulting ~400 bp fragments (as in Fig. 1) also has little effect on transcription but it leaves

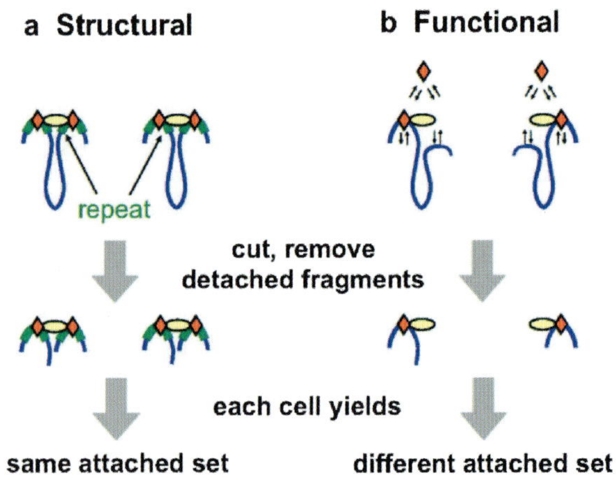

Fig. 1. Molecular ties. **a** Static (structural). DNA repeats (*green*) in two cells bind to the same protein complexes (*ovals*, *diamonds*), looping the fiber. After cutting with a nuclease and removing detached fragments, the same set of repeats from each cell remain bound; when 10% DNA remains, repeats are enriched tenfold. **b** Dynamic (functional). The fiber is looped by attachment to a protein complex, but both attachments and proteins in the complex change from moment to moment. After cutting and removing detached fragments, a different set of fragments remain attached in the two cells; when 10% DNA remains, no fragment is enriched tenfold. This result is obtained if cutting and removal are carried out in isotonic buffers; essentially all active transcription complexes also remain attached. Reprinted with permission from Cook 2003.

fragments from within one or other of the two transcription units. Quantitative analysis shows that each minichromosome is attached at 1–2 points through either one of the two promoters in the plasmid or the body of the transcription units.

Support for the idea that polymerases act as the ties comes from the regeneration of loops from unlooped sperm DNA (Gall and Murphy 1998). When demembranated sperm heads (which are inactive and contain unlooped DNA; above) are injected into the germinal vesicle of amphibian oocytes, the heads swell, accumulate polymerase II, and begin to be transcribed. If the contents of the germinal vesicle are now dispersed in a hypotonic buffer, lampbrush loops derived from both injected sperm and host are seen. The generation of lampbrushes from sperm DNA depends on transcription, as actinomycin D prevents it. Moreover, the active form of polymerase II (marked by hyperphosphorylation of serine 5 in the heptad repeats of the C-terminal domain of its catalytic subunit) becomes concentrated in the lampbrush axis.

Polymerases elongate at $\sim 1.8 \times 10^3$ nucleotides/min, and take ~10 min to transcribe a typical human gene (Kimura et al. 2002). While transcription continues, an active transcription unit will remain attached, and only on termination will it detach to leave the bound polymerase that is now free to exchange with others in the soluble pool.

Other Kinds of Attachment

Transcription factors bound to their (untranscribed) DNA targets probably mediate additional attachments, as those targets resist nucleolytic detachment from the substructure. The first hint that this was so came from studies on nucleoids derived from rat cells transformed by polyoma virus; when most DNA was detached with *Eco*RI, fragments containing viral enhancers remained (Cook et al. 1982). These results were confirmed using isotonic buffers; thus, minichromosomes are attached as much through nontranscribed promoters as through the body of transcription units (above). These kinds of attachment are unlikely to persist for long, as most GFP-tagged transcription factors remain bound to DNA only for seconds (Misteli 2001b; Chen et al. 2002; Hoogstraten et al. 2002). And likely to be established only when assembled into preinitiation complexes, as specific transcription factors like Sp1 and C/EBP are detached with the body of the loop by *Hae*III digestion; while in contrast basal factors like TFIIB and TFIIH remain (Kimura et al. 1999).

Another kind of tie—not to a factory—may be essentially permanent. Histones can carry a "code" (Fischle et al. 2003; Lachner et al. 2003) that ensures they bind tightly to others in heterochromatin or to the lamina (Polioudaki et al. 2001). These are probably the ones that GFP-tagging indicates do not exchange except when DNA is replicated (Kimura et al. 2001), so they can sequester a loop permanently away from a factory. For example, genes involved in immunoglobulin rearrangements are repositioned to centromeric heterochromatin during lymphocyte development coincident with their inactivation (Brown et al. 1999). Other kinds of tie probably exist only fleetingly, and will be best analyzed in vivo (e.g., by fluorescence correlation spectroscopy; Lippincott-Schwartz et al. 2001; Weidemann et al. 2003). These arise because chromatin presents such a huge binding surface to the nucleoplasm; then, many so-called soluble nuclear proteins probably spend much of their time bound transiently to it through low-affinity interactions.

8.4 Clusters of Active Polymerases Organize Rosettes of Loops

A Model for Genome Organization

The principles and results described above lead to a general model for the organization of all genomes (Fig. 2; Cook 1995, 2002). When genes strung along a template are transcribed, active polymerases form into clusters to loop intervening DNA. We call one of these clusters a "factory" as it contains several polymerizing complexes working on different transcription units. Each factory would be surrounded by a rosette (or "cloud") of loops, and strings of nucleosomes and factories (plus surrounding clouds) would constitute the major architectural motifs responsible for organizing the genome. As we have seen, transcription factors in factories would mediate additional attachments. Then, RNA polymerase is not only an enzyme, but a critical structural component that ties the genome in loops. Note that the 3C and RNA TRAP methods provide powerful evidence for the local contact between two active transcription units; for example, the LCR only contacts the β-globin gene that it regulates when both are transcribed (above).

Evidence That Active Polymerases Do Not Track

Textbooks tell us that active RNA polymerases track like locomotives down their templates. In contrast, in our model active polymerases are attached to the substructure, and the immobilized enzyme works by reeling

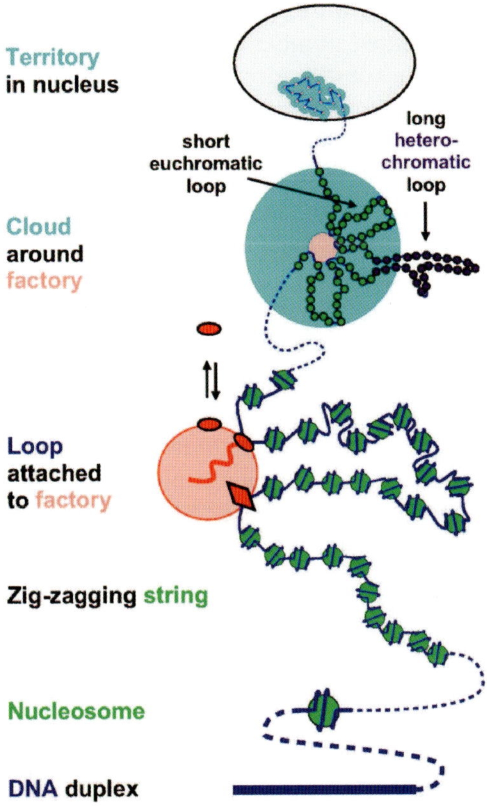

Fig. 2. A model for genome folding. DNA is wound into a nucleosome, and then a zig-zagging string of nucleosomes is tied to a factory through a cluster of transcription factors (*diamond*) or an active polymerase (*oval*). Components of the factory exchange with the soluble pool, and attachments to the factory are made and broken as factors dissociate and transcription terminates. Ten to twenty loops (only three are shown) of 5–200 kbp form a cloud around the factory; long, static, loops are likely to become heterochromatic and attached to the lamina. Fifty to one hundred clouds then form a chromosome territory. Reprinted with permission from Cook 2001 and Cook 2003.

in its template (Cook 1999). Evidence for this is of three general types. First, a tracking polymerase would generate a transcript that is entangled about the template, but this problem does not arise if the polymerase is immobilized. Second, biochemical results indicated that active genes, RNA polymerases, and nascent transcripts are all so closely associated with the underlying structure that they remain when most chromatin is removed with nucleases (as in Fig. 1). Third, we developed a method for localizing nascent transcripts with high resolution, and used it to show that active polymerases engaged on a number of different transcription units are concentrated in a limited number of discrete sites (diameter ~50 nm) firmly associated with the substructure. (Engaged polymerases are allowed to extend transcripts by a few nucleotides in Br-UTP or biotin-CTP before the resulting tagged RNA is immunolabeled [Jackson et al. 1993].) As there turn out to be 8-fold more active molecules of RNA polymerase in a HeLa cell than transcription sites, and as only one polymerase is typically engaged on a transcription unit, each site (diameter ~50 nm) must contain ~8 different polymerases active on ~8 different units.

This suggests active polymerases are immobilized, but can they then work? There is good evidence they can, as tethering them to a slide or plastic bead has no effect on the rate of polymerization of nucleotide triphosphates (Schafer et al. 1991; Cook and Gove 1992). They can be viewed as motors that haul in their templates. They are powerful ones largely due to their low gearing; for each triphosphate hydrolyzed, DNA is reeled in by ~0.34 nm—one-tenth and one-hundredth the step-lengths respectively of kinesin and myosin V (Gelles and Landick 1998).

Specialized Factories

We are all familiar with the prototypic mammalian transcription factory—the nucleolus—which is dedicated to the synthesis of 45S rRNA and the production of ribosomes (Grummt 2003). Active polymerase II and III are also each concentrated in their own discrete factories dedicated to the production of particular transcripts. Thus, in a HeLa nucleus polymerase II transcripts (but not polymerase III, or its transcripts) are concentrated in ~8000 nucleoplasmic sites, while polymerase III transcripts (but not polymerase II, or its transcripts) are found in another ~2000 sites (Pombo et al. 1999).

Some factories specialize even further, and become dedicated to the transcription of specific sets of genes. Examples include: OPT domains (transcribing genes depending on Oct1 and PTF; Pombo et al. 1998), sites containing β-globin and LCR transcripts (above), stress granules (transcribing satellite repeats; Jolly et al. 2004), CBs (transcribing snRNAs; Callan et al. 1991; Frey et al. 1999; Jacobs et al. 1999), and perinucleolar polymerase III factories (transcribing tRNAs; Thompson et al. 2003).

Factories Contain Many Machines Required to Make Mature Transcripts

Active RNA polymerase II is part of a huge complex that carries out many—perhaps all—of the functions (RNA synthesis, capping, splicing, polyadenylation) required to generate a mature message (Maniatis and Reed 2002; Proudfoot et al. 2002). This complex probably also proofreads mRNAs before going on to destroy faulty ones along with any peptides generated during proofreading (Iborra et al. 2001, 2004; Andrulis et al. 2002; Brogna et al. 2002; Libri et al. 2002; Lykke-Andersen 2002). Proofreading involves the nonsense-mediated decay (NMD) pathway (Hilleren and Parker 1999); this probably uses ribosomes to scan mRNAs for inappropriately placed (i.e., premature) termination codons (PTCs), and—if detected—triggers the destruction of those faulty messages (along with any misfolded peptides that might result from proofreading). Nascent peptides—which are presumably made during the scanning by the ribosomes—are found in the nucleoplasmic transcription factories, together with components of the translation and NMD machineries. As inhibiting transcription immediately inhibits this nuclear protein synthesis, the processes must be tightly coupled (Iborra et al. 2001). Moreover, the transcriptional, translational, NMD, and degradative machineries colocalize and coimmunoprecipitate; selected components (translational initiation factor eIF4E, ribosomal subunit S6, NMD factors UPF1/2) also copurify with the catalytic subunit of the polymerase, probably by binding to its C-terminal domain (Iborra et al. 2004).

8.5 Gene Activation

Local Concentrations of Polymerases and Promoters

Nuclei of HeLa cells contain a dispersed pool of RNA polymerase II present at ~1 μM, but this is unlikely to account for much transcription because the local concentration in a factory is ~1000-fold higher (Cook 2001). This is especially so when promoters are tethered close to a factory.

Compare two promoters in a loop, where one lies ~4-fold further away along the DNA from the factory. The distant promoter will be confined to a 64-fold larger volume around the factory (as volume depends on radius^3). This reduces its local concentration by the same amount, and so the frequency with which it contacts the factory. As a result, distant promoters are much less likely to initiate than proximal ones. The long tether has another effect; it buffers the distant promoter from transcription-driven movement, and this immobility makes it likely to acquire the histone code characteristic of "closed" chromatin. The context then becomes self-sustaining: productive collisions of the nearer promoter with the factory attract factors increasing the initiation frequency, and the longer the distant promoter remains inactive the more it is likely to be embedded in heterochromatin.

Gene Regulation and Promoter–Factory Distance

The probability that a promoter collides productively with a factory is increased by increasing promoter mobility (by "opening" chromatin), increasing promoter–factory affinity (through binding of appropriate factors), and reducing promoter–factory distance (by shortening the tether; Iborra et al. 1996; Cook 2003). It will also depend on which other promoters compete for binding sites in nearby factories. For example, a polymerase II unit might be "silenced" by nearby polymerase II units (because they compete too effectively for a polymerizing site) or polymerase III units (because they attach to a remote polymerase III factory distant from any polymerase II factory). Alternatively, adjacent transcription units can stimulate activity. Consider the activation of the human ε-*globin* gene during development. We imagine that it is initially embedded deep in heterochromatin far from a factory. During erythroblast development, the concentration of critical activators rises so the LCR—which would be in open chromatin on the edge of the heterochromatin—now has an increased affinity for the factory. Once the LCR attaches and transcription begins, a polymerase reels in the template. This transiently subdivides the long loop into two smaller ones and the associated movement reverses the histone code, to open ε-*globin* and bring it closer to the factory; now, it is much likelier to attach. Active transcription units can also act as barriers that prevent the spread of heterochromatin down the chromatin fiber; the inevitable movement associated with activity counteracts histone aggregation and the spread of an inactive histone code. Therefore, the pattern of activity of adjacent genes on the chromosome will also determine whether or not a particular gene is active.

8.6 Conclusions

This model illustrated in Fig. 2 has several advantages. First, it is general, and can easily be extended to bacteria. Thus, lysing bacteria in a detergent and 1 M NaCl releases the prototypic factory—a cluster of still-engaged polymerases surrounded by DNA loops. Transcription maintains this structure, as inhibiting it with rifampicin or treatment with ribonuclease (RNase) unfolds it (Pettijohn 1996). Moreover, GFP-tagging reveals that RNA polymerases are concentrated in discrete foci, each of which probably contains several different operons encoding rRNA (Lewis et al. 2000; Cabrera and Jin 2003). Second, it is a minimalist model in which all the basic structural motifs are defined. Third, we have seen it can readily explain how gene activity is regulated. Fourth, it can be extended to mitosis. Then, the contour length of the loops and the basic shape of the chromatin clouds remain unchanged (Jackson et al. 1990; Manders et al. 1999), and decreased transcription coupled to increased cloud:cloud and factory:factory aggregation could drive reassembly into the most compact and stable structure, a cylinder of nucleosomes around an axial core containing the remnants of the factories (Cook 1995). Despite these advantages, many questions remain. For example, we still know very little about the microarchitecture of factories, or what maintains their structure as their constituents exchange continually with others in the surroundings.

Three different kinds of motion of the chromatin loop have been described: (i) the random Brownian motion of DNA segments within the loop, (ii) the directed transcription-dependent reeling-in by immobilized polymerases that continuously changes the contour length of one particular loop, and (iii) the making/breaking of the molecular ties that attach loops to the factory (so adjacent loops split/merge as polymerase initiate/terminate or transcription factors bind/dissociate). These three motions counteract the tendency of chromatin to condense into heterochromatin, with a consequential alteration in histone code.

Acknowledgments

We thank our colleagues for helpful discussions, and the Ministry of Education, Culture, Sports, Science and Technology of Japan, the BBSRC, Cancer Research UK, and The Wellcome Trust for support.

References

Andrulis ED, Werner J, Nazarian A, Erdjument-Bromage H, Tempst P, Lis JT (2002) The RNA processing exosome is linked to elongating RNA polymerase II in Drosophila. Nature 420:837–841

Becker M, Baumann C, John S, Walker DA, Vigneron M, McNally JG, Hager GL (2002) Dynamic behavior of transcription factors on a natural promoter in living cells. EMBO Rep 3:1188–1194

Belmont AS (2002) Mitotic chromosome scaffold structure: new approaches to an old controversy. Proc Natl Acad Sci USA 99:15855–15857

Benyajati C, Worcel A (1976) Isolation, characterization and structure of the folded interphase genome of *Drosophila melanogaster.* Cell 9:393–407

Blanton J, Gaszner M, Schedl P (2003) Protein:protein interactions and the pairing of boundary elements in vivo. Genes Dev 17:664–675

Brogna S, Sato TA, Rosbash M (2002) Ribosome components are associated with sites of transcription. Mol Cell 10:93–104

Brown KE, Baxter J, Graf D, Merkenschlager M, Fisher AG (1999) Dynamic Repositioning of genes in the nucleus of lymphocytes preparing for cell division. Mol Cell 3:207–217.

Cabrera JE, Jin DJ (2003) The distribution of RNA polymerase in Escherichia coli is dynamic and sensitive to environmental cues. Mol Microbiol 50:1493–1505

Callan HG, Gall JG, Murphy C (1991) Histone genes are located at the sphere loci of Xenopus lampbrush chromosomes. Chromosoma 101:245–251

Carter D, Chakalova L, Osborne CS, Dai Y-F, Fraser P (2002) Long-range chromatin regulatory interactions in vivo. Nat Genet 32:623–626

Chen D, Hinkley CS, Henry RW, Huang S (2002) TBP dynamics in living human cells: constitutive association of TBP with mitotic chromosomes. Mol Biol Cell 13:276–284

Chubb JR, Bickmore WA (2003) Considering nuclear compartmentalization in the light of nuclear dynamics. Cell 112:403–406

Chubb JR, Boyle S, Perry P, Bickmore WA (2002) Chromatin motion is constrained by association with nuclear compartments in human cells. Curr Biol 12:439–445

Cook PR (1988) The nucleoskeleton: artefact, passive framework or active site? J Cell Sci 90:1–6

Cook PR (1995) A chromomeric model for nuclear and chromosome structure. J Cell Sci 108:2927–2935

Cook PR (1997) The transcriptional basis of chromosome pairing. J Cell Sci 110:1033–1040

Cook PR (1999) The organization of replication and transcription. Science 284:1790–1795

Cook PR (2001) Principles of nuclear structure and function. Wiley, New York, pp 86–88, 182–184

Cook PR (2002) Predicting three-dimensional genome structure from transcriptional activity. Nat Genet 32:347–352

Cook PR (2003) Nongenic transcription, gene regulation and action at a distance. J Cell Sci 116:4483–4491

Cook PR, Brazell IA (1975) Supercoils in human DNA. J Cell Sci 19:261–279

Cook PR, Gove F (1992) Transcription by an immobilized RNA polymerase from bacteriophage T7 and the topology of transcription. Nucleic Acids Res 20:3591–3598

Cook PR, Lang J, Hayday A, Lania L, Fried M, Chiswell DJ, Wyke JA (1982) Active viral genes in transformed cells lie close to the nuclear cage. EMBO J 1:447–452

Dekker J, Rippe K, Dekker M, Kleckner N (2002) Capturing chromosome conformation. Science 295:1306–1311

Dickinson P, Cook PR, Jackson DA (1990) Active RNA polymerase I is fixed within the nucleus of HeLa cells. EMBO J 9:2207–2214

Dillon N, Trimborn T, Strouboulis J, Fraser P, Grosveld F (1997) The effect of distance on long-range chromatin interactions. Mol Cell 1:131–139

Dundr M, Hoffmann-Rohrer U, Hu Q, Grummt I, Rothblum LI, Phair RD, Misteli T (2002) A kinetic framework for a mammalian RNA polymerase in vivo. Science 298:1623–1626

Fischle W, Wang Y, Allis CD (2003) Binary switches and modification cassettes in histone biology and beyond. Nature 425:475–479

Frey MR, Bailey AD, Weiner AM, Matera AG (1999) Association of snRNA genes with coiled bodies is mediated by nascent snRNA transcripts. Curr Biol 9:126–135

Gall JG, Murphy C (1998) Assembly of lampbrush chromosomes from sperm chromatin. Mol Biol Cell 9:733–747

Gasser SM (2002) Visualizing chromatin dynamics in interphase nuclei. Science 296:1412–1416

Gelles J, Landick R (1998) RNA polymerase as a molecular motor. Cell 93:13–16

Grummt I (2003) Life on a planet of its own: regulation of RNA polymerase I transcription in the nucleolus. Genes Dev 17:1691–1702

Hanscombe O, Whyatt D, Fraser P, Yannoutsos N, Greaves D, Dillon N, Grosveld F (1991) Importance of globin gene order for correct developmental expression. Genes Dev 5:1387–1394

Heun P, Laroche T, Shimada K, Furrer P, Gasser SM (2001) Chromosome dynamics in the yeast interphase nucleus. Science 294:2181–2186

Hilleren P, Parker R (1999) Mechanisms of mRNA surveillance in eukaryotes. Annu Rev Genet 33:229–260

Hoogstraten D, Nigg AL, Heath H, Mullenders LH, van Driel R, Hoeijmakers JH, Vermeulen W, Houtsmuller AB (2002) Rapid switching of TFIIH between RNA polymerase I and II transcription and DNA repair in vivo. Mol Cell 10:1163–1174

Iborra FJ, Pombo A, McManus J, Jackson DA, Cook PR (1996) The topology of transcription by immobilized polymerases. Exp Cell Res 229:167–173

Iborra FJ, Jackson DA, Cook PR (2001) Coupled transcription and translation within nuclei of mammalian cells. Science 293:1139–1142

Iborra FJ, Escargueil AE, Kwek KY, Akoulitchev A, Cook PR (2004) Molecular cross-talk between the transcription, translation, and nonsense-mediated decay machineries. J Cell Sci 117:899–906

Ivarie RD, Schacter BS, O'Farrell PH (1983) The level of expression of the rat growth hormone gene in liver tumour cells is at least eight orders of magnitude less than that in anterior pituitary cells. Mol Cell Biol 3:1460–1467

Jackson DA (2003) The principles of nuclear structure. Chromosome Res 11:387–401

Jackson DA, Cook PR (1985) Transcription occurs at a nucleoskeleton. EMBO J 4:919–925

Jackson DA, Cook PR (1993) Transcriptionally-active minichromosomes are attached transiently in nuclei through transcription units. J Cell Sci 105:1143–1150

Jackson DA, McCready SJ, Cook PR (1984a) Replication and transcription depend on attachment of DNA to the nuclear cage. J Cell Sci Suppl 1:59–79

Jackson DA, Patel SB, Cook PR (1984b) Attachment of repeated sequences to the nuclear cage. Nucleic Acids Res 12:6709–6726

Jackson DA, Dickinson P, Cook PR (1990) The size of chromatin loops in HeLa cells. EMBO J 9:567–571

Jackson DA, Hassan AB, Errington RJ, Cook PR (1993) Visualization of focal sites of transcription within human nuclei. EMBO J 12:1059–1065

Jackson DA, Bartlett J, Cook PR (1996) Sequences attaching loops of nuclear and mitochondrial DNA to underlying structures in human cells: the role of transcription units. Nucleic Acids Res 24:1212–1219

Jacobs EY, Frey MR, Wu W, Ingledue TC, Gebuhr TC, Gao L, Marzluff WF, Matera AG (1999) Coiled bodies preferentially associate with U4, U11, and U12 small nuclear RNA genes in interphase HeLa cells but not with U6 and U7 genes. Mol Biol Cell 10:1653–1663

Jolly C, Metz A, Govin J, Vigneron M, Turner BM, Khochbin S, Vourc'h C (2004) Stress-induced transcription of satellite III repeats. J Cell Biol 164:25–33

Kimura H, Cook PR (2001) Kinetics of core histones in living human cells: little exchange of H3 and H4 and some rapid exchange of H2B. J Cell Biol 153:1341–1353

Kimura H, Tao Y, Roeder RG, Cook PR (1999) Quantitation of RNA polymerase II and its transcription factors in an HeLa cell: little soluble holoenzyme but significant amounts of polymerases attached to the nuclear substructure. Mol Cell Biol 19:5383–5392

Kimura H, Sugaya K, Cook PR (2002) The transcription cycle of RNA polymerase II in living cells. J Cell Biol 159:777–782

Lachner M, O'Sullivan RJ, Jenuwein T (2003) An epigenetic road map for histone lysine methylation. J Cell Sci 116:2117–2124

Lewis PJ, Thaker SD, Errington J (2000) Compartmentalization of transcription and translation in *Bacillus subtilis*. EMBO J 19:710–718

Li G, Sudlow G, Belmont AS (1998) Interphase cell cycle dynamics of a late-replicating, heterochromatic homogeneously staining region: precise choreography of condensation/decondensation and nuclear positioning. J Cell Biol 140:975–989

Libri D, Dower K, Boulay J, Thomsen R, Rosbash M, Jensen TH (2002) Interactions between mRNA export commitment, 3′-end quality control, and nuclear degradation. Mol Cell Biol 22:8254–8266

Lippincott-Schwartz J, Snapp E, Kenworthy A (2001) Studying protein dynamics in living cells. Nat Rev Mol Cell Biol 2:444–456

Lykke-Andersen J (2002) Identification of a human decapping complex associated with hUpf proteins in nonsense-mediated decay. Mol Cell Biol 22:8114–8121

Manders EMM, Kimura H, Cook PR (1999) Direct imaging of DNA in living cells reveals the dynamics of chromosome formation. J Cell Biol 144:813–822

Maniatis T, Reed R (2002) An extensive network of coupling among gene expression machines. Nature 416:499–506

Manuelidis L (1990) A view of interphase chromosomes. Science 250:1533–1540

Marshall WF, Straight A, Marko JF, Swedlow J, Dernburg A, Belmont A, Murray AW, Agard DA, Sedat JW (1997) Interphase chromosomes undergo constrained diffusional motion in living cells. Curr Biol 7:930–939

Misteli T (2001a) The concept of self-organization in cellular architecture. J Cell Biol 155:181–185

Misteli T (2001b) Protein dynamics: implications for nuclear architecture and gene expression. Science 291:843–847

Mott MR, Callan HG (1975) An electron microscope study of the lampbrush chromosomes of the newt *Triturus cristatus*. J Cell Sci 17:241–261

Mueller-Storm HP, Sogo JM, Schaffner W (1989) An enhancer stimulates transcription in trans when attached to the promoter via a protein bridge. Cell 58:767–777

Palstra RJ, Tolhuis B, Splinter E, Nijmeijer R, Grosveld F, de Laat W (2003) The beta-globin nuclear compartment in development and erythroid differentiation. Nat Genet 35:190–194

Pederson T (2000) Half a century of “the nuclear matrix”. Mol Biol Cell 11:799–805

Pettijohn DE (1996) The nucleoid. In: Neidhardt FC, Curtiss R, Ingraham JL, Lin ECC, Brooks Low K, Magasanik B, Reznifoff WS, Riley M, Schaechter M, Umbarger HE (eds) *Escherichia coli* and *Salmonella typhimurium*: cellular and molecular biology. ASM Press, Washington, pp 158–166

Polioudaki H, Kourmouli N, Drosou V, Bakou A, Theodoropoulos PA, Singh PB, Giannakouros T, Georgatos SD (2001) Histones H3/H4 form a tight complex with the inner nuclear membrane protein LBR and heterochromatin protein 1. EMBO Rep 2:920–925

Pombo A, Cuello P, Schul W, Yoon J-B, Roeder RG, Cook PR, Murphy S (1998) Regional and temporal specialization in the nucleus: a transcriptionally-active nuclear domain rich in PTF, Oct1 and PIKA antigens associates with specific chromosomes early in the cell cycle. EMBO J 17:1768–1778

Pombo A, Jackson DA, Hollinshead M, Wang Z, Roeder RG, Cook PR (1999) Regional specialization in human nuclei: visualization of discrete sites of transcription by RNA polymerase III. EMBO J 18:2241–2253

Proudfoot NJ, Furger A, Dye MJ (2002) Integrating mRNA processing with transcription. Cell 108:501–512

Ptashne M (1986) Gene regulation by proteins acting nearby and at a distance. Nature 322:697–701

Robinett CC, Straight A, Li G, Willhelm C, Sudlow G, Murray A, Belmont AS (1996) In vivo localization of DNA sequences and visualization of large-scale chromatin organization using lac operator/repressor recognition. J Cell Biol 135:1685–1700

Sachs RK, van den Engh G, Trask B, Yokota H, Hearst JE (1995) A random-walk/giant-loop model for interphase chromosomes. Proc Natl Acad Sci USA 92:2710–2714

Saitoh Y, Laemmli UK (1978) From the chromosomal loops and the scaffold to the classic bands of metaphase chromosomes. Cold Spring Harbor Symp Quant Biol 58:755–765

Schafer DA, Gelles J, Sheetz MP, Landick R (1991) Transcription by single molecules of RNA polymerase observed by light microscopy. Nature 352:444–448

Sedat J, Manuelidis L (1978) A direct approach to the structure of eukaryotic chromosomes. Cold Spring Harbor Symp Quant Biol 42:331–350

Snow MHL, Callan HG (1969) Evidence for a polarized movement of the lateral loops of newt lampbrush chromosomes during oogenesis. J Cell Sci 5:1–25

Spector DL (2003) The dynamics of chromosome organization and gene regulation. Annu Rev Biochem 72:573–608

Stenoien DL, Patel K, Mancini MG, Dutertre M, Smith CL, O'Malley BW, Mancini MA (2001) FRAP reveals that mobility of oestrogen receptor-alpha is ligand- and proteasome-dependent. Nat Cell Biol 3:15–23

Strukov, YG, Wang Y, Belmont AS (2003) Engineered chromosome regions with altered sequence composition demonstrate hierarchical large-scale folding within metaphase chromosomes. J Cell Biol 162:23–35

Thompson M, Haeusler RA, Good PD, Engelke DR (2003) Nucleolar clustering of dispersed tRNA genes. Science 302:1399–1401

Tolhuis B, Palstra RJ, Splinter E, Grosveld F, de Laat W (2002) Looping and interaction between hypersensitive sites in the active beta-globin locus. Mol Cell 10:1453–1465

Vazquez J, Belmont AS, Sedat JW (2001) Multiple regimes of constrained chromosome motion are regulated in the interphase Drosophila nucleus. Curr Biol 11:1227–1239

Weidemann, T, Wachsmuth, M, Knoch, TA, Müller G, Waldeck W, Langowski J (2003) Counting nucleosomes in living cells with a combination of fluorescence correlation spectroscopy and confocal imaging. J Mol Biol 339, 229–240.

Wu C (1993) Transvection, nuclear structure, and chromatin proteins. J Cell Biol 120:587–590

9. Nuclear Architecture: Topology and Function of Chromatin- and Non-Chromatin Nuclear Domains

Satoshi Tashiro[1], Marion Cremer[2], Irina Solovei[2], and Thomas Cremer[2]

[1]Department of Cellular Biology, Research Institute for Radiation Biology and medicine, Hiroshima University, Kasumi 1-2-3, Hiroshima 734-8553, Japan

[2]Department of Biology II, Ludwig Maximilians University (LMU), Grosshadernerstr. 2, D-82152 Martinsried, Germany

9.1 Introduction

The driving force behind studies on nuclear architecture is based on the assumption that nuclear architecture is an integrated part of the complex epigenetic regulatory mechanisms which control cell type specific gene expression patterns. Epigenetic mechanisms comprise the chromatin level, including DNA methylation, histone modifications and chromatin remodeling factors, and the nuclear level, which includes the dynamics and three-dimensional (3D) spatial higher-order organization of the genome inside the cell nucleus. There is increasing evidence that such a higher-order organization of chromatin arrangement contributes essentially to the regulation of gene expression and other nuclear functions (for review see Spector 2003; van Driel et al. 2003).

The nucleus appears to be a compartmentalized organelle, but the mechanisms involved in nuclear compartmentalization must be different from the mechanisms involved in the compartmentalization of the cytoplasm. In contrast to the organelles in the cytoplasm there are no membrane structures in the nucleoplasm, which could separate different nuclear compartments.

Understanding nuclear architecture in space and time, as we argue, is indispensable to understand the changes of nuclear functions during cell

cycle, cell differentiation, cellular senescence, programmed cell death, as well as the reprogramming of nuclei in animal cloning experiments. In this chapter we describe what is presently known or argued about the structure and arrangements of chromosome territories (CTs), variedly sized chromatin domains, as well as non-chromatin domains.

While it is no longer controversial to say that changes of gene activity are correlated with changes of chromatin structure, it is still a matter of controversy as to what extent changes in higher-order chromatin architecture and arrangements are either the cause or the consequence of changes in gene expression (Singer and Green 1997; Parada et al. 2004). We must therefore study epigenetics as a problem of complex interactions of several hierarchical levels ranging from DNA methylation to histone modifications and chromatin remodeling, to higher-order chromatin arrangements and nuclear architecture at large. These different levels are most likely dependent on each other in complex and unpredictable ways (Geiman and Robertson 2002; van Driel et al. 2003).

Throughout this chapter we attempt to integrate presently known features of the nuclear architecture into the framework of the CT–interchromatin compartment (IC) model of the cell nucleus (Cremer and Cremer 2001; Cremer et al. 1993). At the end we will try to present a view of the functional nuclear architecture, which brings together our present knowledge of nonrandom, yet probabilistic features of CT arrangements with possibly deterministic features of the positioning of active genes in euchromatic nuclear zones and of silent genes in heterochromatin nuclear zones. This model presented is necessarily premature in the light of the facts established to date.

About a century ago Carl Rabl and Theodor Boveri suggested that each chromosome in the cell nucleus of animal cells occupies a distinct territory (Boveri 1909). This concept was abandoned when EM studies performed in the 1950s and 1960s revealed areas of compacted and dispersed chromatin, called hetero- and euchromatin, but failed to distinguish distinct CTs (for review see Wischnitzer 1973). During the 1960s and 1970s the view of the cell nucleus as a bag, where extended chromatin fibers float around in the nuclear sap with little geometrical constraints became popular. At this time only a minority of nuclear researchers argued for a functionally relevant nonrandom higher-order chromatin architecture (for reviews see Comings 1968; Okada and Comings 1979; Vogel and Schroeder 1974). In the late 1970s and 1980s the Rabl–Boveri view of CTs was resurrected, when firm evidence for the territorial organization of interphase chromosomes, CTs was obtained for a variety of species from the animal and plant kingdoms (Zorn et al. 1976; Cremer et al. 1982, 1988;

Manuelidis 1985; Schardin et al. 1985; Lichter et al. 1988; Leitch et al. 1990) (Fig. 1A,B).

9.2 Topology and Function of Chromosome Territories and Chromatin Domains

Structure of Chromosome Territories and Chromatin Domains

Methodological advancements since the 1980s have made it possible to study nuclear architecture from the level of chromatin and non-chromatin domains to the level of the higher-order organization of these components at large. The development of multicolor fluorescence in situ hybridization (M-FISH) protocols, under conditions which preserve the 3D nuclear shape and higher-order chromatin arrangements of nuclei as much as possible (3D-FISH) (Solovei et al. 2002), has allowed studies of chromatin topology from the level of single genes to entire CTs and their suprachromosomal arrangements (Fig. 1C) (for review see Spector 2003; Parada et al. 2004; Pederson 2004).

Pulse-labeling of DNA with thymidine analogues in living mammalian cells during the replication phase has revealed the presence of focal chromatin aggregates or chromatin domains with a DNA content in the order of several hundred kilobases (kb) to several megabases (Mb), termed ~1 Mb chromatin domains (Ma et al. 1998). Early and mid-replicating ~1 Mb chromatin domains can be visualized in different colors by pulse-chase-pulse experiments with CldU and IdU. The mid-replicating domains are gene poor and predominantly arranged at the nuclear periphery and around nucleoli, while early replicating domains are gene rich, and localized in the nuclear interior between the perinuclear and perinucleolar gene-poor compartments. These ~1-Mb domains persist through interphase and we consider them as a major substructure of CTs. Neither the structure of these chromatin domains nor the way in which they are connected with each other has been elucidated to date. When cells with pulse-chase-pulse labeled DNA grow for several additional cell cycles labeled and unlabeled chromatids segregate, giving rise to cells with fewer and fewer labeled chromosome territories. Each labeled CT, however, orientates itself in early G1 such that the typical distribution pattern of early and mid replicating chromatin domains is conserved (Fig. 1D) (Schermelleh et al. 2001).

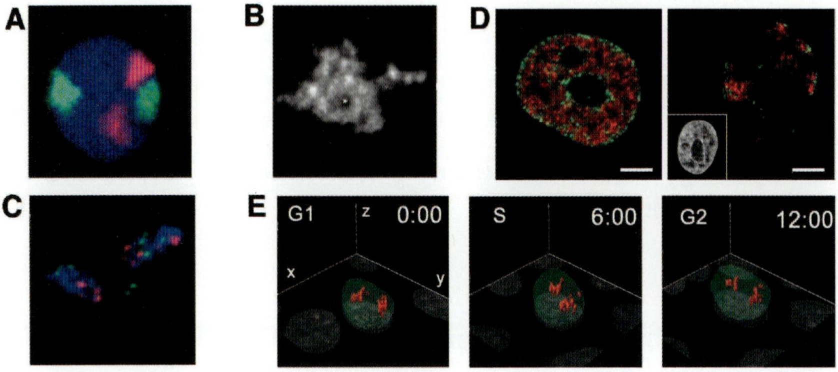

Fig. 1. A Visualization of chromosome territories (CTs) X (*green*) and 8 (*red*) after fluorescence in situ hybridization (FISH) with chromosome-specific painting probes on light optical section of a human female lymphocyte nucleus (*blue*). Note the relatively compact structure of territories in this nucleus. **B** Light optical section through painted CT #18 in a human fibroblast nucleus. Note the irregular shape and focal substructure of the CT showing regions with different chromatin density. **C** Hybridization of two differently labeled pools of BAC probes, assigned to G-dark (*red*) or G-light bands (*green*) to CT #12 (*blue*) in human lymphocyte nucleus. Majority of BAC signals is found within the CT; occasionally signals (*arrow*) extend from CTs (image courtesy of D. Biener, University of Munich). **D** Early- and mid-replicating chromatin domains. *Left*: Typical distribution of early replicating chromatin (*red*) throughout the nuclear interior and mid-replicating chromatin (*green*) at the nuclear and nucleoli periphery on the optical section through a nucleus of neuroblastoma cell after double replication labeling with Cy3- and Cy5-conjugated nucleotides. *Right*: Optical mid-section through nucleus of a cell double replication labeled as on the Left. Note that though only a few CTs are labeled in this nucleus (as a result of sister chromatid segregation during few postlabeling mitotic divisions), the topology of mid- and early-replicating chromatin is maintained through several cell cycles. The inset shows the same nucleus after counterstaining (image courtesy of L. Schermelleh, University of Munich). **E** Three-dimensional (3D) reconstructions of a HeLa cell nucleus with replication labeled chromosome territories (*red*) performed from three time points (0, 6, and 12 h) of a 4D time lapse series covering cell cycle from G1 to G2. Note that the position of the labeled CTs remains almost unchanged during this time interval (image courtesy of L. Schermelleh, University of Munich).

Recently, in vivo labeling approaches have provided insight into the dynamics of CTs, chromatin domains, genes, as well as non-chromatin domains in the nucleus of living cells (for review see Belmont 2003). These approaches have made it possible to perform 4D (space coordinates plus time) studies of the topology of chromatin and non-chromatin nuclear domains in nuclei of cultured, living cells (Fig. 1E). The perspectives of future studies become even brighter in the light of dramatic improvements of laser fluorescence microscopy beyond the Abbe limit of conventional light microscopic resolution (Failla et al. 2002), advancements in bioinformatics for quantitative 3D and 4D evaluations (Swedlow 2003) and higher-order chromatin and chromosome computer modeling (Kreth et al. 2004).

Random or Nonrandom Arrangements of CTs?

The question as to whether CT arrangements are fixed or variable was of concern to Theodor Boveri (Boveri 1909). In his studies of embryos of the *Ascaris equorum*, Boveri argued that the neighborhood of CTs is stably maintained during the resting stage (interphase), but that new neighborhoods can be established in mitosis, when chromosomes move to form the metaphase plate. Accordingly, he assumed that the neighborhood of CTs can differ from cell cycle to cell cycle. Since then scientists have defended both the assumption of highly ordered CT arrangements (Nagele et al. 1995) and the assumption of highly variable, likely random arrangements (Lesko et al. 1995). It has recently become possible to visualize all pairs of autosome territories, the X and the Y territory, in different colors in postmitotic (G0) fibroblast nuclei. This approach allowed to measure their distances from the center of the nucleus (assessment of the radial CT arrangements) and the distances and angles between homologous and heterologous CTs (assessment of the side-by-side arrangements). The results demonstrate a profound variability of *side-by-side* arrangements (Fig. 1F) (Bolzer et al. 2004). In contrast, the *radial* distribution of CTs in human fibroblast nuclei was highly nonrandom: the territories of larger chromosomes were positioned closer to the nuclear rim, while the territories of small chromosomes were located close to the nuclear center.

In nuclei of other cell types than fibroblasts, such as human lymphocytes and a number of epithelial normal and malignant cell types, a highly nonrandom radial distribution of CTs was also observed, but here the distribution was not correlated with size but with *gene density* of the respective chromosomes. Whereas, for example, in human fibroblasts the CTs of the small gene dense chromosomes 18, 19 and Y (all with a similar DNA content) were centrally positioned together, the arrangements of these territories were distinctly different in human lymphocyte nuclei. Here, the CTs #19 were observed in the nuclear center, while the CTs #18 and Y were positioned at the nuclear periphery (Fig. 1G) (Croft et al. 1999; Boyle et al. 2001; Cremer et al. 2001b, 2003). This gene density related chromatin arrangement was also reflected by the polarized distribution of subchromosomal domains within a CT representing either gene dense or gene-poor regions (Fig. 1H). The reason(s) for these cell type specific differences of CT arrangements remain speculative, but nuclear shape may have an important impact. Fibroblast nuclei are elliptically shaped and rather flat, while, e.g., lymphocyte nuclei and other cell types investigated are rather spherical. As a consequence, a gene-poor CT #18 or Y, located in the center of a flat fibroblast nucleus, still allows a close association to the nuclear envelope, whereas the association to the nuclear envelope requires a peripheral position for these CTs in a spherical nucleus. This attachment can be triggered, among others, by HP1, a protein recruited at specific transcriptionally inactive chromatin sites, which interacts with the LBR, a component of the inner nuclear membrane. Although differences in nuclear shape may affect the distribution of CTs, they are unlikely to be sufficient to explain the profound differences between the size-dependent radial order found in fibroblast nuclei and the gene density-dependent order found in lymphocyte nuclei.

Computer simulations of CT arrangements were performed with the intention to model CT arrangements under the condition that these arrangements were enforced by geometrical constraints. Interestingly, these simulations indicated that in both elliptically shaped flat nuclei and spherically shaped nuclei, geometrical constraints would favor the positioning of small CTs at the nuclear periphery and the positioning of large CTs in the nuclear interior. This difference between the predicted CT distribution due to geometrical constraints and the experimentally observed CT distributions suggests that other conditions than geometrical constraints play a decisive role in the formation of cell type specific, nonrandom, radial CT patterns.

Changes of CT Positions During Interphase

To follow up positional changes of entire CTs during different cell-cycle stages of the interphase nucleus, cycling cells were replication-labeled with fluorochrome-labeled nucleotides during S-phase. During the second and later postlabeling mitoses, labeled and unlabeled chromatids segregated randomly, resulting in cells exhibiting nuclei with few labeled CTs located at different nuclear sites. Those CTs can be followed by live cell microscopy through a subsequent interphase and mitosis (Walter et al. 2003). While "large-scale" positional changes (>1 μm) of the intensity gravity centers of CTs were observed in early G1, the mobility of CTs was strongly constrained from mid G1 to late G2. We never observed, for example, movements that led to a positional change of a CT from one side of the nucleus to the opposite site. The constrained mobility of entire CTs, however, is consistent with the possibility that certain subchromosomal regions carry out larger, even directed movements, as described below.

A comparison of CT arrangements between mother and daughter cells in different cell types showed an obvious symmetry in most pairs of daughter cells (Fig 1I), whereas different mitotic cells produced daughter nuclei with largely different CT arrangements (Walter et al. 2003; Gao et al. 2004).

Dynamics of Centromere Positions in Cycling and Postmitotic Cells During Terminal Cell Differentiation

Changes of centromere positions from the nuclear periphery into the nuclear interior and vice versa, as well as movements which result in the formation of centromere clustering (chromocenters) or in their dissolution, provide important examples for dynamic, large-scale changes of chromatin arrangements during the cell cycle and during the terminal differentiation of postmitotic cells (Fig. 1J) (Solovei et al. 2004a and references therein). Impressive changes of the chromatin pattern during the cell cycle can also be observed by the visualization of methylated histones (Fig. 1K). In general, clustering of centromeres into a few large chromocenters after exit of the cell cycle and/or terminal differentiation (see below) has been described for different cell types and may be a common finding for different cell types (Cerda et al. 1999; Beil et al. 2002; Cremer et al. 2004).

Recent studies of centromere positions in Purkinje cell nuclei from mouse cerebellum provide an example for positional changes during terminal cell differentiation (Solovei et al. 2004b and references therein). At birth (P0), mouse Purkinje cells are already in a postmitotic state but the terminal differentiation, including the formation of the impressive dendritic tree of these cells, occurs during the postnatal period. Studies of centromere positions during postnatal days P0 to P28 revealed that centromeres at all stages were typically located close to the nuclear envelope or close to a nucleolus. The fractions of centromeres observed either at the nuclear envelope or at the nucleoli, however, differed profoundly during terminal differentiation (Fig. 1L). Whereas at P0 Purkinje cell nuclei typically showed 2–4 nucleoli and 7–8 centromere clusters mostly at the nuclear periphery, cells at P28 typically showed large clusters of centromeric heterochromatin around the single centrally located nucleolus. It is not yet known whether the moving fraction of centromeres signified a random collection of all centromeres or changes of the orientation of specific CTs. It seems unlikely, however, that these highly reproducible changes of higher-order chromatin occur without a functional necessity.

Gene Topology in Correlation to Their Transcriptional Activity

At present there is an ongoing debate regarding how transcriptionally competent and silent genes are topologically arranged with respect to the

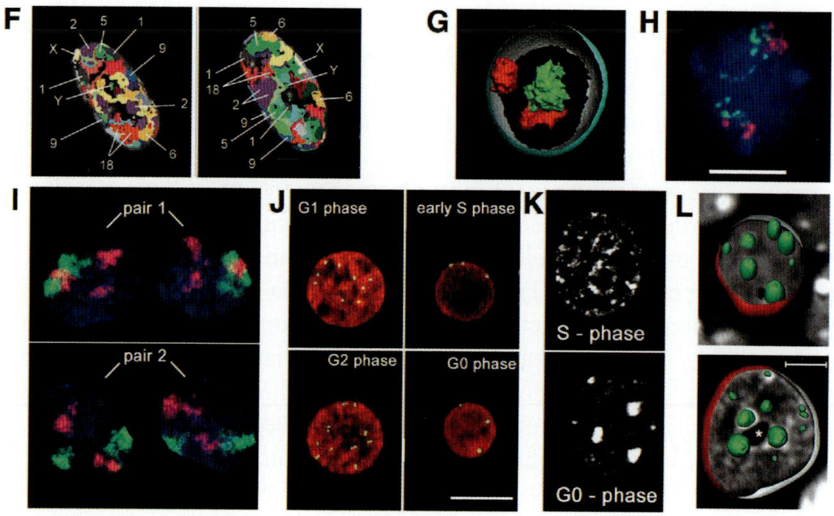

Fig.1. ***(continued)*** **F** Identification of all 24 CTs using a 24-color code on a single optical section through a human male fibroblast nucleus after multi-color FISH. In both shown nuclei CT neighborhood is different: e.g., compare positions of CTs #1, 2, 9 (image courtesy of A. Bolzer, University of Munich). **G** Visualization of CTs #18 (*red*) and #19 (*green*) in a 3D reconstruction of a nucleus of human hemopoietic stem cell. Gene-dense CTs #19 are located in the nuclear interior and often associated, whereas the gene-poor CTs #18 are typically found at the nuclear periphery. **H** Maximum intensity projection of an image stack through a human lymphocyte nucleus (*blue*) after hybridization with pools of differentially labeled BAC probes for gene-rich (*green*) and gene-poor (*red*) regions of chromosome 12. Note the orientation of the gene-dense regions towards the nuclear interior and of the gene-poor regions towards the nuclear periphery (Image courtesy of D. Biener, University of Munich).

Fig. 1. (*continued*) I Visualization of CT #7 (*green*) and of CT #10 (*red*) in nuclei of two sister HeLa cells (there were three copies of chr. #10 in this cell line). Sister nuclei show a clear symmetry of the relative positions of painted CTs. **J** Light optical sections through nuclei of human lymphocytes at different cell cycle stages after kinetochores (*green*) immunostaining. Note different spatial distribution and clustering of kinetochores depending on the cell-cycle stage. **K** Pattern of heterochromatin in nuclei of human breast cancer cell line (MCF-7) after immunodetection of methylated lysines of histone H3. The distribution of chromatin visualized by this antibody changes significantly during the cell cycle: in particular, in S-phase heterochromatin is distributed through a nucleus with few distinct larger clusters adjacent to the nucleoli and nuclear periphery. Upon exit from the cell cycle, heterochromatin clusters significantly grow in size while small signals disappear (image courtesy of Roman Zinner, University of Munich). **L** Three-dimensional reconstructions of nuclei of murine Purkinje cells at birth (P0, *top*) and in the adult animal (P28, *bottom*). Chromocenters (clusters of centromeres, *green*) are visualized by FISH with probe for pericentromere heterochromatin (major satellite repeat). At birth most of the chromocenters are of similar size and equally distributed over the nucleus; during 4 weeks of postnatal development chromocenters undergo some intensive spatial rearrangements and finally form a typical pattern for adult animals with two to three large chromocenters adjacent to the nucleoli and a few small chromocenters adjacent to the nuclear periphery

outside and inside of CTs and chromatin domains, and with respect to the IC. Although there is strong experimental evidence for an interaction between the spatial organization of genes and their activation or repression, data thus far provide a complex and inhomogeneous picture.

The 3D positioning of two X-linked genes, ANT2 and ANT3, were compared in the active and inactive X chromosome territories of female human amniotic fluid cell nuclei. ANT2 is transcriptionally active on Xa, but inactive on Xi. ANT3 is located in the pseudoautosomal region and escapes X-inactivation. It was shown that transcriptionally active ANT2 and ANT3 genes were positioned more peripheral within their CT than the inactive ANT2 gene (Dietzel et al. 1999). In contrast, PcG proteins, which are involved in gene silencing, are localized in the periphery of chromatin domains suggesting that silent genes may be positioned at this periphery as well (Cmarko et al. 2003). For the human MHC locus on chromosome 6 and the epidermal differentiation complex (EDC) on chromosome 1 it was shown that large chromatin loops can protrude away from a chromosome territory at a scale of several microns in cells with a high transcriptional activity of these gene clusters (Volpi et al. 2000; Williams et al. 2002). An analysis of the spatial organization of the gene-rich subtelomeric region 11p15.5 containing genes that are not coordinately expressed also showed that this region frequently extended from the visible bulk DNA of the corresponding chromosome territory (Mahy et al. 2002a). As a consequence it was suggested that local gene density and transcription influences the organization of chromosomes in the nucleus. On the other hand the investigation of the spatial organization of active and inactive genes and noncoding DNA of a ~1 Mb stretch of the WAGR region on 11p13, performed by the same group, showed that both coding and noncoding sequences from this region were similarly located inside of the 11p territory. Accordingly, both ubiquitously expressed genes and genes with a tissue-specific expression can be transcribed from the territory interior, suggesting that large-scale chromatin remodeling to position genes on the surface of a CT is not inevitably required for transcription (Mahy et al. 2002b). The observations of chromatin loops with (active) genes expanding from neighboring CTs and intermingling by chance within a space between more compact central CT portions led the group of Bickmore to challenge the organization of chromosomes as distinct chromosome territories in the interphase nucleus (Chubb and Bickmore 2003).

9.3 Interchromatin Compartment and Non-Chromatin Nuclear Domains

Ultrathin sections of typical mammalian cell nuclei viewed in the transmission electron microscope (TEM) revealed chromatin clumps of various sizes, surrounded by a largely chromatin-free interchromatin space (Visser et al. 2000). The latter contains numerous multiprotein complexes, further referred to as non-chromatin domains. The interchromatin space and its content is called the IC. To some extent these non-chromatin nuclear domains represent sites for the maturing and storage of functional proteins and protein complexes, which can be released from these domains to their sites of action. It is still not clear, however, to which extent such domains have storage functions and to what extent they are directly involved in nuclear functions, such as transcription control, splicing, replication, and repair. Transcription and cotranscriptional splicing takes place in perichromatin fibrils, which expand from the surface of chromatin domain into a boundary zone between chromatin domains and the IC, called the perichromatin region (PR). This evidence clearly supports a nonrandom organization of chromatin domains, which must expose active genes at their surface. Whether silent genes are located in the interior of chromatin domains is not proven. Interestingly, PcG proteins, which are involved in gene silencing, are localized in the periphery of CDs indicating that silent genes may be positioned at the chromatin domain periphery as well. The IC contains a variety of non-chromatin domains, which take up and release functionally important proteins roaming the nuclear space. Below we review a few examples, including Cajal bodies, PML bodies, splicing speckles, and Rad51 foci. Two conceptionally simple, elegant, and widely used live cell approaches, termed FRAP and FLIP, have made it possible to study the dynamic behavior of specific proteins, which roam the nucleus, and their interactions with chromatin and non-chromatin nuclear domains. Fluorescence recovery after photobleaching (FRAP) experiments require that a nuclear protein present in a given chromatin or non-chromatin domain is fused to a blue-, green-, yellow-, or red-fluorescent marker protein (BFP, GFP, YFP, RFP). The fluorescence protein-marked proteins in the given domain are selectively and irreversibly bleached with a laser microbeam, employing conditions which do not impair the function of the marked nuclear protein. The kinetics of fluorescence recovery in the bleached domain provide information on the mobility and binding kinetics of this nuclear protein. In FLIP (fluorescence loss in photobleaching) experiments the laser microbeam is focused to a nuclear region remote from FP-tagged nuclear domains. In the case that the FP marked proteins are not firmly bound to a given nuclear substructure, but roam the entire nuclear space, repeated bleaching of a single nuclear site leads to a loss of fluorescence everywhere in the nucleus. FRAP and FLIP experiments demonstrated a high mobility of many proteins in the nucleoplasm and also showed that the dynamics of individual proteins or protein complexes can differ greatly.

PML Bodies

Approximately 10–20 PML bodies, also termed ND10 or PODs, can be observed in the cell nucleus (Fig. 2A). PML bodies are nuclear matrix (NM)-associated structures of 200–1000 nm in diameter. Their number and size varies during the cell cycle and in different cell types. Electron spectroscopic imaging demonstrates that the core of the PML body is a dense, protein-based structure which does not contain nucleic acids.

The PML body comprises PML, a core component of the nuclear body, and other proteins such as SUMO-1, pRB, p53, BLM, and Sp100 (Sternsdorf et al. 1997; Alcalay et al. 1998; Zhong et al. 1999; Guo et al. 2000). Among these proteins PML is essential for the formation of PML bodies (Ishov et al. 1999). In order to be localized in the nuclear body

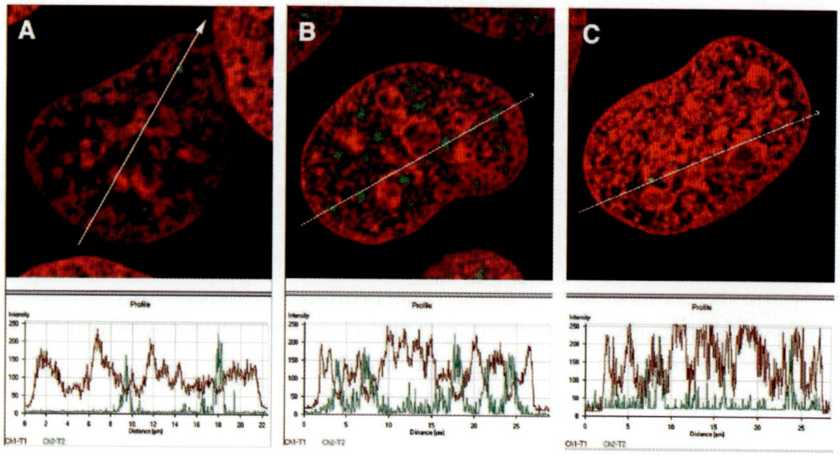

Fig. 2. Non-chromatin nuclear domains and chromosome DNA. Nuclear speckles (A), PML body (B), and Rad51 nuclear foci (C) in a human fibroblast cell line are detected by immunofluorescence staining (*green*) together with chromosome DNA (*red*) by propidium iodide (PI) staining. Profile plots of signal intensity along the arrows in the light optical confocal sections show the segregation of non-chromatin nuclear domains and chromosome DNA

PML must be covalently bound to a ubiquitin-like protein SUMO-1 (PIC1 or sentrin), or sumoylated, at three lysine residues (Sternsdorf et al. 1997; Muller et al. 1998; Zhong et al. 2000a,b,c). Unmodified PML is associated with the soluble nucleoplasmic fraction, and the SUMOylated PML fraction is tightly associated with the NM. PML plays several physiological roles such as a mediator of interferon function (Lavau et al. 1995; Regad et al. 2001), a proapoptotic factor (Quignon et al. 1998; Wang et al. 1998), and a tumor suppressor (Mu 1994; Le et al. 1996). Furthermore, PML might work as a transcription regulator by interacting with p53 and pRB (Alcalay 1998; Guo et al. 2000).

A variety of functions have been suggested for PML bodies (Zhong et al. 2000a,b,c), although the main biological function of this nuclear domain is still unclear. PML is associated in vivo with several DNA repair proteins, including Mre11, Rad51, BLM, TopBP1, and H2AX (Mizoeva and Petrini 2000; Bischof et al. 2001; Xu et al. 2003), suggesting a role of PML body in DNA repair. The number of PML bodies increases after treatment of cells with interferon (Lavau et al. 1995), and a variety of viruses, such as Herpes simplex virus-1 and human cytomegalovirus target PML body (Everett et al. 2004). These observations suggest a role of PML bodies in the antiviral defense. Interaction of PML with the transcription coactivators CBP and p300 (von Mikecz et al. 2000), and inhibition of the transactivation activity of Spl (Vallian et al. 1998) and pRB (Alcalay et al. 1998), as well as transcription repression of Tax (Desbois et al. 1996) and Daxx (Zhong et al. 2000a,b,c) by PML suggest a role of PML bodies in transcription regulation. Indeed, newly synthesized RNA is found in the chromatin-depleted region of the nucleoplasm immediately surrounding the core of the PML body (Boisvert et al. 2000).

PML bodies are classified into three groups according to the dynamics of PML body in living cells visualized by EYFP-Sp100 (Muratani et al. 2002). The first type exhibits a constrained, non-energy-dependent movement. The second type shows very localized movements. These two types of PML bodies are thought to be stably associated with an underlying nuclear structure or with a specific nucleic acid or protein complexes. The third type of PML body exhibits rapid (4–7 μm/min) and more extended movements in the nucleoplasm by a metabolic-energy-dependent mechanism. Another report claims that PML bodies exhibit little movement while small foci that contain Sp100 but not PML are dynamic and fuse with PML bodies (Wiesmeijer et al. 2002). Studies of the dynamics of the components of PML bodies using FRAP and FLIP techniques showed that CBP moves very rapidly into and out of PML bodies, while PML and Sp100 are relatively immobile in these bodies (von Mikecz et al. 2000). Further studies are required to clarify the dynamics of the components that are taken up and released from these bodies.

Cajal Bodies

"Cajal bodies" (CBs) or "coiled bodies" appear as a tangle of coiled, electron-dense threads by EM, and are conserved from plants to animals (Matera 1998). The number and size of CBs varies among cell types and also shows relationships with cell cycle within cell types. Up to 10 CBs per nucleus, ranging from 0.1 to 2 μm in diameter, are typically observed in mammalian cells. Cajal bodies comprise numerous nuclear and nucleolar factors involved in transcription, splicing, snRNA processing and signaling, such as snRNPs, survival of motor neuron protein (SMN), nucleolar snoRNPs, splicing snRNPs, NOPP140, fibrillarin, RNA polymerase I and II, TFIIF, TFIIH, and the TATA binding protein (TBP) (Matera 1999). Since CBs do not contain nascent RNA, the nuclear domains could play a role in coordinating the assembly and maturation of nuclear RNPs and possibly other macromolecular complexes (Jady et al. 2003). Among the components of CBs, p80 coilin, discovered as a human autoantigen, is used as a marker of CBs (Andrade et al. 1991). However, not all coilin is concentrated in CBs. Coilin also shows numerous smaller foci together with a diffuse nuclear staining pattern (Matera 1998). Overexpression of coilin increased the intensity of diffuse nuclear staining, but not the number of CBs, indicating that an excess of this protein does not help to increase the assembly of CBs (Platani et al. 2000).

Cajal bodies are dynamic nuclear domains. Live imaging experiments using plant cells expressing GFP-U2 snRNP B fusion protein and human cells expressing fibrillarin-GFP or GFP-tagged coilin showed that all CBs move within the nucleoplasm (Boudonck et al. 1999; Platani et al. 2000, 2002; Snaar et al. 2000). Movements included translocations through the nucleoplasm, joining of CBs to form larger bodies, and separation of smaller bodies from larger CBs. Mobility of CBs increases after adenosine triphosphate (ATP) depletion or inhibition of transcription, indicating that the association of CBs with chromatin requires ATP and active transcription (Platani et al. 2002). GFP-tagged fibrillarin in CBs and nucleoli of human cells shows high mobility of the protein in both organelles, and has a significantly larger mobile fraction in CBs than in nucleoli (Snaar et al. 2000). A recent study of the dynamics of CBs components in the *Xenopus* germinal vesicle using fluorescence-labeled U7 small nuclear RNA, coilin, and TATA-binding protein (TBP) revealed the slower recovery of fluorescence of GFP-tagged proteins after photobleaching in CBs than in nucleoplasm (Handwerger et al. 2003). These observations suggest that CB components bind to other molecules which could provide a common scaffold for the formation of CBs. Interestingly, a fraction of CBs themselves move to and from nucleoli, suggesting the tethering of CBs to a specific nuclear structure. In fixed somatic cells and oocytes, CBs associate with the histone and snRNA genes, suggesting the direct tethering of CBs to specific gene loci or frequent association of CBs with these loci (Frey and Matera 1995, 2001). Further studies are required to identify these molecules and clarify the mechanisms of the formation and dynamics of CBs.

Splicing Speckles

Electron microscopy studies have identified nuclear domains in mammalian cell nucleus which are built up from numerous small electron dense bodies of 20–25 nm connected by a thin fibril, interchromatin granule clusters (IGCs). Almost all essential pre-mRNA splicing factors are concentrated in the nuclear domains, “splicing speckles,” also known as “SC35 domains" or "IGCs" with irregular shapes in various sizes (for a review, see Lamond and Spector 2003) (Fig. 2B).

Splicing speckles are composed of U1, U2, U4/U6, and U5 snRNPs, arginine/serine-rich (SR) splicing factors, and the hyperphosphorylated large subunit of RNA polymerase II. The arginine/serine-rich domain of a fraction of SR splicing factors is required for targeting protein to nuclear speckles. In addition to these proteins associated with splicing or transcription, kinases, such as CLK (Colwill et al. 1996), PRP4 (Kojima et al. 2001), and PSKH1 (Brede et al. 2002), and phosphatases (protein phosphatase 1) (Trinkle-Mulcahy et al. 1999) that can phosphorylate/dephosphorylate proteins of the splicing machinery have been identified in speckles. Furthermore, 3′-end RNA-processing factors, the eukaryotic translation-initiation factor eIF4E (Dostie et al. 2000), and structural proteins like lamin A (Jagatheesan et al. 1999) are located in speckles.

Studies of the dynamics of GFP-tagged ASF, a splicing factor, have revealed the recruitment of the protein from speckles to sites of active transcription (Misteli et al. 1997). Cells with active transcription exhibit more widespread nucleoplasmic localization of RNA processing factors and a smaller number of speckles, whereas inhibition of transcription by α-amanitin treatment induces accumulation of splicing factors into the nuclear compartments resulting in the formation of enlarged, round-shaped speckles (Misteli et al. 1997). Pulse labeling of transcription sites using bromouridine triphosphate (BrUTP) has revealed no transcription activity within nuclear speckles (Cmarko et al. 1999). In contrast, perichromatin fibrils, which are observed adjacent to IGCs by EM, show incorporation of BrUTP, suggesting that they are formed at transcription sites. These observations indicate that speckles serve as storage or assembly sites of splicing factors, but not compartments of transcription and splicing.

Speckles visualized by the ASF–GFP fusion protein change shape over time and subdomains are capable of both blebbing and fusing with individual speckles (Misteli et al. 1997). But they keep a relative position in interphase nuclei, suggesting that they are mildly tethered to some immobile nuclear structure. The motion of ASF–GFP examined by fluorescence FRAP experiments is at rates up to 100 times slower than free diffusion when it is associated with speckles and, interestingly, also in the nucleoplasm (Kruhlak et al. 2000). These results suggest that ASF interact transiently and frequently with relatively immobile nuclear components in and outside speckles. Recent study of the relationships between Speckles and higher-order chromosome structure showed that speckles were rarely contacted by gene-poor later-replicating (G-band) DNA, and were typically surrounded by a subset of gene-rich or early-replicating (R-band)

DNA (Shopland et al. 2003). Furthermore, two muscle genes became localized to the periphery of speckles during terminal differentiation from proliferative myoblasts to muscle cells (Moen et al. 2004). Speckles could serve as a center of a subset of active genes which form a euchromatin area around the nuclear domains. Speckles could be nuclear domains which are formed in the vicinity or at the center of a group of functionally related active genes located in the same euchromatin area.

It is still unclear how the formation and dynamics of speckles is regulated. Recently, the "Regulated-Exchange" model has been proposed by Lamond and Spector (2003). According to this model, speckles would be formed through a process of self-assembly without an underlying scaffold structure. The respective rates of association and disassociation of individual speckle components define their exchange rates and the sizes of their bound and soluble pools in the nucleus. The association and/or disassociation rates could be controlled in response to specific cellular signals through the regulatory mechanisms, or phosphorylation/dephosphorylation of the RS domain of SR splicing factors. Further studies are required to clarify the mechanism of speckles formation and dynamics.

9.4 Nuclear Domains Associated with DNA Repair

Mre11/Rad50/NBS1 Complexes

The MRE11-RAD50-NBS1 (MRN) complex has been thought to function in diverse aspects of the cellular response to DNA damage, sensing, processing, and repair (for reviews see (D'Amours and Jackson 2002; Petrini and Stracker 2003; van den Bosch et al. 2003). The original components of the MRN complex Mre11, Rad50, and Xrs2 (NBS1 is the vertebrate homologue of Xrs2), were identified by genetic screening for *Saccharomyces cerevisiae* mutants that were hypersensitive to DNA damage or were defective in meiotic recombination. Mre11 and Rad50 are highly conserved in *Schizosaccharomyces pombe*, mouse, and human cells, whereas Xrs2 is not. Mre11 is a core component of the MRN complex and has DNA nuclease, strand-dissociation and strand-annealing activities (for a review see D'Amours and Jackson 2002)). Rad50 is thought to play an important role in binding DNA ends to hold them in close proximity and in controlling the nuclease activity of Mre11 (Paull and Gellert 1999; Hopfner et al. 2001). NBS1 is reported to be a regulator of some biochemical activity of the MRN complex, such as ATP-dependent DNA unwinding and nuclease activity (Carney et al. 1998; Matsuura et al. 1998; Varon et al. 1998).

Mre11 is thought to be distributed homogeneously within the nucleoplasm in undamaged cells, and to be recruited to sites containing double strand breaks (DSBs) after induction of DNA damage in a time- and dose-dependent fashion but not in cells that are treated with other types of the DNA-damage induction like UV irradiation (Maser et al. 1997; Nelms et al. 1998). Mre11 foci colocalize with phosphorylated histone H2AX after induction of DNA damage (Paull et al. 2000), and in cells that undergo V(D)J recombination (Chen et al. 2000) or class switch recombination (Petersen et al. 2001). Rad50 and NBS1 have been shown to accumulate at laser-UVA-microirradiated sites in human breast tumor cells. MRE11 starts to form nuclear foci at sites containing DNA damage 10 min after induction of DNA damage, indicating the role of MRN complex at the early stage of DSB repair (Mirzoeva and Petrini 2001). DSB-induced Mre11 foci have been classified into two groups: small and many foci, which are the earliest DNA-damage-dependent foci observed in damaged cells from 10 min to 8 h after induction of DNA damage, and large and less in number than small foci, appearing 4 h post-treatment (Mirzoeva and Petrini 2001). Another study reported that after partial volume irradiation with ultrasoft X-rays, Mre11 located to the irradiated sites in human fibroblasts, but did not find distinct Rad51 accumulation at these sites (Nelms et al. 1998). These findings strongly suggest the involvement of MRN complex or Mre11 foci in the DSBs repair system of in human cells.

Introduction of detergent treatment before fixation of cells in immunofluorescence staining revealed that Mre11 foci show colocalization with PML bodies in untreated cells (Mirzoeva and Petrini 2001). In mammals, the Mre11 complex binds to Trf2, an important regulator of telomere function (Zhu et al. 2000). The Trf2–Mre11 complex interacts with telomeres specifically in S phase, indicating the involvement of Mre11 complex in telomere replication. Recently, dynamic association of telomere and PML body was reported (Molenaar et al. 2003). These findings suggest that Mre11 foci could play an important role in telomere maintenance in normal cells.

Rad51 Foci

Among genes of the Rad52 epistasis group in *Saccharomyces cerevisiae*, Rad51 plays a central role in this process and also in mitotic recombination (Shinohara et al. 1992). Rad51 forms helical filaments on both single- and double-stranded DNA, promoting homologous pairing and strand exchange (Shinohara et al. 1992, 1993; Ogawa et al. 1993; Baumann et al. 1996;). Mouse Rad51 genes are expressed at high level in thymus, spleen, ovary, and testis and at a low level in brain and other tissues (Shinohara et al. 1993). Both yeast and human Rad51 appear to function as a part of a larger recombination complex that includes other members of the Rad52-epistasis group. Human Rad51 protein interacts with Rad52, replication protein A, and the tumor suppressors p53, BRCA1, BRCA2, Xrcc2, and Xrcc3 (Shinohara et al. 1992; Buchhop et al. 1997; Scully et al. 1997; Sharan et al. 1997; Golub et al. 1998). Rad51-deficient DT40 chicken lymphocytes show increases in chromosome breaks and cell death (Sonoda et al. 1998). This evidence suggests that Rad51-mediated homologous recombination may play an essential role in DNA repair of mammalian cells.

Rad51 protein forms multiple discrete foci in the nucleoplasm, and the percentage of cells with Rad51 foci increase after induction of DNA damage (Haaf et al. 1995) (Fig. 2C). Rad51 foci are detected in a fraction of normal human cells, typically during S phase (Tashiro et al. 1996). It has been shown that defects in Rad54, Xrcc2, Xrcc3 and some other genes, which are required for assembly and stabilization of multimeric Rad51 protein complexes, interfere with Rad51 foci formation (Bishop et al. 1998; Tan et al. 1999; O'Regan et al. 2001). After induction of DSBs, single-stranded DNA (ssDNA), which could be an intermediate of recombinational repair, is detected within Rad51 foci (Raderschall et al. 1999). Furthermore, UVA microirradiation and whole cell UVC irradiation experiments confirmed the recruitment of Rad51 at site of DNA damage (Tashiro et al. 2000). Rad51 also accumulates at discrete foci on chromosomal DNA during meiotic prophase in yeast, mice and human (Plug et al. 1996; Scully et al. 1997). These findings obtained by the recently developed biological techniques indicate the in vivo involvement of Rad51 in the homologous recombination.

Dynamics of Rad52 group proteins in response to DNA damages studied with FRAP experiments using Rad51, Rad52, or Rad54–GFP fusion proteins revealed that about half of the Rad51 proteins, while all of Rad52 and Rad54 molecules, in the nucleoplasm were mobile (Essers et al. 2002). Furthermore, Rad51 proteins showed longer residence time in the nuclear foci than Rad52 or Rad54, indicating that Rad51 is stably associated with damaged DNA induced nuclear foci.

9.5 The Matrix View of Nuclear Architecture

Evidence for a constrained mobility of chromatin and non-chromatin domains (Vazquez et al. 2001; Chubb et al. 2002; Muratani et al. 2002; Platani et al. 2002; Walter et al. 2003) raises the question of a mechanism that enforces such constraints. A possible explanation, which needs to be discussed here, is a NM or scaffold, which may provide attachment sites for DNA and proteins. Since Berezney and Coffey (1974) published a protocol for its preparation (Berezney and Coffey 1974), the biochemically defined NM, enriched in hnRNP proteins and protein complexes associated with virtually any nuclear function (Berezney et al. 1995) has triggered much experimental work and a seemingly endless controversy. From the very beginning proponents of the NM have emphasized its functional importance as a truly existing, in vivo structure. Its opponents have suspected that the insoluble NM is an experimental artifact due to high salt treatments and/or other steps that favor the aggregation of proteins, which in the living cell freely roam the nuclear sap. Electron microscopy studies provided evidence for an in situ NM with a 3D network of branching 10 nm core filaments (Nickerson et al. 1997). These core filaments may provide attachment sites for (ribonucleo-) protein complexes with distinct functional properties, but are still biochemically ill defined. It has not been possible to date to raise antibodies that specifically and exclusively stain this network of core filaments.

At present, few scientists still subscribe to the view that a permanent nuclear skeleton exists in the living cell nucleus, which organizes its chromatin. It has been argued that NM preparations may comprise the partially intact and functional content of the IC in the living cell nucleus (Cremer et al. 2000). This proposal implicates that matrix filaments, if they should finally emerge as truly existing in vivo structures, are located within the IC and not in the interior of higher-order chromatin domains or chromatin fibers. The major functional purpose of matrix filaments may be a role to regulate the access of genes to preformed functional protein complexes contained in the IC.

9.6 Functional Importance of Cell Type-Specific Higher-Order Chromatin Arrangements: A Model View

The CT–IC model has provided a set of hypotheses in favor of a functional nuclear compartmentalization, which can be experimentally tested (Cremer and Cremer 2001a). It proposes that ~1-Mb chromatin domains are major constituents of CTs. Further, it assumes that each ~1-Mb chromatin domain is built up from a series of smaller chromatin loop domains with a DNA content in the order of ~100-kb chromatin domains. Computer simulations of CTs based on the assumption that each ~1-Mb chromatin domain is a distinct structural unit separated from its next two neighbors by chromatin linkers were consistent with experimental observations (Münkel et al. 1999). Clusters of ~1-Mb chromatin domains form chromosome band domains, which in turn are assembled to chromosome arm domains and entire CTs. The CT–IC model takes into account EM evidence that the nucleus contains a largely chromatin-free space lined by chromatin domain surfaces, called the IC. According to the CT–IC model, transcriptionally active genes are exposed at the surface of chromatin domains, and transcription and cotranscriptional splicing occurs in perichromatin filbrils located in the PR, which can be considered as the most peripheral part of chromatin domains or as a part of the IC or as an extra compartment, expanding between chromatin domains and the IC. An important, unresolved question pertains to the location of permanently silent genes. Are they also exposed at the periphery of chromatin domains in connection with the PR and the IC? Or are they hidden in the interior of chromatin domains? The latter possibility would make sense in the case where transcription complexes are preformed like modules which can be put together at the site of transcription. In the case where the size of such transcription modules and the compaction of chromatin domains prevents their diffusion into the chromatin domain interior, silent genes could be safeguarded from an undesirable transcriptional activation by their spatial separation from the IC, where such transcription modules are formed. However, such a mechanistic explanation is most likely insufficient, since it has been shown that the compaction of heterochromatin still allows for the diffusion of individual proteins into its interior (Cheutin et al. 2003) and even fluorescein isothiocyanate-conjugated dextran molecules up to about 70 kDa seem to penetrate freely into heterochromatin chromatin (Verschure et al. 2003) . It seems therefore wise to keep in mind that in addition to size other factors, such as electric charge, may influence the diffusion of proteins and protein complexes from the IC into the interior of chromatin domains.

The following model is an attempt to provide a functional interpretation of the data brought together in this chapter, in particular of the finding that radial CT arrangements are not deterministic, but probabilistic and that side-by-side arrangements of CTs are highly variable. A basic principle of this model is the assumption that the active or silent state of a gene depends on its chromatin environment. Active genes are supposedly embedded in an euchromatic environment, while permanently silent genes are embedded in facultative heterochromatin or associated with constitutive heterochromatin. We consider two possibilities for a gene to switch between different chromatin environments: One possibility is that a so far heterochromatic nuclear domain is remodeled into an euchromatic domain. In this case the gene or a cluster of genes can maintain its 3D nuclear position. The other possibility takes into account that a gene or a cluster of genes moves from a heterochromatic nuclear zone into an already existing euchromatic nuclear zone. The same considerations in reverse order apply to the silencing of so far active genes.

Berezney and Wei (1998) have introduced the concept of nuclear zones as a structural precondition of the topological, nuclear coordination of replication and transcription programs. Here, we broaden this concept and predict that all genes, which are active in a given cell type during a given period of interphase or during a given state of differentiation, must be located in euchromatic nuclear zones, whereas all permanently silent genes in such cells must be located in heterochromatic nuclear zones. These zones comprise more than a single ~1 Mbp chromatin domain and can be built up either by *cis*-interactions between functionally equivalent chromatin domains of a given CT or by *trans*-interactions between functionally equivalent chromatin domains from several neighboring CTs. These deterministic requirements can be fulfilled for all genes involved in a given active or shut-down biochemical or signaling pathway even when these genes are located on different CTs. We consider it functionally irrelevant whether such a set of genes forms a 3D cluster in a given nucleus or whether they are scattered at different nuclear sites, as long as each gene is located in its appropriate nuclear zone. Accordingly, the model is compatible with the experimentally observed probabilistic neighborhood of CTs during clonal cell growth. The nonrandom distribution of gene-dense bands and of gene-poor bands along mitotic chromosomes (containing the transcriptionally active housekeeping genes and the cell type-specific and in most cell types silent genes, respectively), the "polarized", radial orientation of CTs, as well as directed chromatin movements may all have evolved as adaptations to secure a functionally important nuclear architecture that allows the cell type-specific allocation of genes into the respective nuclear zones when a fertilized egg develops into an organism.

References

Alcalay M, Tomassoni L, Colombo E, Stoldt S, Grignani F, Fagioli M, Szekely L, Helin K, Pelicci PG (1998) The promyelocytic leukemia gene product (PML) forms stable complexes with the retinoblastoma protein. Mol Cell Biol 18:1084–1093

Andrade LE, Chan EK, Raska I, Peebles CL, Roos G, Tan EM (1991) Human autoantibody to a novel protein of the nuclear coiled body: immunological characterization and cDNA cloning of p80-coilin. J Exp Med 173:1407–1419

Baumann P, Benson FE, West SC (1996) Human Rad51 protein promotes ATP-dependent homologous pairing and strand transfer reactions in vitro. Cell 87:757–766

Beil M, Durschmied D, Paschke S, Schreiner B, Nolte U, Bruel A, Irinopoulou T (2002) Spatial distribution patterns of interphase centromeres during retinoic acid-induced differentiation of promyelocytic leukemia cells. Cytometry 47:217–225

Belmont A (2003) Dynamics of chromatin, proteins, and bodies within the cell nucleus. Curr Opin Cell Biol 15:304–310

Berezney R, Coffey DS (1974) Identification of a nuclear protein matrix. Biochem Biophys Res Commun 60:1410–1417

Berezney R, Wei X (1998) The new paradigm: integrating genomic function and nuclear architecture. J Cell Biochem Suppl 30–31:238–242

Berezney R, Mortillaro MJ, Ma H, Wei X, Samarabandu J (1995) The nuclear matrix: a structural milieu for genomic function. Int Rev Cytol 162A:1–65

Bischof O, Kim SH, Irving J, Beresten S, Ellis NA, Campisi J (2001) Regulation and localization of the Bloom syndrome protein in response to DNA damage. J Cell Biol 153:367–380

Bishop DK, Ear U, Bhattacharyya A, Calderone C, Beckett M, Weichselbaum RR, Shinohara A (1998) Xrcc3 is required for assembly of Rad51 complexes in vivo. J Biol Chem 273:21482–21488

Boisvert FM, Hendzel MJ, Bazett-Jones DP (2000) Promyelocytic leukemia (PML) nuclear bodies are protein structures that do not accumulate RNA. J Cell Biol 148:283–292

Bolzer, A. Kreth, G,. Solovei, I., Koehler, D., Saracoglu, K., Fauth, C., Muller, S., Eils, R., Cremer, C., Speicher, M.R., Cremer, T. (2005) Twenty-four color chromosome painting in three-dimensionally preserved human fibroblast nuclei and prometaphase rosettes demonstrates the probabilistic nature of nonrandom chromosome arrangements. PLoS Biology 3: 0826-0842

Boudonck K, Dolan L, Shaw PJ (1999) The movement of coiled bodies visualized in living plant cells by the green fluorescent protein. Mol Biol Cell 10:2297–2307

Boveri T (1909) Die Blastomerenkerne von Ascaris megalocephala und die Theorie der Chromosomenindividualität. Arch Zellforsch 3:181–268

Boyle S, Gilchrist S, Bridger JM, Mahy NL, Ellis JA, Bickmore WA (2001) The spatial organization of human chromosomes within the nuclei of normal and emerin-mutant cells. Hum Mol Genet 10:211–219.

Brede G, Solheim J, Prydz H (2002) PSKH1, a novel splice factor compartment-associated serine kinase. Nucleic Acids Res 30:5301–5309

Buchhop S, Gibson MK, Wang XW, Wagner P, Sturzbecher HW, Harris CC (1997) Interaction of p53 with the human Rad51 protein. Nucleic Acids Res 25:3868–3874

Carney JP, Maser RS, Olivares H, Davis EM, Le Beau M, Yates JR, 3rd, Hays L, Morgan WF, Petrini JH (1998) The hMre11/hRad50 protein complex, Nijmegen breakage syndrome: linkage of double-strand break repair to the cellular DNA damage response. Cell 93:477–486

Cerda MC, Berrios S, Fernandez-Donoso R, Garagna S, Redi C (1999) Organisation of complex nuclear domains in somatic mouse cells. Biol Cell 91:55–65

Chen HT, Bhandoola A, Difilippantonio MJ, Zhu J, Brown MJ, Tai X, Rogakou EP, Brotz TM, Bonner WM, Ried T, Nussenzweig A (2000) Response to RAG-mediated VDJ cleavage by NBS1 and gamma-H2AX. Science 290:1962–1965

Cheutin T, McNairn AJ, Jenuwein T, Gilbert DM, Singh PB, Misteli T (2003) Maintenance of stable heterochromatin domains by dynamic HP1 binding. Science 299:721–725

Chubb JR, Bickmore WA (2003) Considering nuclear compartmentalization in the light of nuclear dynamics. Cell 112:403–406

Chubb JR, Boyle S, Perry P, Bickmore WA (2002) Chromatin motion is constrained by association with nuclear compartments in human cells. Curr Biol 12:439–445

Cmarko D, Verschure PJ, Martin TE, Dahmus ME, Krause S, Fu XD, van Driel R, Fakan S (1999) Ultrastructural analysis of transcription and splicing in the cell nucleus after bromo-UTP microinjection. Mol Biol Cell 10:211–223

Cmarko D, Verschure PJ, Otte AP, van Driel R, Fakan S (2003) Polycomb group gene silencing proteins are concentrated in the perichromatin compartment of the mammalian nucleus. J Cell Sci 116:335–343

Colwill K, Pawson T, Andrews B, Prasad J, Manley JL, Bell JC, Duncan PI (1996) The Clk/Sty protein kinase phosphorylates SR splicing factors and regulates their intranuclear distribution. EMBO J 15:265–275

Comings DE (1968) The rationale for an ordered arrangement of chromatin in the interphase nucleus. Am J Hum Genet 20:440–460

Cremer T, Cremer C (2001) Chromosome territories, nuclear architecture and gene regulation in mammalian cells. Nat Rev Genet 2:292–301.

Cremer T, Cremer C, Schneider T, Baumann H, Hens L, Kirsch-Volders M (1982) Analysis of chromosome positions in the interphase nucleus of Chinese hamster cells by laser-UV-microirradiation experiments. Hum Genet 62:201–209

Cremer T, Lichter P, Borden J, Ward DC, Manuelidis L (1988) Detection of chromosome aberrations in metaphase and interphase tumor cells by in situ hybridization using chromosome-specific library probes. Hum Genet 80:235–246

Cremer T, Kurz A, Zirbel R, Dietzel S, Rinke B, Schrock E, Speicher MR, Mathieu U, Jauch A, Emmerich P, et al (1993) Role of chromosome territories in the functional compartmentalization of the cell nucleus. Cold Spring Harbor Symp Quant Biol 58:777–792

Cremer T, Kreth G, Koester H, Fink RHA, Heintzmann R, Cremer M, Solovei I, Zink D, Cremer C (2000) Chromosome territories, interchromatin domain compartment, and nuclear matrix: An integrated view of the functional nuclear architecture. Crit Rev Eukaryotic Gene Expr 12:179–212

Cremer M, von Hase J, Volm T, Brero A, Kreth G, Walter J, Fischer C, Solovei I, Cremer C, Cremer T (2001) Non-random radial higher-order chromatin arrangements in nuclei of diploid human cells. Chromosome Res 9:541–567

Cremer M, Kupper K, Wagler B, Wizelman L, von Hase J, Weiland Y, Kreja L, Diebold J, Speicher MR, Cremer T (2003) Inheritance of gene density-related higher order chromatin arrangements in normal and tumor cell nuclei. J Cell Biol 162:809–820

Cremer M, Zinner R, Stein S, Albiez H, Wagler B, Cremer C, Cremer T (2004) Three dimensional analysis of histone methylation patterns in normal and tumor cell nuclei. Eur J Histochem 48:15–28

Croft JA, Bridger JM, Boyle S, Perry P, Teague P, Bickmore WA (1999) Differences in the localization and morphology of chromosomes in the human nucleus. J Cell Biol 145:1119–1131

D'Amours D, Jackson SP (2002) The Mre11 complex: at the crossroads of DNA repair and checkpoint signalling. Nat Rev Mol Cell Biol 3:317–327

Desbois C, Rousset R, Bantignies F, Jalinot P (1996) Exclusion of Int-6 from PML nuclear bodies by binding to the HTLV-I Tax oncoprotein. Science 273:951–953

Dietzel S, Schiebel K, Little G, Edelmann P, Rappold GA, Eils R, Cremer C, Cremer T (1999) The 3D positioning of ANT2, ANT3 genes within female X chromosome territories correlates with gene activity. Exp Cell Res 252:363–375

Dostie J, Lejbkowicz F, Sonenberg N (2000) Nuclear eukaryotic initiation factor 4E (eIF4E) colocalizes with splicing factors in speckles. J Cell Biol 148:239–247

Essers J, Houtsmuller AB, van Veelen L, Paulusma C, Nigg AL, Pastink A, Vermeulen W, Hoeijmakers JH, Kanaar R (2002) Nuclear dynamics of RAD52 group homologous recombination proteins in response to DNA damage. EMBO J 21:2030–2037

Everett RD, Sourvinos G, Leiper C, Clements JB, Orr A (2004) Formation of nuclear foci of the herpes simplex virus type 1 regulatory protein ICP4 at early times of infection: localization, dynamics, recruitment of ICP27, and evidence for the de novo induction of ND10-like complexes. J Virol 78:1903–1917

Failla AV, Cavallo A, Cremer C (2002) Subwavelength size determination by spatially modulated illumination virtual microscopy. Appl Opt 41:6651–6659

Frey MR, Matera AG (1995) Coiled bodies contain U7 small nuclear RNA and associate with specific DNA sequences in interphase human cells. Proc Natl Acad Sci USA 92:5915–5919

Frey MR, Matera AG (2001) RNA-mediated interaction of Cajal bodies, U2 snRNA genes. J Cell Biol 154:499–509

Gao J, Köhler D, Solovei I, Cremer T, Eils R, Mattes J (2004) Assessing the similarity of spatial configurations using distance differences and bending energy:

Application to chromosomal interphase arrangements in HeLa cell clones. In: ISBI 2004, Arlington, VA, USA, pp 1400–1403

Geiman TM, Robertson KD (2002) Chromatin remodeling, histone modifications,, DNA methylation-how does it all fit together? J Cell Biochem 87:117–125

Golub EI, Gupta RC, Haaf T, Wold MS, Radding CM (1998) Interaction of human rad51 recombination protein with single-stranded DNA binding protein, RPA. Nucleic Acids Res 26:5388–5393

Guo A, Salomoni P, Luo J, Shih A, Zhong S, Gu W, Paolo Pandolfi P (2000) The function of PML in p53-dependent apoptosis. Nat Cell Biol 2:730–736

Haaf T, Golub EI, Reddy G, Radding CM, Ward DC (1995) Nuclear foci of mammalian Rad51 recombination protein in somatic cells after DNA damage and its localization in synaptonemal complexes. Proc Natl Acad Sci USA 92:2298–2302

Handwerger KE, Murphy C, Gall JG (2003) Steady-state dynamics of Cajal body components in the Xenopus germinal vesicle. J Cell Biol 160:495–504

Hopfner KP, Karcher A, Craig L, Woo TT, Carney JP, Tainer JA (2001) Structural biochemistry and interaction architecture of the DNA double-strand break repair Mre11 nuclease, Rad50-ATPase. Cell 105:473–485

Ishov AM, Sotnikov AG, Negorev D, Vladimirova OV, Neff N, Kamitani T, Yeh ET, Strauss JF 3rd, Maul GG (1999) PML is critical for ND10 formation and recruits the PML-interacting protein daxx to this nuclear structure when modified by SUMO-1. J Cell Biol 147:221–234.

Jady BE, Darzacq X, Tucker KE, Matera AG, Bertrand E, Kiss T (2003) Modification of Sm small nuclear RNAs occurs in the nucleoplasmic Cajal body following import from the cytoplasm. EMBO J 22:1878–1888

Jagatheesan G, Thanumalayan S, Muralikrishna B, Rangaraj N, Karande AA, Parnaik VK (1999) Colocalization of intranuclear lamin foci with RNA splicing factors. J Cell Sci 112 (Pt 24):4651–4661

Kojima T, Zama T, Wada K, Onogi H, Hagiwara M (2001) Cloning of human PRP4 reveals interaction with Clk1. J Biol Chem 276:32247–32256

Kreth G, Finsterle J, von Hase J, Cremer M, Cremer C (2004) Radial arrangement of chromosome territories in human cell nuclei: a computer model approach based on gene density indicates a probabilistic global positioning code. Biophys J 86:1–10

Kruhlak MJ, Lever MA, Fischle W, Verdin E, Bazett-Jones DP, Hendzel MJ (2000) Reduced mobility of the alternate splicing factor (ASF) through the nucleoplasm and steady state speckle compartments. J Cell Biol 150:41–51

Lamond AI, Spector DL (2003) Nuclear speckles: a model for nuclear organelles. Nat Rev Mol Cell Biol 4:605–612

Lavau C, Marchio A, Fagioli M, Jansen J, Falini B, Lebon P, Grosveld F, Pandolfi PP, Pelicci PG, Dejean A (1995) The acute promyelocytic leukaemia-associated PML gene is induced by interferon. Oncogene 11:871–876

Le XF, Yang P, Chang KS (1996) Analysis of the growth and transformation suppressor domains of promyelocytic leukemia gene, PML. J Biol Chem 271:130–135

Leitch AR, Mosgoller W, Schwarzacher T, Bennett MD, Heslop-Harrison JS (1990) Genomic in situ hybridization to sectioned nuclei shows chromosome domains in grass hybrids. J Cell Sci 95:335–341

Lesko SA, Callahan DE, LaVilla ME, Wang ZP, Ts'o PO (1995) The experimental homologous and heterologous separation distance histograms for the centromeres of chromosomes 7, 11, and 17 in interphase human T-lymphocytes. Exp Cell Res 219:499–506

Lichter P, Cremer T, Borden J, Manuelidis L, Ward DC (1988) Delineation of individual human chromosomes in metaphase and interphase cells by in situ suppression hybridization using recombinant DNA libraries. Hum Genet 80:224–234

Ma H, Samarabandu J, Devdhar RS, Acharya R, Cheng PC, Meng C, Berezney R (1998) Spatial and temporal dynamics of DNA replication sites in mammalian cells. J Cell Biol 143:1415–1425

Mahy NL, Perry PE, Bickmore WA (2002a) Gene density and transcription influence the localization of chromatin outside of chromosome territories detectable by FISH. J Cell Biol 159:753–763

Mahy NL, Perry PE, Gilchrist S, Baldock RA, Bickmore WA (2002b) Spatial organization of active and inactive genes and noncoding DNA within chromosome territories. J Cell Biol 157:579–589

Manuelidis L (1985) Individual interphase chromosome domains revealed by in situ hybridization. Hum Genet 71:288–293

Maser RS, Monsen KJ, Nelms BE, Petrini JH (1997) hMre11 and hRad50 nuclear foci are induced during the normal cellular response to DNA double-strand breaks. Mol Cell Biol 17:6087–6096

Matera AG (1998) Of coiled bodies, gems, and salmon. J Cell Biochem 70:181–192

Matera AG (1999) Nuclear bodies: multifaceted subdomains of the interchromatin space. Trends Cell Biol 9:302–309

Matsuura S, Tauchi H, Nakamura A, Kondo N, Sakamoto S, Endo S, Smeets D, Solder B, Belohradsky BH, Der Kaloustian VM, Oshimura M, Isomura M, Nakamura Y, Komatsu K (1998) Positional cloning of the gene for Nijmegen breakage syndrome. Nat Genet 19:179–181

Mirzoeva OK, Petrini JH (2001) DNA damage-dependent nuclear dynamics of the Mre11 complex. Mol Cell Biol 21:281–288

Misteli T, Caceres JF, Spector DL (1997) The dynamics of a pre-mRNA splicing factor in living cells. Nature 387:523–527

Moen PT Jr, Johnson CV, Byron M, Shopland LS, de la Serna IL, Imbalzano AN, Lawrence JB (2004) Repositioning of muscle-specific genes relative to the periphery of SC-35 domains during skeletal myogenesis. Mol Biol Cell 15:197–206

Molenaar C, Wiesmeijer K, Verwoerd NP, Khazen S, Eils R, Tanke HJ, Dirks RW (2003) Visualizing telomere dynamics in living mammalian cells using PNA probes. EMBO J 22:6631–6641

Muller S, Matunis MJ, Dejean A (1998) Conjugation with the ubiquitin-related modifier SUMO-1 regulates the partitioning of PML within the nucleus. EMBO J 17:61–70.

Muratani M, Gerlich D, Janicki SM, Gebhard M, Eils R, Spector DL (2002) Metabolic-energy-dependent movement of PML bodies within the mammalian cell nucleus. Nat Cell Biol 4:106–110.

Münkel C, Eils R, Dietzel S, Zink D, Mehring C, Wedemann G, Cremer T, Langowski J (1999) Compartmentalization of interphase chromosomes observed in simulation and experiment. J Mol Biol 285:1053–1065.

Nagele R, Freeman T, McMorrow L, Lee HY (1995) Precise spatial positioning of chromosomes during prometaphase: evidence for chromosomal order. Science 270:1831–1835

Nelms BE, Maser RS, MacKay JF, Lagally MG, Petrini JH (1998) In situ visualization of DNA double-strand break repair in human fibroblasts. Science 280:590–592

Nickerson JA, Krockmalnic G, Wan KM, Penman S (1997) The nuclear matrix revealed by eluting chromatin from a cross-linked nucleus. Proc Natl Acad Sci USA 94:4446–4450

O'Regan P, Wilson C, Townsend S, Thacker J (2001) XRCC2 is a nuclear RAD51-like protein required for damage-dependent RAD51 focus formation without the need for ATP binding. J Biol Chem 276:22148–22153

Ogawa T, Yu X, Shinohara A, Egelman EH (1993) Similarity of the yeast RAD51 filament to the bacterial RecA filament. Science 259:1896–1899

Okada TA, Comings DE (1979) Higher order structure of chromosomes. Chromosoma 72:1–14

Parada LA, Sotiriou S, Misteli T (2004) Spatial genome organization. Exp Cell Res 296:64–70

Paull TT, Gellert M (1999) Nbs1 potentiates ATP-driven DNA unwinding and endonuclease cleavage by the Mre11/Rad50 complex. Genes Dev 13:1276–1288

Paull TT, Rogakou EP, Yamazaki V, Kirchgessner CU, Gellert M, Bonner WM (2000) A critical role for histone H2AX in recruitment of repair factors to nuclear foci after DNA damage. Curr Biol 10:886–895

Pederson T (2004) The spatial organization of the genome in mammalian cells. Curr Opin Genet Dev 14:203–209

Petersen S, Casellas R, Reina-San-Martin B, Chen HT, Difilippantonio MJ, Wilson PC, Hanitsch L, Celeste A, Muramatsu M, Pilch DR, Redon C, Ried T, Bonner WM, Honjo T, Nussenzweig MC, Nussenzweig A (2001) AID is required to initiate Nbs1/gamma-H2AX focus formation and mutations at sites of class switching. Nature 414:660–665

Petrini JH, Stracker TH (2003) The cellular response to DNA double-strand breaks: defining the sensors and mediators. Trends Cell Biol 13:458–462

Platani M, Goldberg I, Swedlow JR, Lamond AI (2000) In vivo analysis of Cajal body movement, separation, and joining in live human cells. J Cell Biol 151:1561–1574

Platani M, Goldberg I, Lamond AI, Swedlow JR (2002) Cajal body dynamics and association with chromatin are ATP-dependent. Nat Cell Biol 4:502–508

Plug AW, Xu J, Reddy G, Golub EI, Ashley T (1996) Presynaptic association of Rad51 protein with selected sites in meiotic chromatin. Proc Natl Acad Sci U S A 93:5920–5924

Quignon F, De Bels F, Koken M, Feunteun J, Ameisen JC, de The H (1998) PML induces a novel caspase-independent death process. Nat Genet 20:259–265.

Raderschall E, Golub EI, Haaf T (1999) Nuclear foci of mammalian recombination proteins are located at single-stranded DNA regions formed after DNA damage. Proc Natl Acad Sci USA 96:1921–1926

Regad T, Saib A, Lallemand-Breitenbach V, Pandolfi PP, de The H, Chelbi-Alix MK (2001) PML mediates the interferon-induced antiviral state against a complex retrovirus via its association with the viral transactivator. EMBO J 20:3495–3505.

Schardin M, Cremer T, Hager HD, Lang M (1985) Specific staining of human chromosomes in Chinese Hamster x man hybrid cell lines demonstrates interphase chromosome territories. Hum Genet 71:281–287

Schermelleh L, Solovei I, Zink D, Cremer T (2001) Two-color fluorescence labeling of early and mid-to-late replicating chromatin in living cells. Chromosome Res 9:77–80

Scully R, Chen J, Plug A, Xiao Y, Weaver D, Feunteun J, Ashley T, Livingston DM (1997) Association of BRCA1 with Rad51 in mitotic and meiotic cells. Cell 88:265–275

Sharan SK, Morimatsu M, Albrecht U, Lim DS, Regel E, Dinh C, Sands A, Eichele G, Hasty P, Bradley A (1997) Embryonic lethality and radiation hypersensitivity mediated by Rad51 in mice lacking Brca2. Nature 386:804–810

Shinohara A, Ogawa H, Ogawa T (1992) Rad51 protein involved in repair and recombination in S. cerevisiae is a RecA-like protein. Cell 69:457–470

Shinohara A, Ogawa H, Matsuda Y, Ushio N, Ikeo K, Ogawa T (1993) Cloning of human, mouse and fission yeast recombination genes homologous to RAD51 and recA. Nat Genet 4:239–243

Shopland LS, Johnson CV, Byron M, McNeil J, Lawrence JB (2003) Clustering of multiple specific genes and gene-rich R-bands around SC-35 domains: evidence for local euchromatic neighborhoods. J Cell Biol 162:981–990

Singer RH, Green MR (1997) Compartmentalization of eukaryotic gene expression: causes and effects. Cell 91:291–294

Snaar S, Wiesmeijer K, Jochemsen AG, Tanke HJ, Dirks RW (2000) Mutational analysis of fibrillarin and its mobility in living human cells. J Cell Biol 151:653–662

Solovei I, Cavallo A, Schermelleh L, Jaunin F, Scasselati C, Cmarko D, Cremer C, Fakan S, Cremer T (2002) Spatial preservation of nuclear chromatin architecture during three-dimensional fluorescence in situ hybridization (3D-FISH). Exp Cell Res 276:10–23

Solovei I, Schermelleh L, Düring K, Engelhardt A, Stein S, von Hase J, Cremer C, Cremer T (2004a) Differences in centromere positioning of cycling and postmitotic human cell types. Chromosoma 112: 410-423

Solovei I, Grandi N, Knoth R, Volk B, Cremer T (2004b) Positional changes of pericentromeric heterochromatin and nucleoli in postmitotic Purkinje cells during murine cerebellum development. Cytogenet Genome Res 105: 302-310

Sonoda E, Sasaki MS, Buerstedde JM, Bezzubova O, Shinohara A, Ogawa H, Takata M, Yamaguchi-Iwai Y, Takeda S (1998) Rad51-deficient vertebrate cells accumulate chromosomal breaks prior to cell death. EMBO J 17:598–608

Spector DL (2003) The dynamics of chromosome organization and gene regulation. Annu Rev Biochem 72:573–608

Sternsdorf T, Jensen K, Will H (1997) Evidence for covalent modification of the nuclear dot-associated proteins PML, Sp100 by PIC1/SUMO-1. J Cell Biol 139:1621–1634.

Swedlow JR (2003) Quantitative fluorescence microscopy and image deconvolution. Methods Cell Biol 72:349–367

Tan TL, Essers J, Citterio E, Swagemakers SM, de Wit J, Benson FE, Hoeijmakers JH, Kanaar R (1999) Mouse Rad54 affects DNA conformation, DNA-damage-induced Rad51 foci formation. Curr Biol 9:325–328

Tashiro S, Kotomura N, Shinohara A, Tanaka K, Ueda K, Kamada N (1996) S phase specific formation of the human Rad51 protein nuclear foci in lymphocytes. Oncogene 12:2165–2170

Tashiro S, Walter J, Shinohara A, Kamada N, Cremer T (2000) Rad51 accumulation at sites of DNA damage and in postreplicative chromatin. J Cell Biol 150:283–291

Trinkle-Mulcahy L, Ajuh P, Prescott A, Claverie-Martin F, Cohen S, Lamond AI, Cohen P (1999) Nuclear organisation of NIPP1, a regulatory subunit of protein phosphatase 1 that associates with pre-mRNA splicing factors. J Cell Sci 112 (Pt 2):157–168

Vallian S, Chin KV, Chang KS (1998) The promyelocytic leukemia protein interacts with Sp1 and inhibits its transactivation of the epidermal growth factor receptor promoter. Mol Cell Biol 18:7147–7156

van den Bosch M, Bree RT, Lowndes NF (2003) The MRN complex: coordinating and mediating the response to broken chromosomes. EMBO Rep 4:844–849

van Driel R, Fransz PF, Verschure PJ (2003) The eukaryotic genome: a system regulated at different hierarchical levels. J Cell Sci 116:4067–4075

Varon R, Vissinga C, Platzer M, Cerosaletti KM, Chrzanowska KH, Saar K, Beckmann G, Seemanova E, Cooper PR, Nowak NJ, Stumm M, Weemaes CM, Gatti RA, Wilson RK, Digweed M, Rosenthal A, Sperling K, Concannon P, Reis A (1998) Nibrin, a novel DNA double-strand break repair protein, is mutated in Nijmegen breakage syndrome. Cell 93:467–476

Vazquez J, Belmont AS, Sedat JW (2001) Multiple regimes of constrained chromosome motion are regulated in the interphase Drosophila nucleus. Curr Biol 11:1227–1239

Verschure PJ, van der Kraan I, Manders EM, Hoogstraten D, Houtsmuller AB, van Driel R (2003) Condensed chromatin domains in the mammalian nucleus are accessible to large macromolecules. EMBO Rep 4:861–866

Visser AE, Jaunin F, Fakan S, Aten JA (2000) High resolution analysis of interphase chromosome domains. J Cell Sci 113:2585–2593

Vogel F, Schroeder TM (1974) The internal order of the interphase nucleus. Humangenetik 25:265–297

Volpi EV, Chevret E, Jones T, Vatcheva R, Williamson J, Beck S, Campbell RD, Goldsworthy M, Powis SH, Ragoussis J, Trowsdale J, Sheer D (2000) Large-scale chromatin organization of the major histocompatibility complex and other regions of human chromosome 6 and its response to interferon in interphase nuclei. J Cell Sci 113 (Pt 9):1565–1576

von Mikecz A, Zhang S, Montminy M, Tan EM, Hemmerich P (2000) CREB-binding protein (CBP)/p300, RNA polymerase II colocalize in transcriptionally active domains in the nucleus. J Cell Biol 150:265–273

Walter J, Schermelleh L, Cremer M, Tashiro S, Cremer T (2003) Chromosome order in HeLa cells changes during mitosis and early G1, but is stably maintained during subsequent interphase stages. J Cell Biol 160:685–697

Wang ZG, Ruggero D, Ronchetti S, Zhong S, Gaboli M, Rivi R, Pandolfi PP (1998) PML is essential for multiple apoptotic pathways. Nat Genet 20:266–272

Wiesmeijer K, Molenaar C, Bekeer IM, Tanke HJ, Dirks RW (2002) Mobile foci of Sp100 do not contain PML: PML bodies are immobile but PML, Sp100 proteins are not. J Struct Biol 140:180–188

Williams RR, Broad S, Sheer D, Ragoussis J (2002) Subchromosomal positioning of the epidermal differentiation complex (EDC) in keratinocyte and lymphoblast interphase nuclei. Exp Cell Res 272:163–175

Wischnitzer S (1973) The submicroscopic morphology of the interphase nucleus. Int Rev Cytol 34:1–48

Xu ZX, Timanova-Atanasova A, Zhao RX, Chang KS (2003) PML colocalizes with and stabilizes the DNA damage response protein TopBP1. Mol Cell Biol 23:4247–4256

Zhong S, Hu P, Ye TZ, Stan R, Ellis NA, Pandolfi PP (1999) A role for PML and the nuclear body in genomic stability. Oncogene 18:7941–7947

Zhong S, Muller S, Ronchetti S, Freemont PS, Dejean A, Pandolfi PP (2000a) Role of SUMO-1-modified PML in nuclear body formation. Blood 95:2748–2752

Zhong S, Salomoni P, Pandolfi PP (2000b) The transcriptional role of PML and the nuclear body. Nat Cell Biol 2:E85–90

Zhong S, Salomoni P, Ronchetti S, Guo A, Ruggero D, Pandolfi PP (2000c) Promyelocytic leukemia protein (PML), Daxx participate in a novel nuclear pathway for apoptosis. J Exp Med 191:631–640

Zhu XD, Kuster B, Mann M, Petrini JH and de Lange T (2000) Cell-cycle-regulated association of RAD50/MRE11/NBS1 with TRF2 and human telomeres. Nat Genet 25:347–352

Zorn C, Cremer T, Cremer C, Zimmer J (1976) Laser UV microirradiation of interphase nuclei and post-treatment with caffeine. A new approach to establish the arrangement of interphase chromosomes. Hum. Genet. 35:83–89

10-1. Regulation of Chromatin Structure by Curved DNA: How Activator Binding Sites Become Accessible

Takashi Ohyama

Department of Biology, Faculty of Education and Integrated Arts and Sciences, Major in Integrative Bioscience and Biomedical Engineering, Graduate School of Science and Engineering, Waseda University, 1-6-1, Nishi-Waseda, Shinjuku-ku, Tokyo 169-8050, Japan

10-1.1 Introduction

A single somatic cell of humans contains DNA fibers of a total length of approximately 2 m, which are compacted, without entanglement, into the nucleus of approximately 1×10^{-5} m in diameter. To greater or lesser degrees, all organisms compact their DNA. Biologically important DNA regions, such as the origins of DNA replication, regulatory regions of transcription, and recombination loci, must all be compacted. The tightly constrained DNA, however, presents the appropriate environment for replication, transcription, and recombination to take place.

In eukaryotes, the DNA fiber is packaged into chromatin. In the last decade, much progress has been made in understanding how transcription is initiated in chromatin (Workman and Kingston 1998; Peterson and Workman 2000; Aalfs and Kingston 2000; Vignali et al. 2000; Wu and Grunstein 2000; Becker and Horz 2002; Narlikar et al. 2002). In the first step, transcription activators bind to their target DNA elements and recruit chromatin remodeling or modifying activities, to alter chromatin structure. We know that activators can bind to their targets, even when those DNA elements are adjacent to nucleosomes, or actually within nucleosomes (Almer et al. 1986; Archer et al. 1992; Zhu and Thiele 1996; Wolffe 1998; Nishikawa et al. 2003). Until recently, however, it was unclear what structures of chromatin allow activator binding, or how those structures are constructed. Several recent studies have shed light on the significance of curved DNA structures for this (Blomquist et al. 1999; Ohyama 2001; Nishikawa et al. 2003). This essay describes some recent advances.

10-1.2 Three-Dimensional Architecture of Naturally Occurring Curved DNA Structures

A DNA bend can be generated either by an exterior force such as a protein binding, or by the nucleotide sequence per se. The former is called (protein-) induced DNA bend or simply DNA bending; and the latter is called curved DNA, bent DNA, or intrinsic DNA curvature. This essay concentrates on the latter. Generally, regularly distributed runs of adenines or thymines (A-tracts or T-tracts), with a periodicity of one run per helical repeat, form bent DNA structures (Trifonov and Sussman 1980; Marini et al. 1982; Wu and Crothers 1984; Hagerman 1986; Diekmann 1986; Ulanovsky and Trifonov 1987; Koo and Crothers 1988; Barbic et al. 2003). The periodicity of the tracts relative to the helical repeat length of DNA (about 10.5 bp) determines the DNA's three-dimensional (3D) architecture (Calladine et al. 1988). When the A- or T-tracts occur with a periodicity almost equal to 10.5 bp, the helical trajectory of DNA becomes planar (Fig. 1), that is to say, a flat curved structure (plane curve or 2D curve) is formed. However, if the tract periodicity is other than this, then a 3D curved structure is formed, like a corkscrew, with either a right- or a left-handed writhe. When the periodicity is less than 10.5 bp (e.g., 9 or 10 bp), DNA adopts a left-handed curved structure, and when the periodicity is larger than 10.5 bp (e.g., 11 or 12 bp), it adopts a right-handed curved structure (Calladine et al. 1988; Hirota and Ohyama 1995; Brukner et al. 1997; Nishikawa et al. 2003). These are sometimes called "space curves." Periodicities of 5–7 bp result in a nearly straight (actually a zigzag) trajectory of the helical axis. Naturally occurring DNA adopts various 3D shapes by combining these basic structures.

10-1.3 Curved DNA and Packaging of Genomes into Chromatin

The structure and mechanical properties of DNA influence the formation, stability and positioning of nucleosomes (Zhurkin et al. 1979; Trifonov and Sussman 1980; Satchwell et al. 1986; Shrader and Crothers 1989; Ioshikhes et al. 1992; Sivolob and Khrapunov 1995; Ioshikhes et al. 1996; Olson et al. 1998; Fitzgerald and Anderson 1998). DNA has to be bent around the histone core, and it therefore seems thermodynamically favorable to form a nucleosome on a DNA sequence that is already appropriately curved (Drew and Travers 1985; Zhurkin 1985; Anselmi et al. 1999). Indeed, nucleosomes often preferentially associate with curved DNA

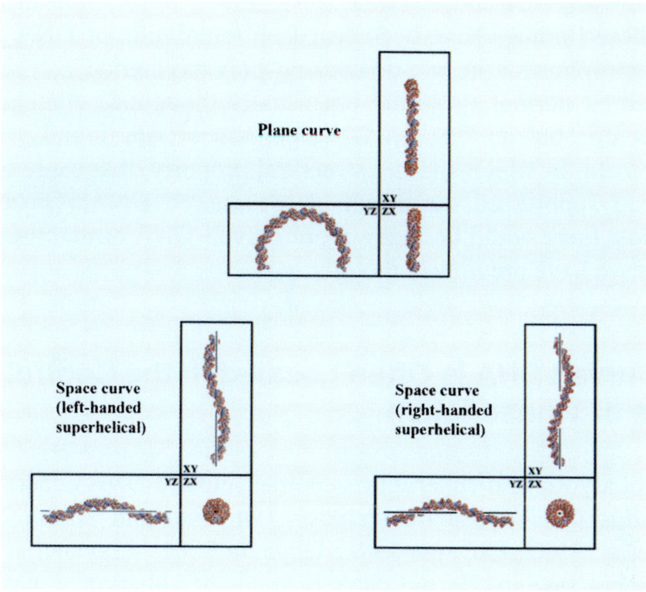

Fig. 1. Intrinsic DNA curvatures with superhelical and circular conformations. The figure shows the structures formed by the nucleotide sequences $(A_5CATG)_8$ (*left*), $(A_5CAGTCA_5CAGTCG)_4$ (*top*), and $(A_5CAGTCAG)_7$ (*right*). They were drawn by a combination of DIAMOD (Dlakic and Harrington 1998) and RASMOL (Sayle and Milner-White 1995).

fragments (Pennings et al. 1989; Costanzo et al. 1990; De Santis et al. 1996; Widlund et al. 1997). By screening a library of DNA fragments from nucleosome cores from the mouse, Widlund et al. (1997) showed that among the fragments that form the most stable nucleosomes, a curved DNA structure is the most common feature.

This suggests that curved DNA structures may frequently occur on eukaryotic genomes in order to package them. Indeed, the repeating units of satellite DNA sequences frequently contain one or more curved DNA structures (Martinez-Balbas et al. 1990; Pasero et al. 1993; Fitzgerald et al. 1994). The satellites are universally associated with regions of constitutive heterochromatin, and comprise anywhere from a few percent to >50% of mammalian genomes (John and Miklos 1979; Singer 1982). Also, curved DNA sites occur repeatedly in human ε-, Gγ-Aγ-ψβ-, δ-, and β-globin, c-*myc*, and immunoglobulin heavy chain μ loci, and in mouse β^{major}-globin locus (Wada-Kiyama and Kiyama 1994, 1995, 1996; Ohki et al. 1998). Most of these findings have been based on retardation of DNA fragments during electrophoresis in non-denaturing polyacrylamide gels. Fragments that migrate at normal speed are not usually thought to contain a curved DNA structure, but an "unseen DNA curvature" was found in one such DNA fragment, in which another structural property, that caused rapid migration, had suppressed the effect of the curved DNA (Ohyama et al. 1998). Thus, there could be many more curved DNA structures in the genome than we have previously thought.

10-1.4 Curved DNA Is Often Located in the Control Regions of Transcription

In both prokaryotic and eukaryotic genomes, curved DNA occurs frequently in regions that control transcription (Ohyama and Hashimoto 1989; Ohyama et al. 1992; Ohyama 1996; Ohyama 2001; Asayama et al. 2002; and references therein). In the class II genes of eukaryotes, both TATA-box-containing and TATA-box-less promoters often contain this structure. Moreover, curved DNA may be common to all class I gene promoters (Marilley and Pasero 1996). Despite the many reports of curved DNA in promoters, the role of curved DNA is not fully understood. As described above, nucleosomes often preferentially associate with curved DNA fragments. However, it is generally thought that if nucleosomes assemble over a promoter region, they would inhibit access and/or assembly of transcription factors. How do eukaryotes circumvent this problem? They appear to use the DNA curvature cleverly.

10-1.5 Chromatin Structure That is Permissive to the Activator Binding

Structures That Encourage Nucleosome Formation

To make target DNA elements accessible, there are logically two options: expose the region toward the environment on the surface of a nucleosome; or make the region free of nucleosomes (Fig. 2). As argued above, it seems thermodynamically favorable to incorporate into a nucleosome a DNA sequence that is already appropriately curved. Thus the

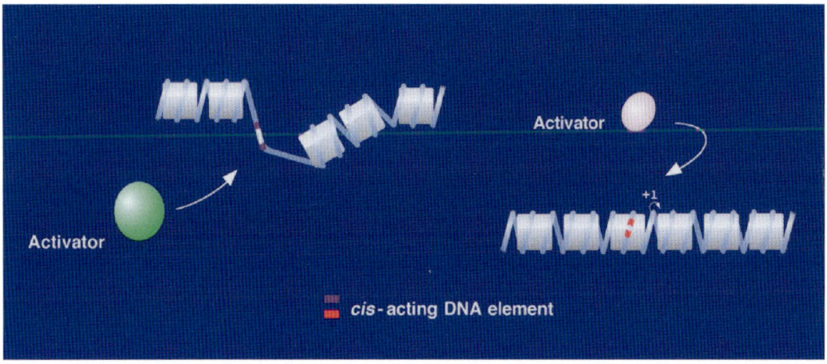

Fig. 2. A chromatin structure that permits activator binding can be formed in two ways: by inhibiting nucleosome assembly on the target element; or by putting the target element on a nucleosome, and displaying it towards the environment.

first option could occur with DNA that has an intrinsic conformation similar to the writhing of DNA in nucleosomes, namely, when it mimics left-handed (negative) supercoils. The second option could occur when the DNA's intrinsic structure is different from this.

Unusual DNA Structures That Inhibit Nucleosome Formation

In *Saccharomyces cerevisiae*, the adenylate kinase gene promoter has a curved DNA structure which is dissimilar to the negative supercoil. This promoter was shown to be free of nucleosomes (Angermayr et al. 2002). Also, in the yeast *GAL80* promoter, intrinsic DNA curvature close to the upstream activator sequence (UAS_{GAL80}) may play the same role. Nucleosomes are not formed on this curved DNA, or on UAS_{GAL80} (Bash et al. 2001). Other unusual DNA structures such as poly (dA•dT) sequences, triple-stranded DNAs, and cruciform DNAs, may also keep DNA free of nucleosomes. Using in vivo UV photo-footprinting and DNA repair by photolyase, Suter et al. (2000) demonstrated that poly (dA•dT) sequences in yeast promoters such as *HIS3*, *URA3*, and *ILV1* were not folded into nucleosomes. Formation of triple-stranded DNA and nucleosome assembly are competing processes. For example, Espinas et al. (1996) studied in vitro assembly of mononucleosomes onto 180 bp DNA fragments containing $(GA•TC)_{22}$, or onto 190 bp fragments with $(GA•TC)_{10}$. Nucleosome assembly was strongly inhibited when triple-stranded DNA was formed at the $(GA \bullet TC)_n$ site, while the formation of triple-stranded DNA was inhibited when the $(GA \bullet TC)_n$ site was incorporated into a nucleosome. The $(GA \bullet TC)_n$ sequences themselves had no influence on nucleosome formation. Similarly, it is known that cruciform structures are unable to associate with core histones, and are located mainly on inter-nucleosomal DNA (Battistoni et al. 1988).

Left-Handedly Curved DNA Can Expose *cis*-DNA Elements in Nucleosomes

When curved DNA structures mimic part of the negative supercoils seen in nucleosomes, they seem to be very effective in recruiting histone cores. Even though the target DNA elements of activators are incorporated into nucleosomes, if they can display their recognition sites on the surface of the nucleosomes, recognition would be facilitated (Ohyama 2001). Curved DNA is implicated in the formation of such structures, and the nucleosome structure formed on the long terminal repeat of the mouse mammary tumor virus (MMTV-LTR) is a good example. Four recognition elements (GRE1–4) of the glucocorticoid receptor, a zinc finger protein, are located within a positioned nucleosome. Two GREs expose their major grooves towards the environment on the surface of this nucleosome, and are recognized by the receptor (Pina et al.1990a; Fletcher et al. 2000; and references therein). An early study revealed the presence of a curved DNA structure between GRE2 and GRE3, and proposed that this may determine the rotational setting of the nucleosomal DNA (Pina et al. 1990b). This curved DNA has a left-handed curved trajectory (Ohyama 2001).

Recently, by creating 35 reporter constructs with the herpes simplex virus thymidine kinase (HSV *tk*) promoter, we studied the relationship between the geometry of DNA upstream of the promoter, nucleosome positioning, and promoter activity (Nishikawa et al. 2003). Left-handed curved, right-handed curved, planar, zigzag, and straight DNA segments were studied. A left-handed curved DNA of about 40 bp activated transcription by about 10-fold when it was linked to the promoter at a specific rotational phase and distance. The other DNA conformations did not have this effect. Transcription was activated by the following mechanism: the histone core was attracted by the left-handed curved DNA; the TATA box was thereby left in the linker DNA with its minor groove facing outwards; this structure enhanced accessibility of the TATA box; and presumably the enhanced interaction between the TATA box and the TFIID activated transcription (Fig. 3).

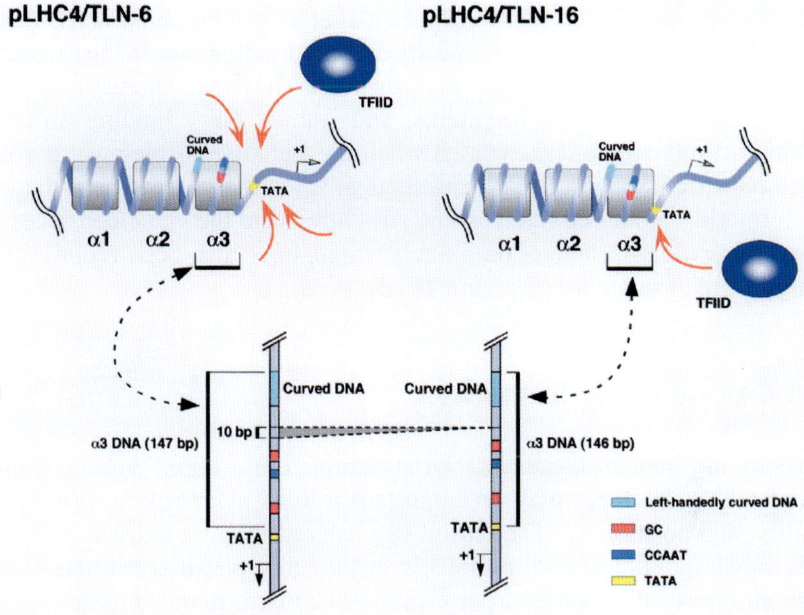

Fig. 3. A left-handed curved DNA structure that is appropriately introduced upstream of a promoter can activate transcription, by modulating local chromatin structure. The figure shows an example using the HSV *tk* promoter as a test system. Curved DNA can attract histone octamers, and depending on the distance between it and the TATA box, can position the box either in the linker DNA region (*pLHC4/TLN-6*) or at the edge of the nucleosome (*pLHC4/TLN-16*) with its minor groove facing outwards. Both make the box more accessible to transcription factors, and activate transcription, although the first structure is more active than the second. The symbols *α1, α2,* and *α3* are nucleosomes formed on the promoter region. Reprinted, with permission, from Nishikawa et al. 2003.

Another Role of DNA Curvature

Even when DNA geometry is dissimilar to the negative supercoils on histone octamers, neighboring DNA sequences may allow a nucleosome to form. In this case, however, the nucleosome structure may be altered. This mechanism also seems to construct nucleosomes that permit activators to bind to DNA. Using the repeated $(A/T)_3NN(G/C)_3NN$ motifs (TG-motifs) that are anisotropically flexible and have a high nucleosome-forming ability (Shrader and Crothers 1989), Blomquist et al. (1999) constructed DNA fragments composed of the TG-motifs and the binding site for the nuclear factor 1 (NF-1) with an A_5-tract on both sides. They then reconstituted nucleosomes on these DNA fragments, and studied NF-1 binding affinity. Binding affinity was more elevated when the flanking A-tracts were positioned out-of-phase with the TG-motifs than when the tracts were in-phase. The formation of altered nucleosome structures, and the enhanced accessibility of a *cis*-DNA element on a nucleosome, have also been reported for a poly (dA•dT) sequence (Zhu and Thiele 1996).

10-1.6 Concluding Remarks

In eukaryotic genomes, many DNA structures are present. Among them, curved DNA structures play an important role in nucleosome formation, stability, and positioning. The reason why curved DNA is frequently located in transcriptional control regions is presumably that it constructs chromatin structures that leave target elements exposed, permitting activators to recognize them and bind. Right-handed curved DNA seems to inhibit nucleosome formation, which is similar to its putative role in prokaryotes, inhibiting assembly of nucleoid structuring proteins (Ohyama 2001). In prokaryotes, the DNA located in the promoter regions frequently has a right-handed curved path of helical axis (Travers 1990; Ohyama et al. 1992; Asayama et al. 1999), which helps open promoter complexes to form (Hirota and Ohyama 1995), and could inhibit association of nucleoid proteins (the effects of left-handed curved DNA structures have not yet been clarified). In eukaryotic chromatin, left-handed curved DNAs ensure the accessibility of target sites of activators, by regulating the rotational setting of neighboring DNA on the surface of histone octamers, or in the close vicinity of the nucleosomes. Although this essay has highlighted the role of DNA curvature in the organization of chromatin infrastructure, DNA curvature may also play other roles, after local chromatin structure has been remodeled.

Acknowledgments

The author would like to acknowledge the contributions of Mr. J. Nishikawa, Mr. Y. Fukue, and Ms. J. Ohyama. This work was supported in part by a Grant-in-Aid from the Ministry of Education, Science, Sports and Culture of Japan.

References

Aalfs JD, Kingston RE (2000) What does "chromatin remodeling" mean? Trends Biochem Sci 25:548–555

Almer A, Rudolph H, Hinnen A, Horz W (1986) Removal of positioned nucleosomes from the yeast *PHO5* promoter upon PHO5 induction releases additional upstream activating DNA elements. EMBO J 5:2689–2696

Angermayr M, Oechsner U, Gregor K, Schroth GP, Bandlow W (2002) Transcription initiation in vivo without classical transactivators: DNA kinks flanking the core promoter of the housekeeping yeast adenylate kinase gene, *AKY2*, position nucleosomes and constitutively activate transcription. Nucleic Acids Res 30:4199–4207

Anselmi C, Bocchinfuso G, De Santis P, Savino M, Scipioni A (1999) Dual role of DNA intrinsic curvature and flexibility in determining nucleosome stability. J Mol Biol 286:1293–1301

Archer TK, Lefebvre P, Wolford RG, Hager GL (1992) Transcription factor loading on the MMTV promoter: a bimodal mechanism for promoter activation. Science 255:1573–1576

Asayama M, Hayasaka Y, Kabasawa M, Shirai M, Ohyama T (1999) An intrinsic DNA curvature found in the cyanobacterium *Microcystis aeruginosa* K-81 affects the promoter activity of *rpoD1* encoding a principal sigma factor. J Biochem (Tokyo) 125:460–468

Asayama M, Kato H, Shibato J, Shirai M, Ohyama T (2002) The curved DNA structure in the 5′-upstream region of the light-responsive genes: its universality, binding factor and function for cyanobacterial *psbA* transcription. Nucleic Acids Res 30:4658–4666

Barbic A, Zimmer DP, Crothers DM (2003) Structural origins of adenine-tract bending. Proc Natl Acad Sci USA 100:2369–2373

Bash RC, Vargason JM, Cornejo S, Ho PS, Lohr D (2001) Intrinsically bent DNA in the promoter regions of the yeast *GAL1-10* and *GAL80* genes. J Biol Chem 276:861–866

Battistoni A, Leoni L, Sampaolese B, Savino M (1988) Kinetic persistence of cruciform structures in reconstituted minichromosomes. Biochim Biophys Acta 950:161–171

Becker PB, Horz W (2002) ATP-dependent nucleosome remodeling. Annu Rev Biochem 271:247–273

Blomquist P, Belikov S, Wrange O (1999) Increased nuclear factor 1 binding to its nucleosomal site mediated by sequence-dependent DNA structure. Nucleic Acids Res 27:517–525

Brukner I, Belmaaza A, Chartrand P (1997) Differential behavior of curved DNA upon untwisting. Proc Natl Acad Sci USA 94:403–406

Calladine CR, Drew HR, McCall MJ (1988) The intrinsic curvature of DNA in solution. J Mol Biol 201:127–137

Costanzo G, di Mauro E, Salina G, Negri R (1990) Attraction, phasing and neighbour effects of histone octamers on curved DNA. J Mol Biol 216:363–374

De Santis P, Kropp B, Leoni L, Sampaolese B, Savino M (1996) Influence of DNA superstructural features and histones aminoterminal domains on mononucleosome and dinucleosome positioning. Biophys Chem 62:47–61

Diekmann S (1986) Sequence specificity of curved DNA. FEBS Lett 195:53–56

Dlakic M, Harrington RE (1998) DIAMOD: display and modeling of DNA bending. Bioinformatics 14:326–331

Drew HR, Travers AA (1985) DNA bending and its relation to nucleosome positioning. J Mol Biol 186:773–790

Espinas ML, Jimenez-Garcia E, Martinez-Balbas A, Azorin F (1996) Formation of triple-stranded DNA at $d(GA•TC)_n$ sequences prevents nucleosome assembly and is hindered by nucleosomes. J Biol Chem 271:31807–31812

Fitzgerald DJ, Anderson JN (1998) Unique translational positioning of nucleosomes on synthetic DNAs. Nucleic Acids Res 26:2526–2535

Fitzgerald DJ, Dryden GL, Bronson EC, Williams JS, Anderson JN (1994) Conserved patterns of bending in satellite and nucleosome positioning DNA. J Biol Chem 269:21303–21314

Fletcher TM, Ryu B-W, Baumann CT, Warren BS, Fragoso G, John S, Hager GL (2000) Structure and dynamic properties of a glucocorticoid receptor-induced chromatin transition. Mol Cell Biol 20:6466–6475

Hagerman PJ (1986) Sequence-directed curvature of DNA. Nature 321:449–450

Hirota Y, Ohyama T (1995) Adjacent upstream superhelical writhe influences an *Escherichia coli* promoter as measured by in vivo strength and in vitro open complex formation. J Mol Biol 254:566–578

Ioshikhes I, Bolshoy A, Trifonov EN (1992) Preferred positions of AA and TT dinucleotides in aligned nucleosomal DNA sequences. J Biomol Struct Dyn 9:1111–1117

Ioshikhes I, Bolshoy A, Derenshteyn K, Borodovsky M, Trifonov EN (1996) Nucleosome DNA sequence pattern revealed by multiple alignment of experimentally mapped sequences. J Mol Biol 262:129–139

John B, Miklos GLG (1979) Functional aspects of satellite DNA and heterochromatin. Int Rev Cytol 58:1–114.

Koo HS, Crothers DM (1988) Calibration of DNA curvature and a unified description of sequence-directed bending. Proc Natl Acad Sci USA 85:1763–1767

Marilley M, Pasero P (1996) Common DNA structural features exhibited by eukaryotic ribosomal gene promoters. Nucleic Acids Res; 24:2204–2211

Marini JC, Levene SD, Crothers DM, Englund PT (1982) Bent helical structure in kinetoplast DNA. Proc Natl Acad Sci USA 79:7664–7668

Martinez-Balbas A, Rodriguez-Campos A, Garcia-Ramirez M, Sainz J, Carrera P, Aymami J, Azorin F (1990) Satelite DNAs contain sequences that induced curvature. Biochemistry 29:2342–234

Narlikar GJ, Fan HY, Kingston RE (2002) Cooperation between complexes that regulate chromatin structure and transcription. Cell 108:475–487

Nishikawa J, Amano M, Fukue Y, Tanaka S, Kishi H, Hirota Y, Yoda K, Ohyama T (2003) Left-handedly curved DNA regulates accessibility to *cis*-DNA elements in chromatin. Nucleic Acids Res 31:6651–6662

Ohki R, Hirota M, Oishi M, Kiyama R (1998) Conservation and continuity of periodic bent DNA in genomic rearrangements between the c-*myc* and immunoglobulin heavy chain μ loci. Nucleic Acids Res 26:3026–3033

Ohyama T (1996) Bent DNA in the human adenovirus type 2 E1A enhancer is an architectural element for transcription stimulation. J Biol Chem 271:27823–27828

Ohyama T (2001) Intrinsic DNA bends: an organizer of local chromatin structure for transcription. BioEssays 23:708–715

Ohyama T, Hashimoto S (1989) Upstream half of adenovirus type 2 enhancer adopts a curved DNA conformation. Nucleic Acids Res 17:3845–3853

Ohyama T, Nagumo M, Hirota Y, Sakuma S (1992) Alteration of the curved helical structure located in the upstream region of the β-lactamase promoter of plasmid pUC19 and its effect on transcription. Nucleic Acids Res 20:1617–1622

Ohyama T, Tsujibayashi H, Tagashira H, Inano K, Ueda T, Hirota Y, Hashimoto K (1998) Suppression of electrophoretic anomaly of bent DNA segments by the structural property that causes rapid migration. Nucleic Acids Res 26:4811–4817

Olson WK, Gorin AA, Lu XJ, Hock LM, Zhurkin VB (1998) DNA sequence-dependent deformability deduced from protein-DNA crystal complexes. Proc Natl Acad Sci USA 95:11163–11168

Pasero P, Sjakste N, Blettry C, Got C, Marilley M (1993) Long-range organization and sequence-directed curvature of *Xenopus laevis* satellite 1 DNA. Nucleic Acids Res 21:4703–4710

Pennings S, Muyldermans S, Meersseman G, Wyns L (1989) Formation, stability and core histone positioning of nucleosomes reassembled on bent and other nucleosome-derived DNA. J Mol Biol 207:183–192

Peterson CL, Workman JL (2000) Promoter targeting and chromatin remodeling by the SWI/SNF complex. Curr Opin Genet Dev 10:187–192

Pina B, Bruggemeier U, Beato M (1990a) Nucleosome positioning modulates accessibility of regulatory proteins to the mouse mammary tumor virus promoter. Cell 60:719–731

Pina B, Barettino D, Truss M, Beato M (1990b) Structural features of a regulatory nucleosome. J Mol Biol 216:975–990

Satchwell SC, Drew HR, Travers AA (1986) Sequence periodicities in chicken nucleosome core DNA. J Mol Biol 191:659–675

Sayle RA, Milner-White EJ (1995) RASMOL: biomolecular graphics for all. Trends Biochem Sci 20:374–376

Shrader TE, Crothers DM (1989) Artificial nucleosome positioning sequences. Proc Natl Acad Sci USA 86:7418–7422

Singer MF (1982) Highly repeated sequences in mammalian genomes. Int Rev Cytol 76:67–112

Sivolob AV, Khrapunov SN (1995) Translational positioning of nucleosomes on DNA: the role of sequence-dependent isotropic DNA bending stiffness. J Mol Biol 247:918–931

Suter B, Schnappauf G, Thoma F (2000) Poly(dA•dT) sequences exist as rigid DNA structures in nucleosome-free yeast promoters in vivo. Nucleic Acids Res 28:4083–4089

Travers AA (1990) Why bend DNA? Cell 60:177–180

Trifonov EN, Sussman JL (1980) The pitch of chromatin DNA is reflected in its nucleotide sequence. Proc Natl Acad Sci USA 77:3816–3820

Ulanovsky LE, Trifonov EN (1987) Estimation of wedge components in curved DNA. Nature 326:720–722

Vignali M, Hassan AH, Neely KE, Workman JL (2000) ATP-dependent chromatin-remodeling complexes. Mol Cell Biol 20:1899–1910

Wada-Kiyama Y, Kiyama R (1994) Periodicity of DNA bend sites in human ε-globin gene region. Possibility of sequence-directed nucleosome phasing. J Biol Chem 269:22238–22244

Wada-Kiyama Y, Kiyama R (1995) Conservation and periodicity of DNA bend sites in the human β-globin gene locus. J Biol Chem 270:12439–12445

Wada-Kiyama Y, Kiyama R (1996) An intrachromosomal repeating unit based on DNA bending. Mol Cell Biol 16:5664–5673

Widlund HR, Cao H, Simonsson S, Magnusson E, Simonsson T, Nielsen PE, Kahn JD, Crothers DM, Kubista M (1997) Identification and characterization of genomic nucleosome-positioning sequences. J Mol Biol 267:807–817

Wolffe AP (1998) Chromatin: structure and function 3rd edn. Academic Press, London

Workman JL, Kingston RE (1998) Alteration of nucleosome structure as a mechanism of transcriptional regulation. Annu Rev Biochem 67:545–579

Wu HM, Crothers DM (1984) The locus of sequence-directed and protein-induced DNA bending. Nature 308:509–513

Wu J, Grunstein M (2000) 25 years after the nucleosome model: chromatin modifications. Trends Biochem Sci 25:619–623

Zhu Z, Thiele DJ (1996) A specialized nucleosome modulates transcription factor access to a *C. glabrata* metal responsive promoter. Cell 87:459–470

Zhurkin VB (1985) Sequence-dependent bending of DNA and phasing of nucleosomes. J Biomol Struct Dyn 2:785–804

Zhurkin VB, Lysov YP, Ivanov VI (1979) Anisotropic flexibility of DNA and the nucleosomal structure. Nucleic Acids Res 6:1081–1096

10-2. Actin-Related Proteins Involved in Nuclear and Chromatin Dynamics

Masahiko Harata

Laboratory of Molecular Biology, Department of Molecular and Cell Biology, Division of Life Science, Graduate School of Agricultural Science, Tohoku University, Tsutsumidori-Amamiyamachi 1-1, Aoba-ku, Sendai 981-8555, Japan

10-2.1 Introduction

Actin plays central roles in the organization and dynamics of the cytoskeleton. Within a few decades of the first isolation of actin from muscle in 1941, it was shown that actin filaments form the main architecture of the cytoskeleton, and that the dynamics of the cytoskeleton are regulated by the assembly/disassembly of the filament, which depends on the adenosine diphosphate/triphosphate (ADP/ATP) exchange and on association with various actin-binding proteins. Because of these characteristics of actin, previous researchers hypothesized that actin and/or its evolutionarily related molecules were involved in the organization and dynamics of the nucleus. However, actin filaments were observed only in the cytoplasm, and no molecule evolutionarily related to actin was identified at the time. The hypothesis was therefore regarded with skepticism for a long time.

In 1992, the first molecule, other than its isoforms, that is evolutionarily related to actin was identified (Lees-Miller et al. 1992; Schwob and Martin 1992). This molecule was called actin-related protein, abbreviated Arp. The discovery of Arp revealed that the actin family consists of actin isoforms and Arps. The first Arp identified was localized in the cytoplasm; however, some Arps were later found to be localized predominantly in the nucleus of a wide variety of organisms (Weber et al. 1995; Harata et al. 1999a), showing the involvement of actin family molecules in the organization and dynamics of the nucleus. In this essay, I will describe the progress of research on Arps localized in the nucleus.

10-2.2 General Aspects of Actin-Related Proteins (Arps)

Arps and actin appear to share a common ancestor, and they show 40%–70% similarity in their amino acids sequences to each other (Shafer and Schroer 1999). Compared to actin, the surface regions of Arp molecules show little conservation; however, Arps are supposed to be highly similar in their 3D structure in the core region including the ATP-binding pocket (Fig. 1) (Harata et al. 1994; Boyer and Peterson 2000). Therefore, the difference in the molecular surface, particularly the presence of specific insertions, most probably contributes to their unique functions, whereas properties based on the similarity in the ATP-binding pocket would be shared. Since the identification of the first Arp, various Arps have been found in a wide range of eukaryotic organisms with the progress of genome-sequencing projects. These comparative studies show that this protein family is much more divergent than previously thought. While analyzed Arps possess important and distinct functions from actin, the function of many of the Arps has not yet been addressed.

Budding yeast, whose entire genome sequence has been determined, is a particularly appropriate organism to classify and analyze the functions of Arps. The ten *ARP* genes of budding yeast have been classified according to their similarity to actin and are designated *ARP1* to *ARP10*, where Arp1 is the most similar and Arp10 is the least similar to actin (Poch and Winsor 1997).

Various Arps have been identified not only in yeast but also in higher eukaryotes. For example, about ten members of the Arp family have been reported so far in humans. The number of the Arps identified so far in humans seems to be small compared to the difference in genome size between yeast and humans. On the other hand, when we searched the human genome sequence with the amino acids sequence of actin or Arps, we identified more than 30 candidate sequences for human genes encoding Arps. While it is unknown whether all these sequences code for and express Arps, it is certain that many Arps still remain to be identified in higher organisms including humans.

10-2.3 Arps Localized in the Nucleus

In 1994, we reported the discovery of a third member of the Arp family in budding yeast, which we named Act3 (Harata et al. 1994). According to the current classification described above, Act3 is called Arp4. At the time, Arps were expected to be predominantly localized in the cytoplasm as

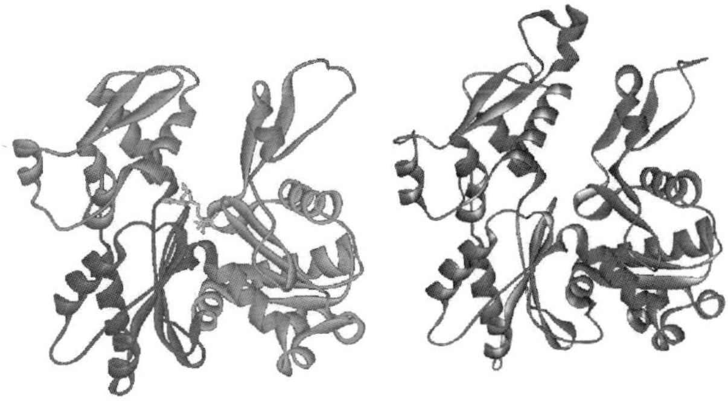

Fig. 1. The deduced three-dimensional (3D) structures of actin and Arp. The 3D structures of hArp6 (*right*) were deduced with the SWISS-Model program, visualized with a Swiss-PdbViewer (Peitsch 1995, 1996; Guex and Peitsch 1997), and compared with that of rabbit skeletal muscle actin (*left*). The adenosine triphosphate (ATP) molecule in the ATP-binding pocket of actin is indicated as a stick model

actin is. We found, however, that Arp4 was localized in the nucleus (Weber et al. 1995). This report was the first to show the presence of an Arp in the nucleus, and the possible involvement of Arps in the function of the chromatin and nucleus.

When the subcellular localizations of the other yeast Arps were investigated, unexpectedly more than half of them, including Arp5, Arp6, Arp7, Arp8, and Arp9, were predominantly localized in the nucleus (Harata et al. 2000), and Arp1, Arp2, Arp3, and Arp10 were observed in the cytoplasm. This suggests that the roles of Arps in the nucleus are not less significant than those in the cytoplasm.

We have also discovered Arps localized in the nucleus in mammals. In mammals, we identified two isoforms of an Arp that are putative orthologues of yeast Arp4, and designated them ArpNα (accession no. AB015906) and ArpNβ (accession no. AB015907), respectively (Harata et al. 1999a). ArpNβ is also called BAF53 as a 53-kDa component of mammalian SWI/SNF (BAF) complex (see the following sections) (Zhao et al.

1998). ArpNα and ArpNβ/BAF53 are extremely similar (97%) except for short amino acid stretches, called the α- and β-specific regions, respectively. Interestingly, while ArpNβ/BAF53 is expressed in all tissues tested and cultured cells, ArpNα was observed only in brain tissues; in addition, the expression of ArpNα was shown to occur during the neural differentiation of embryonic carcinoma (EC) cells (Harata et al. 1999a; Kuroda et al. 2002). In addition to ArpNα and ArpNβ/BAF53, recently we showed that human and chicken orthologues of budding yeast Arp6 (hArp6 and gArp6, respectively) are localized in the nucleus (Kato et al. 2000; Ohfuchi et al.2006).

10-2.4 Involvement of Arps in Nuclear and Chromatin Dynamics

Nuclear Dynamics and Protein Families

Nuclear dynamics are realized by collaboration of various protein families, including the histone and lamin families as well-known examples. ATP-dependent chromatin remodeling enzymes, which belong to the helicase family, are suggested to contribute to nuclear compartmentalization though chromatin modulation (Isogai and Tjian 2003). The finding of an Arp localized in the nucleus shows that the actin family is an attractive candidate for the protein families involved in nuclear dynamics. Recent biochemical approaches reveal that the nuclear Arps are components of complexes involved in chromatin modulation in both yeasts and vertebrates. In addition, in mammals, some Arps are suggested to be included in the nuclear architecture.

Chromatin Modification Complexes Containing Arps

Advances in chromatin research, especially using the budding yeast system, have revealed that at least two classes of protein complexes in the nucleus govern the modulation of chromatin structure: one class is the ATP-dependent chromatin remodeling complexes, the other is the histone modification complexes. These complexes consist of a catalytic subunit and various specific components. The catalytic subunits, the ATPases of the remodeling complexes and the HATs of the modifying complexes, are responsible for the enzymatic reactions. While many of the other components that are required for the regulation of the function of these complexes are diverse, actin-related proteins (Arps) were consistently found in multiple chromatin remodeling and HAT complexes from yeast to mammals.

In budding yeast, NuA4 HAT complex contains Arp4 together with actin (Galarneau et al. 2000). In chromatin remodeling complexes, Ino80 complex contains Arp4, Arp5, and Arp8 together with actin (Shen et al. 2000), and both SWI/SNF and the RSC complexes contain Arp7 and Arp9 (Cairns et al. 1998; Peterson et al. 1998). Swrl complex, which replaces histone H2A with the variant H2AZ, was shown to contain Arp4 and Arp6 together with actin (Mizuguchi et al. 2004). Importantly, in most cases examined, Arps are required for the activity of these complexes (Galarneau et al. 2000; Shen et al. 2003; Görzer et al. 2003).

In mammals, ArpNβ/BAF53 is a component of various chromatin remodeling and related complexes, including SWI/SNF (BAF) (Zhao et al. 1998), SWI/SNF-B (PBAF) (Nie et al. 2000), p400 (Fuchs et al. 2001), cMyc-associated (Park et al. 2002), and WINAC complexes (Kitagawa et al. 2003). TIP60 HAT complex also contains ArpNβ/BAF53 (Ikura et al. 2000). Because of its brain-specific expression (Harata et al. 1999a), ArpNα was not present in these isolated complexes; however, exogenously expressed ArpNα was co-immunoprecipitated with BRM, an enzymatic component of mammalian SWI/SNF (BAF) complex, as well as with ArpNβ/BAF53 (Kuroda et al. 2002). Recently, a SWI/SNF-like neuron-specific chromatin remodeling complex (bBAF) was identified, and its neuron-specific component was identical to ArpNα (Olave et al. 2002). These results, taken together with their extremely similar amino acid sequences, suggest that both ArpNα and ArpNβ/BAF53 are mutually selective components of various complexes involved in chromatin modulation.

Arps and Nuclear Architecture

Their association with the functions of the actin family in cytoplasm implies that nuclear Arps are involved not only in chromatin modulation but also in nuclear architecture. Human nuclear Arps, hArpNβ/BAF53 and hArp6, were shown to be contained in the nuclear matrix, which is expected to represent a nuclear architecture (Sung et al. 2001). Proteomic analysis revealed that human hArp2 and hArp3 are components organizing the nucleolus (Andersen et al. 2002). However, information concerning the relationship between nuclear architecture and Arps is still limited, probably because of the difficulties in analyzing proteins organizing nuclear architecture. It is expected that progress in the proteomic research of nuclear architecture will provide important information.

10-2.5 Molecular Mechanisms by Which Arps Contribute to Nuclear and Chromatin Dynamics

The molecular mechanisms of the nuclear Arps are still largely unknown. However, interesting reported characters of Arps suggest their possible molecular mechanisms in nuclear and chromatin dynamics.

Intermolecular Interactions

Actin is known to interact with many kinds of proteins including various actin-binding proteins, and Arps are also thought to share this property. However, the nuclear Arps possess specific insertions in the molecular surface, and they are thought to show distinct intermolecular interactions from actin.

Insertion II, one of the specific insertions of the budding yeast Arp4, interacts with core histones (Harata et al. 1999b), and this property is supposed to contribute to the recruitment of the complexes containing Arp4 (for example, NuA4 HAT and Ino80 chromatin remodeling complexes) onto chromatin (Galarneau et al. 2000; Harata et al. 2002). Recently, the budding yeast Arp8 was also shown to interact with core histones (Shen et al. 2003). It seems likely the histone-binding ability is conserved in multiple Arps localized in the nucleus.

In mammals, nuclear Arps were shown to interact with a transcriptional coregulator. We showed that the α-specific region of ArpNα interacted with the transcriptional corepressor CtBP (C-terminal binding protein) (Oma et al. 2003). On the other hand, the β-specific region of ArpNβ/BAF53, the isoform of ArpNα, coprecipitated with HAT activity (Park et al. 2002). Taken together with the brain specific expression of ArpNα, such a difference in the intermolecular interactions of Arp-isoforms would contribute to the function and/or development of the brain in mammals.

Role of the ATP-Binding Pocket in Arps

As described above, actin and Arp share the conserved ATP-binding pocket. While ATP binding ability and its regulatory role in the functions of the respective cytoplasmic complexes have been shown for cytoplasmic Arps, ATP binding activity of nuclear Arps has been controversial. Recently we demonstrated that the budding yeast Arp4 is able to bind ATP. In addition, we showed that the ATP binding dissociated Arp4 from the other components of the nuclear complexes and, on the other hand, that the inhibition of ATP binding increased the amount of Arp4 that is incorporated into the complexes (Sunada et al. 2005). Since the presence of Arp4 is essential for the activities of the complex, the association/dissociation of Arp4 caused by the ATP-binding pocket is expected to be connected with the regulation of the complex function. Taken together with the observation that the assembly/disassembly of actin filaments is regulated by the exchange of bound adenine nucleotide to actin, it is likely that the ATP-binding pocket of the nuclear Arps performs a similar function in the organization and functional regulation of nuclear complexes. This property might explain the presence of Arps in many complexes involved in chromatin modulation, whose activity and organization are dynamically regulated by intracellular signal transduction and cell cycle progression. In addition, Arps are possible components of the nuclear architecture (see above), and therefore, the nuclear Arps might contribute to the regulation of nuclear dynamics through assembly/disassembly of the nuclear architecture.

10-2.6 Perspectives

Recent analyses of nuclear Arps suggest their involvement in chromatin and nuclear dynamics as components of chromatin modulation complexes and the nuclear architecture. Although nuclear dynamics are realized by the collaboration of many molecules, further analyses of nuclear Arps will disclose novel molecular mechanisms in the maintenance and regulation of nuclear dynamics.

References

Andersen JS, Lyon CE, Fox AH, Leung AK, Lam YW, Steen H, Mann M, Lamond AI (2002) Directed proteomic analysis of the human nucleolus. Curr Biol 12:1–11

Boyer LA, Peterson CL (2000) Actin-related proteins (Arps): conformational switches for chromatin-remodeling machines? BioEssays 22:666–672

Cairns BR, Erdjument-Bromage H, Tempst P, Winston F, Kornberg RD (1998) Two actin-related proteins are shared functional components of the chromatin-remodeling complexes RSC and SWI/SNF. Mol Cell 2:639–651

Fuchs M, Gerber J, Drapkin R, Sif S, Ikura T, Ogryzko V, Lane WS, Nakatani Y, Livingston DM (2001) The p400 complex is an essential E1A transformation target. Cell 106:297–307

Galarneau L, Nourani A, Boudreault AA, Zhang Y, Heliot L, Allard S, Savard J, Lane WS, Stillman DJ, Côté J (2000) Multiple links between the NuA4 histone acetyltransferase complex and epigenetic control of transcription. Mol Cell 5:927–937

Görzer I, Schüller C, Heidenreich E, Krupanska L, Kuchler K, Wintersberger U (2003) The nuclear actinß-related protein Act3p-Arp4 of *Saccharomyces cerevisiae* is involved in transcription regulation of stress genes. Mol Microbiol 50:1155–1171

Guex N, Peitsch MC (1997), SWISS-MODEL and the Swiss-PdbViewer: an environment for comparative protein modelling. Electrophoresis 18:2714–2723

Harata M, Karwan A, Wintersberger U (1994) An essential gene of *Saccharomyces cerevisiae* coding for an actin-related protein. Proc Natl Acad Sci USA 91:8258–8262

Harata M, Mochizuki R, Mizuno S (1999a) Two isoforms of a human actin-related protein show nuclear localization and mutually selective expression between brain and other tissues. Biosci Biotechnol Biochem 63:917–923

Harata M, Oma Y, Mizuno S, Jiang YW, Stillman DJ, Wintersberger U (1999b) The nuclear actin-related protein of *Saccharomyces cerevisiae,* Act3p/Arp4, interacts with core histones. Mol Biol Cell 10:2595–2605

Harata M, Oma Y, Tabuchi T, Zhang Y, Stillman DJ, Mizuno S (2000) Multiple actin-related proteins of *Saccharomyces cerevisiae* are present in the nucleus. J Biochem 128:665–671

Harata M, Zhang Y, Stillman DJ, Matsui D, Oma Y, Nishimori K, Mochizuki R (2002) Correlation between chromatin association and transcriptional regulation for the Act3p/Arp4 nuclear actin-related protein of *Saccharomyces cerevisiae*. Nucleic Acids Res 30:1743–1750

Ikura T, Ogryzko VV, Grigoriev M, Groisman R, Wang J, Horikoshi M, Scully R, Qin J, Nakatani Y (2000) Involvement of the TIP60 histone acetylase complex in DNA repair and apoptosis. Cell 102:463–473

Isogai Y, Tjian R (2003) Targeting genes and transcription factors to segregated nuclear compartments. Curr Opin Cell Biol 15:296–303

Kato M, Sasaki M, Mizuno S, Harata M (2001) Novel actin-related proteins in vertebrates: similarities of structure and expression pattern to Arp6 localized on *Drosophila* heterochromatin. Gene 268:133–140

Kitagawa H, Fujiki R, Yoshimura K, Mezaki Y, Uematsu Y, Matsui D, Ogawa S, Unno K, Okubo M, Tokita A, Nakagawa T, Ito T, Ishimi Y, Nagasawa H, Matsumoto T, Yanagisawa J, Kato S (2003) The chromatin-remodeling complex WINAC targets a nuclear receptor to promoters and is impaired in Williams syndrome. Cell 113:905–917

Kuroda Y, Oma Y, Nishimori K, Ohta T, Harata M (2002) Brain-specific expression of the nuclear actin-related protein ArpNα and its involvement in mammalian SWI/SNF chromatin remodeling complex. Biochem Biophys Res Commun 299:328–334

Lees-Miller JP, Henry G, Helfman DM (1992) Identification of act2, an essential gene in the fission yeast *Schizosaccharomyces pombe* that encodes a protein related to actin. Proc Natl Acad Sci USA 89:80–83

Nie Z, Xue Y, Yang D, Zhou S, Deroo BJ, Archer TK, Wang W (2000) A specificity and targeting subunit of a human SWI/SNF family-related chromatin-remodeling complex. Mol Cell Biol 23:8879–8888

Ohfuchi E, Kato M, Sakaki M, Sugimoto K, Oma Y, Harata M. (2006) Vertebrate Arp6, a novel nuclear actin-related protein, interacts with heterochromatin protein 1. Eur J Cell Biol 85: 411-421

Olave I, Wang W, Xue Y, Kuo A, Crabtree GR (2002) Identification of a polymorphic, neuron-specific chromatin remodeling complex. Genes Dev 16:2509–2517

Oma Y, Nishimori K, Harata M (2003) The brain-specific actin-related protein ArpN alpha interacts with the transcriptional co-repressor CtBP. Biochem Biophys Res Commun 301:521–528

Park J, Wood MA, Cole MD (2002) BAF53 forms distinct nuclear complexes and functions as a critical c-Myc-interacting nuclear cofactor for oncogenic transformation. Mol Cell Biol 22:1307–1316

Peitsch MC (1995) Protein modeling by E-mail. Bio/Technology 13:658–660

Peitsch MC (1996) ProMod and Swiss-Model: Internet-based tools for automated comparative protein modeling. Biochem Soc Trans 24:274–279

Peterson CL, Zhao Y. Chait BT (1998) Subunits of the yeast SWI/SNF complex are members of the actin-related protein (ARP) family. J Biol Chem 273:23641–23644

Poch O, Winsor B (1997) Who's who among the *Saccharomyces cerevisiae* actin-related proteins? A classification and nomenclature proposal for a large family. Yeast 13:1053–1058

Schafer DA, Schroer TA (1999) Actin-related proteins. Annu Rev Cell Dev Biol 15:341–363

Schwob E, Martin RP (1992) New yeast actin-like gene required late in the cell cycle. Nature 355:179–182

Shen X, Mizuguchi G, Hamiche A, Wu C (2000) A chromatin remodelling complex involved in transcription and DNA processing. Nature 406:541–544

Shen X, Ranallo R, Choi E, Wu C (2003) Involvement of actin-related proteins in ATP-dependent chromatin remodeling. Mol Cell 12:147–155

Sunada R, Gorzer I, Oma Y, Yoshida Y, Suka N, Wintersberger U, Harata M. (2005) The nuclear actin-related protein Act3p/Arp4p is involved in the dynamics of chromatin-modulating complexes. Yeast 22:753-768

Weber V, Harata M, Hauser H, Wintersberger U (1995) The actin-related protein Act3p of *Saccharomyces cerevisiae* is located in the nucleus. Mol Biol Cell 6:1263–1270

Sung YH, Choi EY, Kwon H (2001) Identification of a nuclear ArpN as a component of human SWI/SNF complex and its selective association with a subset of active genes. Mol Cells 11:75–81

Zhao K, Wang W, Rando OJ, Xue Y, Swiderek K, Kuo A, Crabtree GR (1998) Rapid and phosphoinositol-dependent binding of the SWI/SNF-like BAF complex to chromatin after T lymphocyte receptor signaling. Cell 95:625–636

10-3. Effects of 5-Bromodeoxyuridine on Chromatin Structure

Dai Ayusawa

Kihara Institute for Biological Research, Yokohama City University, Mai-oka-cho 641-12, Totsuka-ku, Yokohama 244-0813, Japan

10-3.1 Introduction

Normal human cells in culture enter senescence upon passages or treatment with various agents. Among the treatments, 5-bromodeoxyuridine (BrdU) is unique in that it most clearly induces a senescence-like phenomenon in every type of mammalian cells (Michishita et al. 1999). This system ensures genetic and biochemical approaches to study cellular senescence. In contrast, normal cells stop proliferation by stochastic accumulation of senescent cells over a long period of culture. 5-Bromodeoxyuridine is well known as a modulator of cellular differentiation. Therefore, studies of the molecular basis for BrdU facilitate understanding of chromatin-mediated cellular differentiation and cellular senescence.

We have extensively characterized BrdU-responsive genes using a polymerase chain reaction (PCR)-based subtractive hybridization (Suzuki et al. 2001) and DNA microarrays (Minagawa et al. 2005) in HeLa and SUSM-1. As expected, expression of the genes behaved similarly in normal senescent cells although far less pronounced than in BrdU-treated cells. These analyses give a whole picture of time-dependent expression of senescence-associated genes. Since molecular genetic analysis is possible with the BrdU-inducible system, it will provide a clue to solve the new and old question why BrdU modulates expression of particular genes.

10-3.2 Distamycin A Potentiates the Actions of BrdU

Distamycin A, which specifically binds to the narrow minor grooves of AT-tracts, dramatically potentiates the effects of BrdU. Distamycin A alone has only a marginal effect on cellular senescence. Netropsin, Hoechst 33258, and DAPI show similar effects whereas chromomycin A3, which binds to GC-rich sequences, has no effect. The high-mobility group protein HMGI/Y binds to the minor groove of AT-tracts through a conserved DNA-binding peptide motif called AT hook (Reeves 2001), and AT-hook domains potentiate the effects of BrdU (Satou et al. 2004). HMGI/Y induces structural changes in chromatin and formation of stereospecific complexes called enhanceosomes, but C-terminal or linker domains involved in assembly of enhanceosome or protein–protein interaction are not required for the above effect.

10-3.3 5-Bromouracil Substitutions Change Topology of MAR/SAR DNA

We hypothesize that BrdU targets a certain type of AT-rich sequences, such as matrix/scaffold attached regions (MAR/SARs). MAR/SARs are very AT-rich sequences of several hundred base pairs, preferentially located on G-bands and heterochromatin, and specifically associated with the nuclear matrix (NM). MAR/SAR sequences bend along the axis of their double-stranded DNA helices and present specific binding sites for AT-minor groove binding ligands. They strongly affect expression of genes and chromosome dynamics with DNA-binding proteins such as histone H1 and HMG-I/Y.

We examined some properties of MAR/SAR DNA of rat satellite DNA origin (Nakamura et al. 1991) upon substitution of thymine with 5-bromouracil by PCR with bromodeoxyuridine triphosphate (BrdUTP). DNase I footprinting analysis revealed two classes of binding sites to distamycin A. One class contained members with high affinity to distamycin A, and their binding capacity to distamycin A was not affected by 5-bromouracil substitutions. The other class contained ones with low affinity to distamycin A, and their binding capacity to distamycin A was weakened by 5-bromouracil substitutions.

The MAR/SAR DNA was subjected to conventional polyacrylamide gel electrophoresis (Fig. 1). When run at 4°C, the MAR/SAR DNA migrated slower depending on the extent of 5-bromouracil substitutions. At 52°C, no difference was observed. When incubated with distamycin A before electrophoresis, the thymine-containing DNA migrated faster as its concentrations increased. In contrast, the 5-bromouracil-containing DNA was less affected by distamycin A. In agreement with the results of DNase

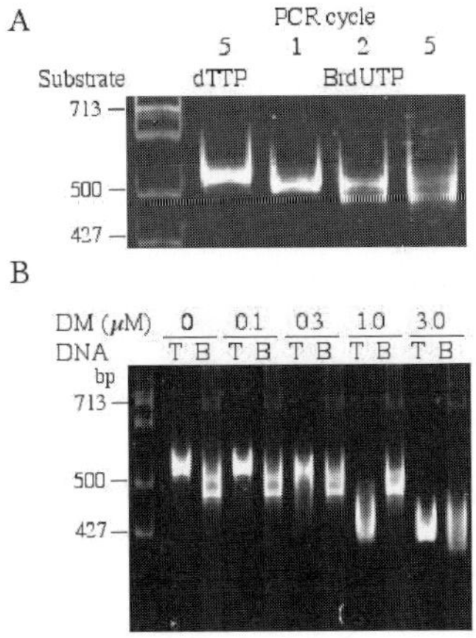

Fig. 1. Effects of 5-bromouracil substitutions on the topology of matrix/scaffold attached region (MAR/SAR) DNA. Polyacrylamide gel electrophoresis of MAR/SAR DNA. Rat MAR/SAR DNA (Nakamura et al. 1991) was amplified by polymerase chain reaction (*PCR*) with deoxythymidine triphosphate (*dTTP*) or bromodeoxyuridine triphosphate (*BrdUTP*) for the cycles indicated, and run on polyacrylamide gel at 4°C. Size marker, *Hin*fI digest of ϕX174 DNA. Polyacrylamide gel electrophoresis of MAR/SAR DNA after incubation with distamycin A. The same DNA was amplified by five cycles of PCR with dTTP (*T*) or BrdUTP (*B*), incubated with increasing concentrations of distamycin A (*DM*), and run on polyacrylamide gel at 4°C

I footprinting analysis, the 5-bromouracil-containing DNA behaved differently from its thymine-containing counterpart.

10-3.4 Interaction of MAR/SAR DNA with the NM

The NM plays an important role in higher-order organization of chromatin structure to tether actively transcribed genes via MAR/SARs. Any alteration in the interaction between MAR/SAR s and the NM will lead to altered expression of linked genes.

We examined interaction between nuclear halos (NM) of HeLa cells and an intronic MAR/SAR sequence of the IgH gene enhancer (Ogino et al. 2000). The end-labeled MAR/SAR DNA was mixed with the halos, incubated with an excess amount of carrier DNA, and then centrifuged to yield pellet and supernatant fractions. DNA was purified and electrophoresed on a polyacrylamide gel. Upon substitution of thymine with 5-

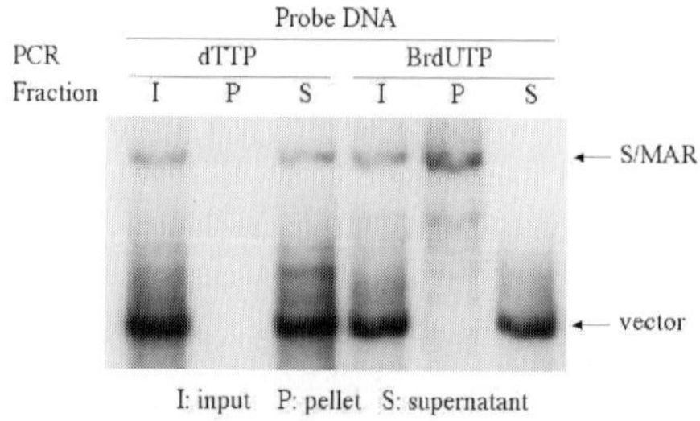

Fig. 2. Binding of MAR/SAR DNA to nuclear halos. Duplex oligonucleotide (5′-TCTTTAATTTCTAATATATTTAGAA-3′) derived from an intronic MAR/SAR of the IgH gene enhancer was ligated to heptamer, and cloned into plasmid to serve as a template. DNA was amplified by PCR with dTTP or BrdUTP using T3 and T7 primers. Equal amounts of radiolabeled DNA (*I*) were incubated with nuclear halos (7×10^6 cell equivalents) prepared from HeLa cells, and centrifuged to yield pellet (*P*) and supernatant (*S*) fractions. Equal counts (1000 cpm) of purified DNA were run on 5% polyacrylamide gel, and an autoradiogram was prepared

bromouracil, the proportion of the MAR/SAR DNA in the pellet fractions markedly increased (Fig. 2). Control DNA did not bind to the nuclear halos under any conditions.

Then, we examined affinity of the end-labeled rat MAR/SAR DNA to nuclear scaffold proteins by filter binding assay (Suzuki et al. 2003). When *Escherichia coli* DNA was used as a competitor, the thymine- and 5-bromouracil-containing DNA bound to the proteins similarly (Fig. 3). In contrast, when thymine- or 5-bromouracil-containing HeLa DNA was used as a competitor, the 5-bromouracil-containing MAR/SAR DNA bound more strongly to the proteins than the thymine-containing counterpart. Distamycin A inhibited the binding dose-dependently when *E. coli* DNA was used as a competitor. However, when HeLa DNA was used as a competitor, distamycin A did not affect the binding of the 5-bromouracil-containing MAR/SAR DNA whereas high concentrations of distamycin A inhibited the binding of the thymine-containing counterpart. When 5-bromouracil-containing HeLa DNA was used as a competitor, distamycin A did not inhibit the binding of the 5-bromouracil-containing MAR/SAR DNA, but did significantly the thymine-containing counterpart. *Escherichia coli* and HeLa DNA differ greatly in GC contents and/or distribution of AT-tracts. These results suggest that 5-bormouracil substitutions change the binding properties of

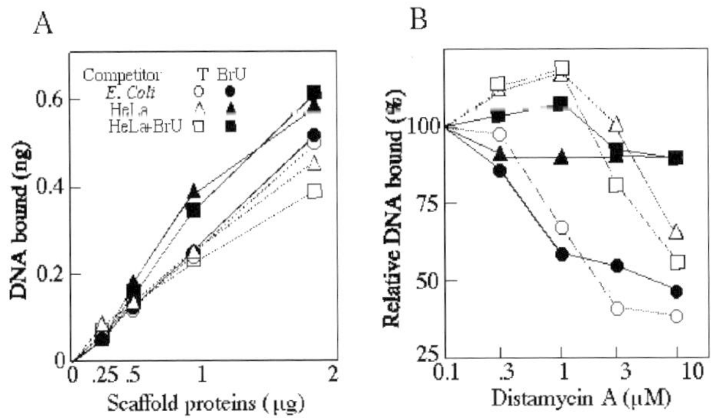

Fig. 3A,B. Binding properties of MAR/SAR DNA to nuclear scaffold proteins. **A** Effects of 5-bromouracil substitutions on the binding of rat MAR/SAR DNA to nuclear scaffold proteins. The rat MAR/SAR DNA containing thymine or 5-bromouracil as in Fig. 1B was end-labeled with ^{32}P and incubated with increasing amounts of scaffold proteins in the presence of competitor DNA. HeLa DNA containing 5-bromouracil (*HeLa+BrU*) was prepared by culturing the cells with 50 μM BrdU for 1 week. Radioactivity bound to nuclear scaffold proteins was determined by an imaging analyzer. The values are averages of triplicate assays. **B** Effects of distamycin A on the binding of rat MAR/SAR DNA to nuclear scaffold proteins. The DNA was incubated with a fixed amount of scaffold proteins and competitor DNA in the presence of increasing concentrations of distamycin A, and processed as in **A**

MAR/SAR DNA to the NM both in the presence and absence of distamycin A.

10-3.5 Discussion

AT-binding ligands induce chromosome decondensation and position effect variegation, and inhibit DNA bending, nucleosome assembly, and binding of MAR/SAR DNA to the NM. 5-Bromodeoxyuridine shares the above phenomena with AT-binding ligands except that the DNA binding to the NM is strengthened. This increased and persistent interaction between MAR/SAR DNA and the NM under the conditions favoring chromosome decondensation may account for the synergism by BrdU and AT-binding ligands.

The presence of a bromine atom at the 5′ position of uracil in double-stranded DNA may provide an altered environment for DNA binding proteins. Because bromine is more electronegative than a methyl group, enforced coulombic attraction might generally stabilize interaction between DNA and proteins. Alternatively, bromine substitution might induce a change in DNA topology due to increase in neighboring base–base interaction or stacking of base pairs. Recently, DNA topology is shown to involve nucleosome assembly and positioning.

References

Girard F, Bello B, Laemmli UK, Gehring WJ (1998) In vivo analysis of scaffold-associated regions in Drosophila: a synthetic high-affinity SAR binding protein suppresses position effect variegation. EMBO J 17:2079–2085

Michishita E, Nakabayashi K, Suzuki T, Kaul SC, Ogino H, Fujii M, Mitsui Y, Ayusawa D (1999) 5-Bromodeoxyuridine induces senescence-like phenomena in mammalian cells regardless of cell type or species. J Biochem 126:1052–1059

Minagawa S, Nakabayashi K, Fujii M, Scherer SW, Ayusawa D (2004) Functional and chromosomal clustering of genes responsive to 5-bromodeoxyuridine in human cells. Exp Gerontol 39: 1069-1078

Nakamura K, Ikeda Y, Iwakami N, Hibino Y, Sugano N (1991) Bending of a highly repetitive component in rat nuclear DNA. Biochem Int 25:355–362

Ogino H, Fujii M, Satou W, Suzuki T, Michishita E, Ayusawa D (2002) Binding of 5-bromouracil-containing S/MAR DNA to the nuclear matrix. DNA Res 9:25–29

Reeves R (2001) Molecular biology of HMGA proteins: hubs of nuclear function. Gene 277:63–81

Satou W, Suzuki T, Noguchi T, Ogino H, Fujii M, Ayusawa D (2004) AT-hook proteins stimulate induction of senescence markers triggered by 5-bromodeoxyuridine in mammalian cells. Exp Gerontol 39:173–179

Suzuki T, Minagawa S, Michishita E, Ogino H, Fujii M, Mitsui Y, Ayusawa D (2001) Induction of senescence-associated genes by 5-bromodeoxyuridine in HeLa cells. Exp Gerontol 36:465–474

Suzuki T, Michishita E, Ogino H, Fujii M, Ayusawa D (2002) Synergistic induction of the senescence-associated genes by 5-bromodeoxyuridine and AT-binding ligands in HeLa cells. Exp Cell Res 276:174–184

10-4. Transcriptional Modulation by Nuclear Matrix Protein P130/MAT3 Associated with MAR/SAR

Yasuhide Hibino[1], Tatsuhiro Usui[1], and Koichi Hiraga[2]

[1]Department of Clinical Dietetics and Human Nutrition, Faculty of Pharmaceutical Sciences, Josai University, 1-1 Keyakidai, Sakado, Saitama 350-0295, Japan

[2]Department of Biochemistry, School of Medicine, Toyama University, 2630 Sugitani, Toyama 930-0194, Japan

10-4.1 Nuclear Architecture/Genomic DNA

Various types of proteins are required for nuclear organization, which confers a variety of nuclear functions in eukaryotic cells. A set of proteins forms the chromosomal backbone (Paulson and Laemmli 1977) and different sets of proteins including chromatin-remodeling factors regulate the utilization of genetic code in chromosomal DNA (Boulikas 1995; Moazed 2001; Muchardt and Yaniv 2001). Eukaryotic chromosomes are topologically attached to the nuclear matrix (NM) or scaffold, a network of protein fibers referred to as the skeletal framework of the nucleus. The NM or scaffold is operationally defined as the residual structures that remain insoluble after extraction of nuclei with a high concentration of either salt or detergent. Several members of the group of proteins classified as components of the NM can directly bind to particular segments of chromosomal DNA. A DNA segment to which matrix and scaffold proteins can bind is termed a matrix or scaffold attachment region (MAR/SAR) (Cockerill and Garrard 1986; Gasser and Laemmli 1986; Cockerill et al. 1987; Jarman and Higgs 1988). One significant role of MAR/SARs that was revealed by structural analyses of the nucleus and chromosomes is the matrix- or scaffold-mediated stabilization of chromosomal structure. Chromosomal loops with an approximate length of ~60 kilobases yield compact configurations of chromosomes. In the interphase cell nucleus, chromatin is arranged in spatially separate, chromosome-specific territories, which are maintained intact on the NM as a three-dimensional (3D) organization of chromosomes (Zink et al. 1998; Ma et al. 1999). The organized structure of chromosomes is likely modulated in association with transcription, replication, and RNA splicing, although mechanisms of the preferential alteration of chromosomal structure are largely unknown at present.

Properties of Highly Repetitive DNA and MAR/SAR

Highly repetitive DNA components have been shown to be located preferentially in the heterochromatic region of the interphase nucleus and are localized at the centromere or telomere of metaphase chromosomes (Singer 1982). Such repetitive components have been shown to have some properties characteristic of bent DNA (Martínez-Balbás, et al. 1990; Hibino et al. 1993) and are known to be nuclear MAR/SAR. In fact, these components are enriched in the NM or nuclear scaffold (Razin et al. 1979; Small, et al. 1982). Several investigators have demonstrated that A runs or AT tracts in DNA are capable of bending the helix axis (Hagerman 1990). Radic et al. (1987) showed that mouse satellite DNA (AvaII repeat) consisting of many A runs is located throughout the centromeric region of the metaphase chromosome and contains a stable curvature, which can be alleviated by distamycin A specific for AT-rich DNA. Accordingly, the bend in the repetitive component is recognized by a non-histone nuclear protein, which may be directly involved in centromeric heterochromatin condensation. In addition, Shrader and Crothers (1990) have suggested that the sequence-directed bending of the DNA helix axis plays a role in nucleosome formation.

DNA Methylation

In the animal genome, cytosine residues in the sequence 5′CpG are often post-synthetically methylated. Cytosine methylation is the most common modification of DNA found in nature and has been implicated in DNA repair, chromatin organization, transcription, X-chromosome inactivation, transposition, recombination, mutagenesis, replication, and genomic imprinting. Differential methylation of Alu repeats has been associated with genomic imprinting in germ line cells or reflect differences in Alu transcriptional activity. In vitro, sperm Alu-binding protein has been shown to be sufficiently specific to selectively protect Alu CpGs from methylation and to be responsible for the unmethylated state of Alu sequences in the male germ line (Chesnokov and Schmid 1995). Moreover, it has been reported that the methyl-CpG-binding proteins interact specifically with methylated DNA and mediate transcriptional repression (Boyes and Bird 1991; Nan et al. 1997).

10-4.2 Properties of MAR/SAR-Binding Protein, P130/MAT3

It has generally been accepted that nuclear MARs/SARs partition the chromatin structure into functional loop domains and that some non-histone proteins are nuclear elements, essential for organizing the loops. Such MAR/SAR-binding proteins have already been found in several eukaryotic cells. Romig et al. (1992) showed that a nuclear scaffold protein from HeLa cells, SAF-A (scaffold attachment factor A), which has an apparent molecular weight of 120 kDa, binds to AT-rich sites in several SARs from vertebrate cells and that this protein is capable of forming large aggregates which mediate construction of looped DNA. It has also been shown that SAF-A is identical to a component of hnRNP particles (hnRNP-U), and the suggestion made that hnRNP-U/SAF-A functions in higher-order chromatin organization and hnRNA metabolism (Fackelmayer et al. 1994).

Hibino et al. (1993) purified P130 from P fraction, which is a residual fraction prepared from rat liver nuclei by the removal of proteins extractable with digitonin and lithium salicylate, and of DNA readily hydrolyzable with DNase, according to specific binding of P130 to a highly repetitive DNA component, the *Xmn*I fragment, which is also cloned from DNA included in the rat liver nuclear scaffold (Ikeda et al. 1990). We found that P130 is homologous to matrin 3 which is abundant in the nuclear scaffold fraction and has two zinc finger motifs and two RNA-binding domains (Hibino et al. 1998a). These results indicated that P130/matrin 3 (P130/MAT3) might be involved in the transcription and/or RNA process. Hibino et al. (2000) also suggested by Southwestern analysis that in addition to binding to the *Xmn*I fragment, P130/MAT3 bound to various types of MAR/SAR segments by recognizing an ATATAT sequence, defined as a base-unpairing region (Dickinson et al. 1992), as a binding site. P130/MAT3 is a different phosphorylated form of P123 with an apparent size of 123 kDa, and is more readily bound to the *Xmn*I fragment than P123 (Hibino et al. 1998b). The *Xmn*I fragment can be methylated both in vivo and in vitro on cytosine of the unique *Hin*dIII site. In a

Southwestern analysis, the methylated *Xmn*I fragment did not bind to P130/MAT3, and instead bound to an 83-kDa polypeptide. Therefore, the interaction of the *Xmn*I segments in the genome with a nuclear scaffold appeared to be regulated in two distinct fashions: the phosphorylation state of P130/MAT3, and the methylation and demethylation of the *Hin*dIII site of the *Xmn*I segments as MAR/SARs. On the other hand, animal and plant genomes carry MAR/SARs, which also interact with NM. A MAR/SAR positioned in the vicinity of a promoter can reportedly augment transcription of a reported gene, when the recombinant construct is integrated into the genome of the host animal, suggesting that the MAR/SAR plays a role in modulating transcription of nuclear genes (Xu et al. 1989; Forrester et al. 1994; Jenuwein et al. 1997). In addition, methylation and demethylation of MAR/SARs appeared to regulate chromatin remodeling in transcription regulation of nuclear genes (Poljak et al. 1994). Therefore, we examined whether, like the MAR/SAR, the native and methylated *Xmn*I fragment can modulate the activity of transcription through interaction with proteins. Hibino et al. (2000) demonstrated that the *Xmn*I fragment augmented the SV40 promoter-mediated luciferase gene transcription determined by transiently expressed luciferase activity in various types of cells. Moreover, the *Xmn*I fragment with the methylated *Hin*dIII site did not augment this transcription. As the binding of the *Xmn*I fragment and P130/MAT3 in vitro is inhibited by methylation of the cytosine base at this *Hin*dIII site, the suppression of the transcription augmentation by methylation indicates the requirement of P130/MAT3 for processes of transcription augmentation. Further, as the *Xmn*I fragment bound to P130/MAT3 in Southwestern analysis, various types of MAR/SAR fragments bound to two nuclear scaffold proteins with sizes similar to those of P130/MAT3 and its isoform, P123. MAR/SAR segments in recombinant plasmids, therefore, can interact with P130/MAT3, and the interaction of MAR/SAR with P130/MAT3 appeared to be required for augmentation of luciferase gene transcription. Thus, this study for the first time demonstrated the similarity of a highly repetitive DNA component to a MAR/SAR in interaction with a scaffold protein, P130/MAT3, by which an appropriately positioned promoter is activated. Of the NM proteins, SATB1 (special AT-rich sequence binding protein 1) and SAF-B, both of which can bind to a MAR/SAR, suppressed transcription activity, when bound to elements stably integrated together with a reporter construct into the genome (Kohwi-Shigematsu et al. 1997; Li et al. 1996; Nayler et al. 1998). An AT-rich sequence has been assigned to a base-unpairing region necessary for the interaction of a MAR/SAR fragment with SATB1, in the thymus and testis (Dickinson et al. 1992). Due to, at least in part, this property, SATB1 orchestrates the temporal and spatial expression of genes during T-cell development as a global regulator of cell function in specific cell lineages (Alvarez et al. 2000; Cai et al. 2003). P130/MAT3, present in all tissues examined including the thymus, bound to DNA having the synthetic ATATAT sequence, but not to variant DNA, indicating that P130/MAT3, distinct from SATB1, likewise requires base unpairing to recognize its binding site in genomic DNA. The binding of P130/MAT3 to the *Xmn*I and MAR/SAR fragments might result in circumstances under which proteins required to form the structural and functional basis of chromatins are more efficiently assembled. If this is the case, then the transcription augmentation observed is considered to reflect the binding efficacy of P130/MAT3 and the highly repetitive DNA component or MAR/SAR on assembly of transcription machinery.

Methylation of the *Hin*dIII site in the *Xmn*I fragment appears likely to regulate, with the aid of demethylation, the reversible appearance and disappearance of the binding sites of P130/MAT3 on chromosomal DNA. Therefore, methylation of the *Hin*dIII site of highly repetitive DNA components might be one of the underlying mechanisms of chromatin remodeling in which P130/MAT3 can play a role, presumably together with the methyl-CpG-binding protein (MeCP2) with a size of 83 kDa. Three global mechanisms of gene regulation, DNA methylation, histone deacetylation and histone methylation, are postulated to be linked by MeCP2 (Nan et al. 1998; Fuks et al. 2003). It has already been demonstrated that chicken erythrocyte histone deacetylase forms a complex with proteins that are components of the NM (Li et al. 1996). Weitzel et al. (1997) reported that ARBP, an abundant NM protein with a high affinity for the MAR/SAR, is homologous to the rat protein MeCP2. ARBP and MeCP2 are localized in a repetitive sequence of pericentromeric heterochromatin in mouse chromosomes. These results indicate that the MAR/SAR and its binding proteins may function in methylation/demethylation of DNA and region-specific histone modification processes and play an important role in the regulation of gene expression by binding to sites for the specific organization of DNA sequences under different conditions.

Detailed characterization of the structure, function, and subcellular localization of P130/MAT3 and of the precise distribution of highly repetitive DNA components over the chromosomal DNA is required to clarify the interesting functions of the nuclear scaffold protein P130/MAT3.

References

Alvarez JD, Yasui DH, Niida H, Joh T, Loh DY, Kohwi-Shigematsu T (2000) The MAR-binding protein SATB1 orchestrates temporal and spatial expression of multiple genes during T-cell development. Genes Dev 14:521–535

Boulikas T (1995) Chromatin domains and prediction of MAR sequences. Int Rev Cytol 162A:279–388

Boyes J, Bird A (1991) DNA methylation inhibits transcription indirectly via a methyl-CpG binding protein. Cell 64:1123–1134

Cai S, Han H-J, Kohwi-Shigematsu T (2003) Tissue-specific nuclear architecture and gene expression regulated by SATB1. Nat Genet 34:42–51

Chesnokov IN, Schmid CW (1995) Specific Alu binding protein from human sperm chromatin prevents DNA methylation. J Biol Chem 270:18539–18542

Cockerill PN, Garrard WT (1986) Chromosomal loop anchorage of the kappa immunoglobulin gene occurs next to the enhancer in a region containing topoisomerase II sites. Cell 44:273–282

Cockerill PN, Yuen M-H, Garrard WT (1987) The enhancer of the immunoglobulin heavy chain locus is flanked by presumptive chromosomal loop anchorage elements. J Biol Chem 262:5394–5497

Dickinson LA, Joh T, Kohwi Y, Kohwi-Shigematsu T (1992) A tissue-specific MAR/SAR DNA-binding protein with unusual binding site recognition. Cell 70:631–645

Fackelmayer FO, Dahm K, Renz A, Ramsperger U, Richter A (1994) Nucleic-acid-binding properties of hnRNP-U/SAF-A, a nuclear-matrix protein which binds DNA and RNA in vivo and in vitro. Eur J Biochem 221:749–757

Forrester WC, van Genderen C, Jenuwein T, Grosschedl R (1994) Dependence of enhancer-mediated transcription of the immunoglobulin mu gene on nuclear matrix attachment regions. Science 265:1221–1225

Fuks, F, Hurd PJ, Wolf D, Nan X, Bird AP, Kouzarides T (2003) The methyl-CpG-binding protein MeCP2 links DNA methylation to histone methylation. J Biol Chem 278:4035–4040

Gasser SM, Laemmli UK (1986) Cohabitation of scaffold binding regions with upstream/enhancer elements of three developmentally regulated genes of *D. melanogaster*. Cell 46:521–530

Hagerman P J (1990) Sequence-directed curvature of DNA. Annu Rev Biochem 59:775–781

Hibino Y, Nakamura K, Tsukada S, Sugano N (1993) Purification and characterization of nuclear scaffold proteins which bind to a highly repetitive bent DNA from rat liver. Biochim Biophys Acta 1174:162–170

Hibino Y, Ohzeki H, Hirose N, Morita Y, Sugano N (1998a) Involvement of DNA methylation in binding of a highly repetitive DNA component to nuclear scaffold proteins from rat liver. Biochem Biophys Res Commun 252:296–301

Hibino Y, Ohzeki H, Hirose N, Sugano N (1998b) Involvement of phosphorylation in binding of nuclear scaffold proteins from rat liver to a highly repetitive DNA component. Biochim Biophys Acta 1396:88–96

Hibino Y, Ohzeki H, Sugano N, Hiraga K (2000) Transcription modulation by a rat nuclear scaffold protein, P130, and a rat highly repetitive DNA component or various types of animal and plant matrix or scaffold attachment regions. Biochem Biophys Res Commun 279:282–287

Ikeda Y, Nakamura K, Iwakami N, Hibino Y, Sugano N (1990) Base sequences of highly repetitive components in nuclear DNAs from rat liver and rat-ascites hepatoma. Cancer Lett 55:201–208

Jarman AP, Higgs DR (1988) Nuclear scaffold attachment sites in the human globin gene complexes. EMBO J 7:3337–3344

Jenuwein T, Forrester WC, Fernández-Herrero LA, Laible G, Dull M, Grosschedl R (1997) Extension of chromatin accessibility by nuclear matrix attachment regions. Nature 385:269–272

Kohwi-Shigematsu T, Maass K, Bode J (1997) A thymocyte factor SATB1 suppresses transcription of stably integrated matrix-attachment region-linked reporter genes. Biochemistry 36:12005–12010

Li W, Chen HY, Davie JR (1996) Properties of chicken erythrocyte histone deacetylase associated with the nuclear matrix. Biochem J 314:631–637

Ma H, Siegel AJ, Berezney R (1999) Association of chromosome territories with the nuclear matrix: disruption of human chromosome territories correlates with the release of a subset of nuclear matrix proteins. J Cell Biol 146:531–542

Martínez-Balbás A, Rodríguez-Campos A, García-Ramírez M, Sainz J, Carrera P, Aymamí J, Azorín F (1990) Satellite DNAs contain sequences that induce curvature. Biochemistry 29:2342–2348

Moazed D (2001) Common themes in mechanisms of gene silencing. Mol Cell 8:489–498

Muchardt C, Yaniv M (2001) When the SWI/SNF complex remodels.the cell cycle. Oncogene 20:3067–3075

Nan X, Campoy FJ, Bird A (1997) MeCP2 is a transcriptional repressor with abundant binding sites in genomic chromatin. Cell 88:471–481

Nan X, Ng H-H, Johnson CA, Laherty CD, Turner BM, Eisenman RN, Bird A (1998) Transcriptional repression by the methyl-CpG-binding protein MeCP2 involves a histone deacetylase complex. Nature 393:386–389

Nayler O, Strätling W, Bourquin JP, Stagljar I, Lindemann L, Jasper H, Hartmann AM, Fackelmayer FO, Ullrich A, Stamm S (1998) SAF-B protein couples transcription and pre-mRNA splicing to SAR/MAR elements. Nucleic Acids Res 26:3542–3549

Paulson JR, Laemmli UK (1977) The structure of histone-depleted metaphase chromosomes. Cell 12:817–828

Poljak L, Seum C, Mattioni T, Laemmli UK (1994) SARs stimulate but do not confer position independent gene expression. Nucleic Acids Res 22:4386–4394

Radic MZ, Lundgren K, Hamkalo BA (1987) Curvature of mouse satellite DNA and condensation of heterochromatin. Cell 50:1101–1108

Razin SV, Mantieva VL, Georgiev GP (1979) The similarity of DNA sequences remaining bound to scaffold upon nuclease treatment of interphase nuclei and metaphase chromosomes. Nucleic Acids Res 7:1713–1735

Romig H, Fackelmayer FO, Renz A, Ramsperger U, Richter A (1992) Characterization of SAF-A, a novel nuclear DNA binding protein from HeLa cells with high affinity for nuclear matrix/scaffold attachment DNA elements. EMBO J 11:3431–3440

Shrader TE, Crothers DM (1990) Effects of DNA sequence and histone-histone interactions on nucleosome placement. J Mol Biol 216:69–84

Singer MF (1982) Highly repeated sequences in mammalian genomes. Int Rev Cytol 76:67–112

Small D, Nelkin B, Vogelstein B (1982) Nonrandom distribution of repeated DNA sequences with respect to supercoiled loops and the nuclear matrix. Proc Natl Acad Sci USA 79:5911–5915

Weitzel JM, Buhrmester H, Strätling WH (1997) Chicken MAR-binding protein ARBP is homologous to rat methyl-CpG-binding protein MeCP2. Mol Cell Biol 17:5656–5666

Xu M, Hammer RE, Blasquez VC, Jones SL, Garrard WT (1989) Immunoglobulin kappa gene expression after stable integration. II. Role of the intronic MAR and enhancer in transgenic mice. J Biol Chem 264:21190–21195

Zink D, Cremer T, Saffrich R, Fischer R, Trendelenburg MF, Ansorge W, Stelzer EH (1998) Structure and dynamics of human interphase chromosome territories in vivo. Hum Genet 102:241–251

10-5. Breaking and Tessellating the Contiguous Nuclear Genome

Kojiro Ishii

PRESTO, Japan Science and Technology Agency,
Division of Cell Biology, Institute of Life Science, Kurume University,
1-1 Hyakunen-koen, Kurume 839-0864, Japan

10-5.1 Preface

"Preformation" and "epigenesis" were the two antithetical concepts theoretical biologists of the early 18th century advocated to explain ontogenesis. Apparently their repercussions still remain and keep vibrating, at least in a subject of the current biological study; i.e. how the usage of "genetic" information, an ultimate "preformed" material, is controlled by the "epigenetics," a derivative of "epigenesis," and conversely, how such "epigenetic" regulations are restricted by the *cis*-acting "genetic" elements. The genetic elements described in the latter issue are experimentally defined as "boundary elements" or "insulators," and have been studied extensively for decades (Gerasimova and Corces 2001). Here I am going to summarize some of the recent advances to unravel the mechanics of boundary elements that hinder spread of epigenetic heterochromatin. They appear to be classified functionally into two groups, the antagonistic chromosomal effect and the structural tethering in the nucleus.

10-5.2 Dynamic Functions That Establish Heterochromatin Boundaries

Heterochromatin and euchromatin represent the discrete domains interspersed throughout the eukaryotic chromosomes in which gene expression is either repressed or facilitated. They can be identified by the distinct condensation level and typical epigenetic hallmarks. For instance, inactive heterochromatin in many organisms were shown to be associated with hypoacetylated and methylated histones, which would call in a series of specific structural proteins, whereas active euchromatin often consists of hyperacetylated histones, exhibiting increased enzyme accessibility (reviewed in Felsenfeld and Groudine 2003).

Recent analyses with the yeast *Saccharomyces cerevisiae* revealed the involvement of histone acetyl-transferases (HATs) in delimiting the transcriptionally silent heterochromatin domains; propagation of silencing at yeast *HM* loci and subtelomeric region was blocked by the chromatin tethering of several HATs with various specificity, that is, GNAT-type HAT Gcn5p or MYST-type HATs Sas2p and Esa1p (Donze and Kamakaka 2001; Chiu et al. 2003; Ishii and Laemmli 2003). This incidence was initially detected as a restoration of reporter gene expression in each context, but it was then carefully proven to be different from direct gene activation (Chiu et al. 2003; Ishii and Laemmli 2003). Chiu et al. (2003) further demonstrated that targeted Gcn5p or Esa1p creates a >2 kb segment of hyperacetylated chromatin within the heterochromatic region. Thus histone hyperacetylation seems to counteract the nature of repressive heterochromatin enzymatically, which leads to stop its propagation. It is noteworthy, however, that the actual boundary does not correspond to the HAT-targeted site, but at some adjacent point where the changeover from hypoacetylation to hyperacetylation occurs, and therefore is not necessarily position-fixed nor confers a sharp transition (Ishii and Laemmli 2003).

Such competition between the opposing effects upon histone acetylation was found to control the natural boundary of yeast telomeric heterochromatin. The MYST-HAT Sas2p was shown to compete the acetylation of histone H4 lysine 16 with Sir2p, a deacetylase required for the hypoacetylation state of yeast heterochromatin, whereby heterochromatin readily spread inward from the telomere ends and repressed adjacent subtelomeric genes in the absence of Sas2p (Kimura et al. 2002; Suka et al. 2002). A similar phenomenon was also observed in cells lacking two GNAT-HATs Gcn5p and Elp3p simultaneously, supporting the idea that heterochromatin blockade can be directed by different acetylation reactions (Kristjuhan et al. 2003). Further support for histone acetylation was brought up by Ladurner et al. (2003), who showed that Bdf1p, a chromosomal protein that binds to acetylated histone H4 through its bromodomain, served for maintaining Sas2p-influencing euchromatin and protecting regions adjacent to telomeres and *HM* loci from deacetylation by Sir2p.

In addition to histone modification, the euchromatic feature is often marked by the histone variants, and Meneghini et al. (2003) also demonstrated that spread of heterochromatin at *HM* loci and subtelomeric region is normally suppressed by the effort of euchromatin-associating Htz1p, a histone H2A variant in yeast. Interestingly, mutations in SWR-C, a chromatin remodeling complex copurified with Htz1p as well as Bdf1p, also resulted in the gain of heterochromatin regions (Krogan et al. 2003); thus all three major epigenetic chromatin determinants, histone modification, chromatin remodeling, and histone variants, are likely to participate together in delimiting heterochromatin.

Finally, not only such epigenetic factors but also many transcription regulators, most of which bind to specific DNA sequences, were demonstrated to harbor potentials to block heterochromatin in yeast (Fourel et al. 2002; Ishii and Laemmli 2003; Yu et al. 2003). Their blocking property was not in good accordance with the gene activation but reminiscent of that of HATs, implying the shared mechanism between them (Ishii and Laemmli 2003). Consistently, suppression of chromosomal silencing by a transcriptional activator was also demonstrated in mammalian cells, but the requirement of the activator was found to be only transient, suggesting the involvement of epigenetic modifications that persist in the absence of activator (Sutter et al. 2003). Taken together, any dynamic activities that confer a euchromatic effect can be recruited to a position defined by DNA-binding transcription factor and fulfill the barrier function toward nearby heterochromatin, possibly as an ensuring mechanism of transcriptional activation.

10-5.3 Structural Functions That Demarcate the Chromatin Domains

Although the above studies stress more on the dynamic, enzymatic aspect of the boundary elements, another independent lines of evidence suggest their mode of action as a physical constraint. Such a structural consequence might be achieved through clustering the elements to form discrete loop domains and/or tethering the elements to the nuclear architecture.

In contrast to the progressive changeover the dynamic functions created in between the heterochromatin and euchromatin, a sharp and distinct boundary element was detected in yeast chromatin by tethering *Drosophila* BEAF protein (Ishii et al. 2002; Ishii and Laemmli 2003). BEAF was originally isolated as an associating factor of a classical fly boundary element, *scs'*, and the yeast analysis revealed its C-terminal oligomerization domain to be sufficient to establish the evolutionarily conserved boundary element. A genetic screen was thus performed to identify yeast proteins that possess similar activity as BEAF, which resulted in the isolation of a variety of proteins involved in the macromolecule transport between nucleus and cytoplasm. Their transport function turned out to be dispensable, however, and subsequent analyses led to the conclusion that such transporters essentially tethered the chromatin loci physically to the nuclear pore complex (NPC) to establish boundary elements (Ishii et al. 2002). Although a direct connection between the NPC and native boundary elements remains elusive and BEAF has been suggested to work independently of NPC, of significance was that the involvement of nuclear architecture was highlighted in the functional context of boundary elements after a nonbiased screen.

Other *Drosophila* boundary proteins Su(Hw) and Mod(mdg4), both of which associate to the *gypsy* retrotransposon and many other genomic sites, were previously shown to form large foci at the periphery of the nucleus and to organize the subnuclear localization of chromatin fiber through their homotypic interaction (Gerasimova et al. 2000). Then Byrd and Corces (2003) demonstrated that they were actually present in the nuclear matrix (NM) fraction and that the chromatin fiber bracketed by the two binding sites was emanating from the matrix to form a loop. Similarly, CTCF protein, which associates universally to mammalian insulators that block the enhancer action, was also incorporated into the NM fraction, and was anchored to another subnuclear structure, the nucleolus, together with its target chromatin (Dunn et al. 2003; Yusufzai et al. 2004). Moreover, an oligomer form of CTCF could be detected analogous to Su(Hw)/Mod(mdg4) and BEAF, implying that CTCF might tie multiple target sites together, just like the proposed model for *gypsy* and *scs'* (Gerasimova et al. 2000; Blanton et al. 2003; Yusufzai et al. 2004).

10-5.4 Interplay Between Dynamic and Structural Functions

Apparently the two functions mentioned above are totally segregated mechanisms. Substantial evidence indicates, however, that they might actually be different sides of the same coin.

BEAF associates to *scs'*, which was postulated to delimit the puff-forming heat shock locus *87A7*. But *scs'* also corresponds to the promoter of two divergent transcripts, and Kuhn et al. (2004) recently argued that it rather resided within the puff, playing a direct role in regulating gene expression at least in the context of polytene chromosome. Interestingly, a central domain of BEAF exhibited a weak transactivation potential in yeast assay and blocked heterochromatin by itself in a similar manner to HATs, although such features were hindered in the full-length construct (Ishii and Laemmli 2003). Considering that *scs'* was originally marked by its nuclease hypersensitivity, the enigmatic link between the structural and dynamic activities of BEAF-mediated boundary element might be the reflection of its complete physiology in interphase nuclei and essential for function.

Not only the middle domain of BEAF but also CTCF exhibited a HAT-type heterochromatin blocking potential in yeast assay (Ishii and Laemmli 2003). This is in line with the recent finding that CTCF contributed to the transcriptional activation at mouse *Igf2/H19* imprinted locus in addition to blocking enhancer (Schoenherr et al. 2003). To our surprise, however, CTCF was demonstrated in its native circumstance to be dispensable for protecting transgene from heterochromatin-mediated silencing (Recillas-Targa et al. 2002). Yet histone hyperacetylation was shown to be the hallmark of CTCF-associating genuine boundary elements, even though its extent correlated rather with the heterochromatin-blocking potential than the amount of CTCF (Litt et al. 2001; Recillas-Targa et al. 2002). The fully potent boundary element for both active and silent chromatin might be achieved through the coalition of a structural contribution exerted by CTCF, which seems inert in yeast, and a dynamic effect created by some protein and/or CTCF itself, presumably in context-dependent manner.

10-5.5 Postface

It is possible that various different functions define the chromatin boundaries according to each local situation. As a unity of "preformed" genetic information, however, chromatin may well confine those "epigenetic" regulations in an autonomous and universal way. Indeed, current variation in the notion of boundary elements can simply be because it is so complex that each study has managed to dissect only a limited aspect of it. Further careful examination should elucidate the nature of the boundary elements.

References

Blanton J, Gaszner M, Schedl P (2003) Protein:protein interactions and the pairing of boundary elements in vivo. Genes Dev 17:664–675

Byrd K, Corces VG (2003) Visualization of chromatin domains created by the gypsy insulator of Drosophila. J Cell Biol 162:565–574

Chiu YH, Yu Q, Sandmeier JJ, Bi X (2003) A targeted histone acetyltransferase can create a sizable region of hyperacetylated chromatin and counteract the propagation of transcriptionally silent chromatin. Genetics 165:115–125

Donze D, Kamakaka RT (2001) RNA polymerase III and RNA polymerase II promoter complexes are heterochromatin barriers in *Saccharomyces cerevisiae*. EMBO J 20:520–531

Dunn KL, Zhao H, Davie JR (2003) The insulator binding protein CTCF associates with the nuclear matrix. Exp Cell Res 288:218–223

Felsenfeld G, Groudine M (2003) Controlling the double helix. Nature 421:448–453

Fourel G, Miyake T, Defossez PA, Li R, Gilson E (2002) General regulatory factors (GRFs) as genome partitioners. J Biol Chem 277:41736–41743

Gerasimova TI, Corces VG (2001) Chromatin insulators and boundaries: effects on transcription and nuclear organization. Annu Rev Genet 35:193–208

Gerasimova TI, Byrd K, Corces VG (2000) A chromatin insulator determines the nuclear localization of DNA. Mol Cell 6:1025–1035

Ishii K, Laemmli UK (2003) Structural and dynamic functions establish chromatin domains. Mol Cell 11:237–248

Ishii K, Arib G, Lin C, Van Houwe G, Laemmli UK (2002) Chromatin boundaries in budding yeast: the nuclear pore connection. Cell 109:551–562

Kimura A, Umehara T, Horikoshi M (2002) Chromosomal gradient of histone acetylation established by Sas2p and Sir2p functions as a shield against gene silencing. Nat Genet 32:370–377

Kristjuhan A, Wittschieben BO, Walker J, Roberts D, Cairns BR, Svejstrup JQ (2003) Spreading of Sir3 protein in cells with severe histone H3 hypoacetylation. Proc Natl Acad Sci USA 100:7551–7556

Krogan NJ, Keogh MC, Datta N, Sawa C, Ryan OW, Ding H, Haw RA, Pootoolal J, Tong A, Canadien V, Richards DP, Wu X, Emili A, Hughes TR, Buratowski S, Greenblatt JF (2003) A Snf2 family ATPase complex required for recruitment of the histone H2A variant Htz1. Mol Cell 12:1565–1576

Kuhn EJ, Hart CM, Geyer PK (2004) Studies of the role of the Drosophila scs and scs′ insulators in defining boundaries of a chromosome puff. Mol Cell Biol 24:1470–1480

Ladurner AG, Inouye C, Jain R, Tjian R (2003) Bromodomains mediate an acetyl-histone encoded antisilencing function at heterochromatin boundaries. Mol Cell 11:365–376

Litt MD, Simpson M, Recillas-Targa F, Prioleau MN, Felsenfeld G (2001) Transitions in histone acetylation reveal boundaries of three separately regulated neighboring loci. EMBO J 20:2224–2235

Meneghini MD, Wu M, Madhani HD (2003) Conserved histone variant H2A.Z protects euchromatin from the ectopic spread of silent heterochromatin. Cell 112:725–736

Recillas-Targa F, Pikaart MJ, Burgess-Beusse B, Bell AC, Litt MD, West AG, Gaszner M, Felsenfeld G (2002) Position-effect protection and enhancer blocking by the chicken beta-globin insulator are separable activities. Proc Natl Acad Sci USA 99:6883–6888

Schoenherr CJ, Levorse JM, Tilghman SM (2003) CTCF maintains differential methylation at the Igf2/H19 locus. Nat Genet 33:66–69

Suka N, Luo K, Grunstein M (2002) Sir2p and Sas2p opposingly regulate acetylation of yeast histone H4 lysine16 and spreading of heterochromatin. Nat Genet 32:378–383

Sutter NB, Scalzo D, Fiering S, Groudine M, Martin DI (2003) Chromatin insulation by a transcriptional activator. Proc Natl Acad Sci USA 100:1105–1110

Yu Q, Qiu R, Foland TB, Griesen D, Galloway CS, Chiu YH, Sandmeier J, Broach JR, Bi X (2003) Rap1p and other transcriptional regulators can function in defining distinct domains of gene expression. Nucleic Acids Res 31:1224–1233

Yusufzai TM, Tagami H, Nakatani Y, Felsenfeld G (2004) CTCF tethers an insulator to subnuclear sites, suggesting shared insulator mechanisms across species. Mol Cell 13:291–298

Perspective—toward understanding the *in situ* genome function

Kyosuke Nagata

Department of Infection Biology, Graduate School of Comprehensive Human Sciences and Institute of Basic Medical Sciences, University of Tsukuba, 1-1-1Tennodai, Tsukuba 305-8575, Japan

During the latest decade, the studies that have attempted to understand the regulatory mechanism of the structure and function of chromatin have had a great research impact with respect to the research in the field related to nuclear functions, including transcription and replication. The analysis of the molecular mechanism of replication and transcription began with the dissection and reconstitution of *cell-free* systems that mimicked the accurate enzymatic processes in test tubes. The *cell-free* replication system of the adenovirus genome DNA replication—the first *cell-free* eukaryotic DNA replication system—was reconstituted with purified viral proteins and host cell factors. Using this system, NF-I was identified as one of the host factors, and it was shown to be a sequence-specific DNA binding protein—one of the first cases of such paradigm proteins. The reconstituted *cell-free* SV40 DNA replication system has provided us with an important basis for understanding cellular genome replication. The *cell-free* transcription system that used nuclear extracts and a DNA fragment corresponding to the adenovirus major late promoter region built a basic sculpture/landscape of the molecular mechanism of transcription. The identification and characterization of TF-IID was followed by identification of TATA-box binding protein followed by the discovery of co-activators followed by the identification of the proteins involved in chromatin regulation. Yes, chromatin became on stage. The tremendous efforts that had been made to clarify transcription-related physiological events, such as epigenetic regulation, encountered a particular limitation when using the *cell-free* systems described above. One of the obvious reasons for this limitation is due to the fact that the system utilizes naked DNA as a template. The actual template within the cell comprises DNA and chromosomal proteins, which together form chromatin. Thus, the age to study the nuclear function at the level of chromatin came. This brings us to a brief discussion on chromatin.

Again, biochemical approaches as well as forward and reverse genetics have been revealing the regulatory mechanism of chromatin. A variety of factors that are involved in the structural, and thereby, the functional changes of chromatin are now known to us. Allow me to elaborate in brief, three major mechanisms (Fig. 1) execute the change that is associated with activation and/or activated states, namely, the ATP-dependent mechanism brought about by chromatin remodeling factors containing ATP-hydrolyzing enzymes, histone modifications, and histone chaperone-mediated reactions. DNA methylation and histone modification together are responsible for gene silencing and heterochromatinization. Noncoding RNAs and siRNA were also found to be important for chromatin regulation. The progresses that were made in research were able to provide answers pertaining to the fundamental mechanisms of chromatin regulation and some parts of epigenetic regulation; however, they did not provide answers to all the questions concerning these processes. For instance, the mechanism that operates events between loci on the same chromosome, such as allelic exclusion, locus control, and insulator-mediated regulation, or the mechanism that is involved in specific chromosomal translocation between chromosomes can be correctly understood only in the context of the nucleus structure. We should know and would like to know the structure and function of the genome in its original site within a particular cell, designated the *in situ* genome, based on the accumulating information that is being derived from studies with naked DNA and chromatin and on a novel study aspect for the nucleus.

The nucleus is the site of operation of the genome function. It is a noteworthy fact that the nucleus is a compartment that is surrounded by the nuclear envelope; thus, the nuclear import and export of biomolecules should be considered with regard to nuclear function. More importantly, the regulated localization of the genome should play key roles in the regulation of the genome and nuclear function. At this point, certain facts remain unclear. These include the manner in which the interphase chromatin occupies the limited space in every chromosome; the reason for the location of heterochromatin along the nuclear periphery; the manner in which and the location where replication foci, transcription foci, and spliceosomes are formed; the manner in which the looped domain structure of the genome is formed; the manner in which genes in the domain are coordinately regulated; and so on. The nucleolus is, for example, considered to be a transcription focus that assembles the ribosomal DNAs that are spread in several chromosomes. However, the detailed mechanism of the assembly and disassembly of the nucleolar organizing region (which comprises ribosomal DNA) and nucleolar factors (nucleolar proteins and RNAs) has not yet been elucidated. Interestingly, some tRNA genes as well as 5S

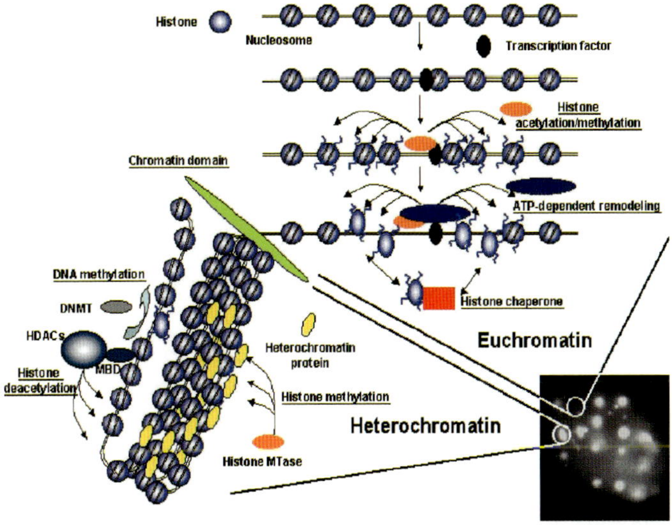

Fig. 1. A model of the formation and change of chromatin. The structural change in euchromatin for transcription begins with binding of transcription factors to their cognate sites followed by histone modification and chromatin remodeling around the transcription start site. In this figure, the histone chaperone is hypothesized to have a function as histone acceptor during the structural and functional change of chromatin. These processes could also be responsible for the maintenance of euchromatin. The formation of heterochromatin implicates DNA methylation by DNMT (DNA methyltransferase), histone modification such as histone deacetylation by HDAC (histone deacetylase) and histone methylation by histone MTase (histone methyltransferase), and binding of heterochromatin-associated proteins

RNA genes are assembled in the nucleolus. This fact raises a question regarding the mechanism by which RNA polymerase II-dependent genes that neighbor the aforementioned RNA genes escape from accumulation into the nucleolus and are organized as domains independent of these genes. In addition, various subnuclear structures such as the PML body and Cajal body exist, and their exact functions have not been adequately clarified. The findings of the studies on the structure and function of the nucleus shall now be discussed.

Instead of summarizing the content of this book, I have attempted to emphasize the content of a recent workshop on nuclear dynamics in order to gain insight into the present status of research in the field of the nuclear structure and function. The EMBO/FEBS Conference on Nuclear Structure and Dynamics held during September 24–28, 2005 in La Grande Motte, near Montpellier in France was possibly one of the first comprehensive meetings in Europe concerning this subject. This meeting witnessed a gathering of individuals who have an interest in this field of research and discussions pertaining to recent progresses in the same.

The Montpellier meeting comprised 13 sessions, each of which reflected the trends in this research field. Session1 entitled “Nuclear compartmentalization, nuclear bodies, and genome function,” session 4 entitled “The nuclear periphery,” and session 11 entitled “Regulation of interphase chromosome architecture and positioning” described the structure/organization of the nucleus and the roles of each nuclear structure in gene regulation. The authentic dynamics of the nuclear structure was discussed at the level of 3 dimensions (D) and 3D + the forth D, that is, the time scale in session 2 entitled “Dynamic 3D chromatin structure” and session 5 entitled “The dynamic nucleus during the cell cycle”. The visualization of molecular behavior, molecular interaction, and the structural change of the nuclear structures as well as their real-time measurement was obviously advantageous for detailed studies on the nuclear dynamics. On these lines, session 3 entitled “Emerging technologies” focused on the recently developed methods. With regard to the nuclear dynamics, of interest were the reports discussing the possibility that the movement of nuclear components and the regulation of the gene activity seem to be under positive contraction between loci and/or chromosomes. Topics concerning chromatin regulation in transcription were discussed in session 7 entitled “Chromatin and epigenetics.” The regulation of genome function with regard to the chromatin level was also discussed in session 8 entitled “DNA repair in situ and chromatin,” session 9 entitled “DNA replication and genome function,” and session 10 entitled “Chromosome segregation and inheritance.” The fact that there were three sessions pertaining to functional RNAs and RNA behavior is noteworthy (session 6 entitled “Noncoding RNA and gene silencing,” session 12 entitled “Small nuclear RNAs,” and session 13 entitled “RNA processing in situ and RNA transport”).

A majority of the authors who have contributed to this book were participants in the International Workshop on the Nuclear Dynamics: Approaches from Biochemistry, Molecular Biology and Visual Biology that was held at Yokohama on December 5, 2002. The authors who did not participate in the Yokohama meeting but participated in the Montpellier meeting as chairpersons/organizers and as coauthors of presenters also lent their efforts to the preparation of this book. The Yokohama meeting was held three years ago and aimed at discussing the leading edge at that time.

Presently, all the authors have summarized their current achievements; therefore, the recent progress made in this research field can be traced.

The development of novel technology has facilitated the development of an understanding of the functions of previously identified molecules. Questions have been raised with respect to the manner in which the unsolved mechanism that governs the establishment and regulation of the nuclear structure and function can be approached. To this end, we require a system(s) to clarify the molecules/molecular functions involved in the nuclear event of interest. We have been learning the utility of the genetic and biochemical approaches from the studies on the molecular mechanism at the level of naked DNA and chromatin. Once a molecule(s) is identified, it would be possible to reveal its function(s), even in an individual. For instance, with regard to the identification of the molecules involved in nuclear import and export, biochemical assays of cellular extracts that were performed using permeable cells and isolated nuclei proved to be quite useful. Thus, one of the more interesting methods with greater potential for use is the system of reconstitution of nuclei in a test tube that leads to the reproduction of a given nuclear event(s) along with the nuclear timetable. Newport and colleague developed a system of reconstitution of nuclei by using *Xenopus* egg extracts and naked DNA. The reconstituted nuclei are capable of replicating DNA. This system still possesses potential in research; for example, it is able to reveal the principle based on which the nuclear architecture is constructed, the manner in which nuclear territory is determined, etc. In addition to reconstitution of nuclei, attempts have been made to alter the property of isolated nuclei and nuclei in permeable cells. Laskey and colleague reported that G1 nuclei initiate cdk/cyclin-dependent DNA replication in the presence of S phase nuclei or S phase nuclear extracts. Trials that are attempting to alter the gene expression pattern of a given nucleus have begun; these trials use a similar experimental procedure in which the nuclei prepared from one cell type are incubated with cellular extracts prepared from another cell type. A trial that attempted to initialize and reprogram the transcription pattern, *i.e.*, to erase the original gene expression pattern of a somatic cell and direct it to that of a different cell type is of great significance. Transplantation of nuclei to another cell type and cell fusion between the different cell types are also used for the reprogramming. However, nuclear reconstitution is of greater advantage to the other methods because it is relatively easier to achieve fractionation of donor extracts and complementation of acceptor nuclei in the reconstitution system. Further, the combination of this system with other cell/individual-based systems shows great promise.

Finally, I wish to thank all the authors and contributors who have jointly published this book. I would particularly like to thank Dr. Tae Shimoyama, who played an important role by painstakingly editing the book with great tenacity, and Dr. Shige H. Yoshimura, who performed the tedious task of being in communication with the authors who have contributed to this book. Special thanks are due to Ms. Haruka Iwano for being a source of constant encouragement and motivation. I reiterate, this book could not have taken form without the help and cooperation of these people.

Index